Methods in Enzymology

Volume 388
PROTEIN ENGINEERING

METHODS IN ENZYMOLOGY

EDITORS-IN-CHIEF

John N. Abelson   Melvin I. Simon

DIVISION OF BIOLOGY
CALIFORNIA INSTITUTE OF TECHNOLOGY
PASADENA, CALIFORNIA

FOUNDING EDITORS

Sidney P. Colowick and Nathan O. Kaplan

Methods in Enzymology

Volume 388

Protein Engineering

EDITED BY

Dan E. Robertson

DIVERSA CORPORATION
SAN DIEGO, CALIFORNIA

Joseph P. Noel

THE SALK INSTITUTE FOR BIOLOGICAL STUDIES
LA JOLLA, CALIFORNIA

ELSEVIER
ACADEMIC
PRESS

AMSTERDAM • BOSTON • HEIDELBERG • LONDON
NEW YORK • OXFORD • PARIS • SAN DIEGO
SAN FRANCISCO • SINGAPORE • SYDNEY • TOKYO

Elsevier Academic Press
525 B Street, Suite 1900, San Diego, California 92101-4495, USA
84 Theobald's Road, London WC1X 8RR, UK

This book is printed on acid-free paper. ∞

For all information on all Academic Press publications
visit our Web site at www.academicpress.com

ISBN: 0-12-182793-3

PRINTED IN THE UNITED STATES OF AMERICA
04   05   06   07   08   9   8   7   6   5   4   3   2   1

Table of Contents

Section I. Methodology

Section II. Applications: Optimization and Screening

Section III. Applications: Directed Evolution of Enzymatic Function

Section IV. Applications: Evolution of Biosynthetic Pathways

Section V. Devices, Antibodies, and Vaccines

Contributors to Volume 388

Article numbers are in parentheses and following the names of contributors. Affiliations listed are current.

OMAR B. ALEXANDER (12), *Department of Biochemistry, Emory University School of Medicine, Rollins Research Center, Atlanta, Georgia 3022*

FRANCES H. ARNOLD (4), *Division of Chemistry and Chemical Engineering, California Institute of Technology, Pasadena, California 91125*

WERNER BASENMATTER (9), *Laboratory of Organic Chemistry, Swiss Federal Institute of Technology, Zurich CH-8093, Switzerland*

MARKUS BAUMANN (18), *Institute of Chemistry and Biochemistry, Department of Technical Chemistry and Biotechnology, Greifswald University, Greifswald D-17487, Germany*

STEPHEN J. BENKOVIC (6), *Department of Chemistry, Pennsylvania State University, University Park, Pennsylvania 16802*

BEN BERKHOUT (28), *Department of Human Retrovirology, Academic Medical Center, 1105-AZ Amsterdam, The Netherlands*

ROBERT R. BIRGE (26), *Departments of Chemistry and Molecular and Cell Biology, University of Connecticut, Storrs, Connecticut 06269*

UWE T. BORNSCHEUER (18), *Institute of Chemistry and Biochemistry, Department of Technical Chemistry and Biotechnology, Greifswald University, Greifswald D-17487, Germany*

JOSEPH CHAPPELL (8), *Department of Agronomy: Plant Physiology, Biochemistry, and Molecular Biology Program, College of Agriculture, University of Kentucky, Lexington, Kentucky 40502*

JOEL R. CHERRY (15), *Novozymes Biotech, Inc., Davis, California 95616*

DAVID W. COLBY (27), *Department of Chemical Engineering and Biological Engineering Division, Massachusetts Institute of Technology, Cambridge, Massachusetts 02139*

ATZE T. DAS (28), *Department of Human Retrovirology, Academic Medical Center, 1105-AZ Amsterdam, The Netherlands*

JONATHAN S. DORDICK (13), *Chemical and Biological Engineering Department, Rensselaer Polytechnic Institute, Troy, New York 12180*

MARK J. DYCAICO (11), *Diversa Corporation, San Diego, California 92121*

JEFFREY ENDELMAN (4), *Division of Chemistry and Chemical Engineering, California Institute of Technology, Pasadena, California 91125*

FRED FENEL (14), *Carbozyme Ltd., 02015 Espoo, Finland*

DIRK FRANKE (20), *Department of Chemistry, The Scripps Research Institute, La Jolla, California 92037*

MELISSA L. GEDDIE (12), *Department of Biochemistry, Emory University School of Medicine, Rollins Research Center, Atlanta, Georgia 3022*

CHRISTILYN P. GRAFF (27), *Department of Chemical Engineering and Biological Engineering Division, Massachusetts Institute of Technology, Cambridge, Massachusetts 02139*

KEVIN A. GRAY (1), *Diversa Corporation San Diego, California 92121*

BRYAN T. GREENHAGEN (8), *College of Agriculture, University of Kentucky, Lexington, Kentucky 40502*

SHIGEAKI HARAYAMA (2), *Biotechnology Development Center, National Institute of Technology and Evaluation, Kisarazu-shi, Chiba 292-0818, Japan*

ERIK HENKE (18), *Institute of Chemistry and Biochemistry, Department of Technical Chemistry and Biotechnology, Greifs-wald University, Greifswald D-17487, Germany*

JASON R. HILLEBRECHT (26), *Departments of Chemistry and Molecular and Cell Biology, University of Connecticut, Storrs, Connecticut 06269*

DONALD HILVERT (9), *Laboratory of Organic Chemistry, Swiss Federal Institute of Technology, Zurich CH-8093, Switzerland*

SUK-BONG HONG (22), *Wyeth Research, Cambridge, Massachusetts 60064*

ALEXANDER R. HORSWILL (6), *Department of Chemistry, Pennsylvania State University, University Park, Pennsylvania 16802*

CHE-CHANG HSU (20), *Department of Chemistry, The Scripps Research Institute, La Jolla, California 92037*

JANNE JÄNIS (14), *Department of Chemistry, University of Joensuu, 80101 Joensuu, Finland*

OSAMU KAGAMI (2), *Marine Biotechnology Institute, Kamaishi Wate 026-0001, Japan*

SATOSHI KANEKO (3), *National Food Research Institute, Biological Function Division, Molecular Function Laboratory, Tsukuba, Ibaraki 305-8642, Japan*

PETER KAST (9), *Laboratory of Organic Chemistry, Swiss Federal Institute of Technology, Zurich CH-8093, Switzerland*

BRENDA A. KELLOGG (27), *Department of Chemical Engineering and Biological Engineering Division, Massachusetts Institute of Technology, Cambridge, Massachusetts 02139*

CHAITAN KHOSLA (23), *Departments of Chemical Engineering, Chemistry, and Biochemistry, Stanford University, Stanford, California 94305*

MIHO KIKUCHI (2), *Marine Biotechnology Institute, Kamaishi Wate 026-0001, Japan*

MASANOBU KITAGAWA (7), *Department of Biosciences and Informatics, Faculty of Science and Technology, Keio University, Yokohama 223-8522, Japan*

KANEHISA KOJOH (7), *Department of Biosciences and Informatics, Faculty of Science and Technology, Keio University, Yokohama 223-8522, Japan*

MONIKA KONARZYCKA-BESSLER (18), *Institute of Chemistry and Biochemistry, Department of Technical Chemistry and Biotechnology, Greifswald University, Greifswald D-17487, Germany*

JEREMY F. KOSCIELECKI (26), *Department of Chemistry, University of Connecticut, Storrs, Connecticut 06269*

KEITH A. KRETZ (1), *Diversa Corporation San Diego, California 92121*

PAWAN KUMAR (23), *Department of Chemical Engineering, Stanford University, Stanford, California 94305*

MIKE LAFFERTY (11), *Diversa Corporation, San Diego, California 92121*

MICHAEL H. LAMSA (15), *Novozymes Biotech, Inc., Davis, California 95616*

PYUNG C. LEE (25), *Department of Biochemistry, University of Minnesota, St. Paul, Minnesota 55126*

MATTI LEISOLA (14), *Laboratory of Bioprocess Engineering, Helsinki University of Technology, 02015-HUT, Finland*

UWE LINNE (24), *Biochemie/Fachbereich Chemie, Philipps-Universität Marburg, Hans Meerwein Strasse, 35032 Marburg, Germany*

PAUL E. O'MAILLE (8), *Structural Biology Laboratory, The Salk Institute for Biological Studies, La Jolla, California 92037*

MOHAMED A. MARAHIEL (24), *Biochemie/Fachbereich Chemie, Philipps-Universität Marburg, Hans-Meerwein Strasse, 35032 Marburg, Germany*

ICHIRO MATSUMURA (12), *Department of Biochemistry, Emory University School of Medicine, Rollins Research Center, Atlanta, Georgia 3022*

BENJAMIN N. MIJTS (25), *Department of Biochemistry, University of Minnesota, St. Paul, Minnesota 55126*

TODD A. NAUMANN (6), *Department of Chemistry, Pennsylvania State University, University Park, Pennsylvania 16802*

JOSEPH P. NOEL (8), *Structural Biology Laboratory, The Salk Institute for Biological Studies, La Jolla, California 92037*

KI-HOON OH (17), *R & D Center of Bioproducts, Institute of Science and Technology, C.J. Corporation, Ichon 467-810, Korea*

MICHIKO ONIMARU (7), *Department of Biosciences and Informatics, Faculty of Science and Technology, Keio University, Yokohama 223-8522, Japan*

MARC OSTERMEIER (10), *Laboratory of Organic Chemistry, Swiss Federal Institute of Technology, Zurich CH-8093, Switzerland*

HEE-SUNG PARK (17), *Department of Biological Studies, Korea Advanced Institute of Science and Technology, Yusung-Gu, Taejon 305-701, Korea*

DAVID E. PASCHON (10), *Department of Chemical and Biomolecular Engineering, Johns Hopkins University, Baltimore, Maryland 21218*

FRANK M. RAUSHEL (22), *Department of Chemistry, Texas A&M University, College Station, Texas 77843*

MANFRED T. REETZ (21), *Max-Planck-Institut für Kohlenforschung, 45470 Mülheim An Der Ruhr, Germany*

TOBY H. RICHARDSON (1), *Diversa Corporation San Diego, California 92121*

DAN E. ROBERTSON (1), *Diversa Corporation San Diego, California 92121*

LORI A. ROWE (12), *Department of Biochemistry, Emory University School of Medicine, Rollins Research Center, Atlanta, Georgia 3022*

MARLEN SCHMIDT (18), *Institute of Chemistry and Biochemistry, Department of Technical Chemistry and Biotechnology, Greifswald University, Greifswald D-17487, Germany*

ROLF D. SCHMID (19), *Institute of Technical Biochemistry, University of Stuttgart, D-70569 Stuttgart, Germany*

CLAUDIA SCHMIDT-DANNERT (25), *Department of Biochemistry, University of Minnesota, St. Paul, Minnesota 55126*

JAY M. SHORT (1), *Diversa Corporation San Diego, California 92121*

JONATHAN J. SILBERG (4), *Division of Chemistry and Chemical Engineering, California Institute of Technology, Pasadena, California 91125*

GRAZYNA E. SROGA (13), *Chemical and Biological Engineering Department, Rensselaer Polytechnic Institute, Troy, New York 12180*

BORIS STEIPE (16), *University of Toronto, Program in Proteomics and Bioinformatics Departments of Biochemistry and Molecular and Medical Genetics, Toronto M5S 1A1, Ontario, Canada*

HAK SUNG-KIM (17), *Department of Biological Studies, Korea Advanced Institute of Science and Technology, Yusung-Gu, Taejon 305-701, Korea*

JEFFREY S. SWERS (27), *Department of Chemical Engineering and Biological Engineering Division, Massachusetts Institute of Technology, Cambridge, Massachusetts 02139*

NORIKO TABATA (7), *Department of Biosciences and Informatics, Faculty of Science and Technology, Keio University, Yokohama 223-8522, Japan*

XUQIU TAN (1), *Diversa Corporation San Diego, California 92121*

YI TANG (23), *Department of Chemical Engineering, Stanford University, Stanford, California 94305**

MING-DAW TSAI (8), *Johnston Laboratory, Departments of Chemistry and Biochemistry, Ohio State University, Columbus, Ohio 43210*

TORU TSUJI (7), *Department of Biosciences and Informatics, Faculty of Science and Technology, Keio University, Yokohama 223-8522, Japan*

OSSI TURUNEN (14), *Laboratory of Bioprocess Engineering, Helsinki University of Technology, 02015-HUT, Finland*

VLADA B. URLACHER (19), *Institute of Technical Biochemistry, University of Stuttgart, D-70569 Stuttgart, Germany*

KOEN VERHOEF (28), *Department of Human Retrovirology, Academic Medical Center, 1105-AZ Amsterdam, The Netherlands*

KEVIN J. WISE (26), *Departments of Chemistry and Molecular and Cell Biology, University of Connecticut, Storrs, Connecticut 06269*

K. DANE WITTRUP (27), *Department of Chemical Engineering and Biological Engineering Division, Massachusetts Institute of Technology, Cambridge, Massachusetts 02139*

CHI-HUEY WONG (20), *Department of Chemistry, The Scripps Research Institute, La Jolla, Californai 92037*

HIROSHI YANAGAWA (7), *Department of Biosciences and Informatics, Faculty of Science and Technology, Keio University, Yokohama 223-8522, Japan*

YIK A. YEUNG (27), *Department of Chemical Engineering and Biological Engineering Division, Massachusetts Institute of Technology, Cambridge, Massachusetts 02139*

HUIMIN ZHAO (5), *Department of Chemical and Biological Engineering, University of Illinois at Urbana, Urbana, Illinois 61801*

*Current address: Department of Chemical Engineering, University of California, Los Angeles, California 90095.

Preface

Based upon a few principle folding paradigms, proteins have evolved to carry out a vast number of highly tuned, interactive functions within the metabolic web. The process of natural selection, using a limited set of amino acids, a few basic mechanisms for mutation and recombination, and the functional output of the resultant three dimensional architecture, is able to parse a nearly infinite variety of amino acid sequences and lengths to optimize phenotypic parameters and protein functions in the context of the cell, the organism, and sometimes the organism's environmental niche. The results of this selection process are a testament to the plasticity of protein folds and the underlying chemistry associated with a particular three dimensional arrangement of amino acids, which together validate the efficiency of Nature's evolutionary process perfected over several billion years.

Within the last decade, a confluence of new technologies in molecular biology, biochemistry, and automation engineering has revolutionized our ability to understand, harness, and ultimately accelerate the evolutionary process. Using comprehensive mutagenesis, large libraries of gene sequences and their associated variants can be rapidly created and, with intelligent, targeted assays, screened for the acquisition and optimization of any intensive or extensive phenotypic variable. Substrate and enantiomeric selectivity, catalytic activity, relief from inhibition, temperature, or oxidant stability, operating optima as well as efficiency of gene expression in native and heterologous hosts, are a few of the parameters that have been optimized in a laboratory setting. Nearly all general protein classes have been targeted for engineering and found to be amenable to the rapid acquisition of adaptive changes necessary for new applications. Those of us who may have thought that Nature had already reached seeming perfection in the form of optimized gene products were surprised by the seemingly unlimited potential for phenotypic enhancement and *ex situ* application of proteins. These new technologies have engendered the exploration of many new medical, agricultural, and industrial opportunities.

We have divided this volume into five sections to best highlight particularly inventive and effective laboratory evolution techniques, some high throughput and ultra high throughput screening technologies, and some results of the application of these technology developments in the fields of industrial catalysis, pathway engineering, protein therapeutics, and device design. We hope to present a balanced picture of the current state of the art as it exists in 2004.

We thank all of the contributing authors for sharing their knowledge with the scientific community. This field is in its infancy and it is by the application and extension of efforts like these that new knowledge and new technologies will be built upon this already firmly rooted foundation.

DAN E. ROBERTSON
JOSEPH P. NOEL

METHODS IN ENZYMOLOGY

VOLUME 262. DNA Replication
Edited by JUDITH L. CAMPBELL

VOLUME 263. Plasma Lipoproteins (Part C: Quantitation)
Edited by WILLIAM A. BRADLEY, SANDRA H. GIANTURCO, AND JERE P. SEGREST

VOLUME 264. Mitochondrial Biogenesis and Genetics (Part B)
Edited by GIUSEPPE M. ATTARDI AND ANNE CHOMYN

VOLUME 265. Cumulative Subject Index Volumes 228, 230–262

VOLUME 266. Computer Methods for Macromolecular Sequence Analysis
Edited by RUSSELL F. DOOLITTLE

VOLUME 267. Combinatorial Chemistry
Edited by JOHN N. ABELSON

VOLUME 268. Nitric Oxide (Part A: Sources and Detection of NO; NO Synthase)
Edited by LESTER PACKER

VOLUME 269. Nitric Oxide (Part B: Physiological and Pathological Processes)
Edited by LESTER PACKER

VOLUME 270. High Resolution Separation and Analysis of Biological Macromolecules (Part A: Fundamentals)
Edited by BARRY L. KARGER AND WILLIAM S. HANCOCK

VOLUME 271. High Resolution Separation and Analysis of Biological Macromolecules (Part B: Applications)
Edited by BARRY L. KARGER AND WILLIAM S. HANCOCK

VOLUME 272. Cytochrome P450 (Part B)
Edited by ERIC F. JOHNSON AND MICHAEL R. WATERMAN

VOLUME 273. RNA Polymerase and Associated Factors (Part A)
Edited by SANKAR ADHYA

VOLUME 274. RNA Polymerase and Associated Factors (Part B)
Edited by SANKAR ADHYA

VOLUME 275. Viral Polymerases and Related Proteins
Edited by LAWRENCE C. KUO, DAVID B. OLSEN, AND STEVEN S. CARROLL

VOLUME 276. Macromolecular Crystallography (Part A)
Edited by CHARLES W. CARTER, JR., AND ROBERT M. SWEET

VOLUME 277. Macromolecular Crystallography (Part B)
Edited by CHARLES W. CARTER, JR., AND ROBERT M. SWEET

VOLUME 278. Fluorescence Spectroscopy
Edited by LUDWIG BRAND AND MICHAEL L. JOHNSON

VOLUME 279. Vitamins and Coenzymes (Part I)
Edited by DONALD B. MCCORMICK, JOHN W. SUTTIE, AND CONRAD WAGNER

Section I

Methodology

[1]   Gene Site Saturation Mutagenesis: A Comprehensive Mutagenesis Approach

By KEITH A. KRETZ, TOBY H. RICHARDSON, KEVIN A. GRAY,
DAN E. ROBERTSON, XUQIU TAN, and JAY M. SHORT

Introduction

Directed protein evolution has been broadly applied to mimic and increase the rate of the natural evolution processes of mutagenesis, recombination, and selection to meet the growing demands for biomolecules tailored to applications in research, chemical, pharmaceutical, food, and other industries.[1–4] Many mutagenesis methods have been developed to introduce one or a few point mutations, including random mutagenesis by error-prone polymerase chain reaction (PCR),[5] chemical treatment,[6,7] mutator strains,[8] and site-directed mutagenesis by PCR amplification using mutagenic oligonucleotides or sequence overlap extension.[9,10] In addition, groups have mutagenized proteins by several different methods, including the introduction of amber mutations in the T4 lysozyme followed by expression in different suppressor strains[11] and cysteine scanning mutagenesis on the metal tetracycline/H$^+$ antiporter.[12] Furthermore, saturation cassette mutagenesis has been employed to study a specific segment of a protein.[13,14] However, saturation mutagenesis at only selected amino acid positions has had limited use for directed evolution.[15–17] Several years ago,

[1] A. Fieck, D. L. Wyborski, and J. M. Short, *Nucleic Acids Res.* **20,** 1785 (1992).

[2] J. M. Short, *Nature Biotechnol.* **15,** 1322 (1997).

[3] R. R. Chirumamilla, R. Muralidhar, R. Marchant, and P. Nigam, *Mol. Cell. Biochem.* **224,** 159 (2001).

[4] M. Chartrain, P. M. Salmon, D. K. Robinson, and B. C. Buckland, *Curr. Opin. Biotechnol.* **11,** 209 (2000).

[5] R. C. Cadwell and G. F. Joyce, *PCR Methods Appl.* **2,** 28 (1992).

[6] S. Ohnuma, T. Nakazawa, H. Hemmi, A. M. Hallberg, T. Koyama, K. Ogura, and T. Nishino, *J. Biol. Chem.* **271,** 10087 (1996).

[7] R. M. Myers, L. S. Lerman, and T. Maniatis, *Science* **229,** 242 (1985).

[8] A. Greener, M. Callahan, and B. Jerpseth, *Methods Mol. Biol.* **57,** 375 (1996).

[9] S. N. Ho, H. D. Hunt, R. M. Horton, J. K. Pullen, and L. R. Pease, *Gene* **77,** 51 (1989).

[10] R. M. Horton, *Mol. Biotechnol.* **3,** 93 (1995).

[11] D. Rennell, S. E. Bouvier, L. W. Hardy, and A. R. Poteete, *J. Mol. Biol.* **222,** 67 (1991).

[12] N. Tamura, S. Konishi, S. Iwaki, T. Kimura-Someya, S. Nada, and A. Yamaguchi, *J. Biol. Chem.* **276,** 20330 (2001).

[13] S. Delagrave and D. C. Youvan, *Biotechnology* **11,** 1548 (1993).

[14] W. Huang, J. Petrosino, M. Hirsch, P. S. Shenkin, and T. Palzkill, *J. Mol. Biol.* **258,** 688 (1996).

METHODS IN ENZYMOLOGY, VOL. 388

A For 32-fold degeneracy

```
5' GAT CAG AAC GCT TTC ATC GAG GGT GTG CTC CCG AAA TTC GTC GTC
    D   Q   N   A   F   I   E   G   V   L   P   K   F   V   V
```

```
5' ATCAGAACGCTTTCATCGAGNNKGTGCTCCCGAAATTCGTCGT   3' coding strand
5' ACGACGAATTTCGGGAGCACMNNCTCGATGAAAGCGTTCTGAT   3' non-coding strand
```

B For 64-fold degeneracy

```
5' GAT CAG AAC GCT TTC ATC GAG GGT GTG CTC CCG AAA TTC GTC GTC
    D   Q   N   A   F   I   E   G   V   L   P   K   F   V   V
```

```
5' ATCAGAACGCTTTCATCGAGNNNGTGCTCCCGAAATTCGTCGT   3' coding strand
5' ACGACGAATTTCGGGAGCACNNNCTCGATGAAAGCGTTCTGAT   3' non-coding strand
```

FIG. 1. Oligonucleotide primer design. (See color insert.)

we described a very non-stochastic approach, Gene Site Saturation Mutagenesis[TM] (GSSM[TM]),[*] which systematically explores minimally all possible single amino acid substitutions along a protein sequence (Fig. 1). [18–21] This comprehensive technique introduces point mutations into every position within a target gene using degenerate primer sets containing 32 or 64 codons to generate a complete library of variants. Unlike rational mutagenesis, GSSM does not require prior knowledge of the structure or mechanism of the target protein due to its ability to generate all mutations at all positions within the protein. Unbiased pools of variants are produced for screening, and the number of amino acid substitutions is greatly expanded as compared to error-prone and chemical-based mutagenesis. The mutagenized library can be screened using a variety of high-throughput assays to identify variants with improved properties. Beneficial mutations can be combined to generate a new library that will contain variants of all possible multiple point mutations or can be added sequentially into a single gene. In either case, combinations of single site mutations can be tested for an optimal protein.

The GSSM approach has been applied successfully to improve a number of different proteins.[21a,21b,22,23] The properties that have been

[15] K. Miyazaki and F. H. Arnold, *J. Mol. Evol.* **49,** 716 (1999).

[16] A. Peterson and B. Seed, *Cell* **54,** 65 (1988).

[17] G. Chen, I. Dubrawsky, P. Mendez, G. Georgiou, and B. L. Iverson, *Protein Eng.* **12,** 349 (1999).

[18] J. M. Short, Patent US 6,171,820 B1 (1999).

[19] J. M. Short, Patent US 6,562,594 B1 (2003).

[20] J. Braman, C. Papworth, and A. Greener, *Methods Mol. Biol.* **57,** 31 (1996).

[21] M. P. Weiner, G. L. Costa, W. Schoettlin, J. Cline, E. Mathur, and J. C. Bauer, *Gene* **151,** 119 (1994).

[*] *Gene Site Saturation Mutagenesis* and *GSSM* are trademarks of Diversa Corporation in the U.S. and/or other countries.

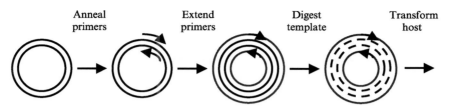

FIG. 2. Reaction mechanism. (See color insert.)

improved include, but are not limited to, stability, substrate specificity, pH performance, enantioselectivity, binding affinity, protein folding, gene expression, and productivity. In many cases, a single amino acid substitution is sufficient to attain the necessary level of performance[22,24]; however, combinations of mutations often provide further enhancement, including both additivity and synergy.[23,25]

Additional information about the protein is also obtained during the screening of the variants. For example, it is possible to map every position within the protein for its sensitivity to mutation and to determine which positions have a positive, negative, or neutral affect on a particular property. Thus, GSSM often provides valuable information concerning the active site, reaction mechanism, and structure–function relationships of an enzyme.

Methods

Overview

The basic procedure utilizes a supercoiled, double-stranded DNA vector with an insert of interest (generally the entire expression construct) and two synthetic primers containing the desired mutation(s) (Fig. 2). The primers, each complementary to opposite strands of the vector, extend during temperature cycling by means of a nonstrand displacing, proofreading

[21a] N. Palackel, Y. Brennan, W. N. Callen, P. Dupree, G. Frey, F. Goubet, G. P. Hazelwood, S. Healy, Y. E. Kang, K. A. Kretz, E. Lee, X. Ian, G. L. Tomlinson, J. Verruto, Y. W. K. Wong, E. J. Mathur, J. M. Short, D. E. Robertson, and B. A. Steer, *Prot. Sci.* **13**, 494 (2004).

[21b] J. B. Garrett, K. A. Kretz, E. O'Donoghue, J. Kerevuo, W. Kim, N. R. Barton, G. P. Hozelwood, J. M. Short, D. E. Robertson, and K. A. Gray, *Appl. Environ. Microbiol.* **70**, 3041 (2004).

[22] G. DeSantis, K. Wong, B. Farwell, K. Chatman, Z. Zhu, G. Tomlinson, H. Huang, X. Tan, L. Bibbs, P. Chen, K. Kretz, and M. J. Burk, *J. Am. Chem. Soc.* **125**, 11476 (2003).

[23] K. A. Gray, T. H. Richardson, K. Kretz, J. M. Short, J. P. F. Bartnek, R. Knowles, L. Kan, P. E. Swanson, and D. E. Robertson, *Adv. Synth. Catalysis* **343**, 607 (2001).

[24] S. L. Marcus, R. Polakowski, N. O. Seto, E. Leinala, S. Borisova, A. Blancher, F. Roubinet, S. V. Evans, and M. M. Palcic, *J. Biol. Chem.* **278**, 12403 (2003).

[25] B. Arezi, C. J. Hansen, and H. H. Hogrefe, *J. Mol. Biol.* **322**, 719 (2002).

DNA polymerase. On incorporation of the oligonucleotide primers, a mutated plasmid containing staggered nicks is generated. The entire plasmid, with insert, is amplified. Following temperature cycling, the product is treated with *Dpn*1. DNA isolated from almost all *Escherichia coli* strains is *dam* methylated and therefore susceptible to *Dpn*1 digestion. The *Dpn*1 endonuclease is specific for methylated and hemimethylated DNA and is used to digest the parental DNA template, leaving the mutation-containing synthesized DNA. *E. coli* is then transformed with the nicked vector DNA incorporating the desired mutation.

Oligonucleotide Primers

Two oligonucleotide primers are used for each position within a protein to be mutagenized (see Fig. 1). Both primers contain the mutation of interest and anneal to the same sequence on opposite strands of the template. The mutation is typically in the middle of the primer with 20 bases of correct sequence on each side. The coding strand primer generally contains NNK at the position to be mutagenized whereas the noncoding strand contains MNN. N is an equal mixture of all four nucleotides; K is an equal mixture of G and T; and M is an equal mixture of A and C. Mutagenesis with this primer design yields a mixture of 32 different sequences coding for all 20 amino acids and one stop codon. If all possible codon substitutions are desired, NNN should be substituted for the codon of interest on both strands, resulting in a mixture of 64 different sequences. Other mixtures of nucleotides at the site of the mutation may be used to code for subsets of amino acids or for special circumstances, such as increasing or decreasing GC content.

Amplification

Reagents

Plasmid template isolated from a dam$^+$ host
Pfu Turbo DNA polymerase and buffer (recommend Stratagene)
10 mM dNTP mix (recommend Applied Biosystems)
Mutation inducing oligonucleotide primers
*Dpn*1 restriction enzyme and buffer (recommend NEB)

Template Preparation. The template is prepared from a dam$^+$ host so that it is susceptible to *Dpn*1 digestion at the end of the protocol. Most common *E. coli* strains are dam$^+$. Both mini- and midipreps (Qiagen) of template DNA have worked well in this protocol. If an A_{260} measurement is used to determine DNA concentration, it is strongly suggested that it be confirmed by agarose gel analysis.

Protocol. For each individual reaction, make up the following reaction mix. Make a negative control reaction where the primers are replaced with an equal volume of water or 5T0.1E.

Buffer	5 μl	10× *PfuTurbo* DNA polymerase buffer
dNTP mix	4 μl	10 mM
Template	X μl	50 ng
Enzyme	X μl	2.5 units *Pfu Turbo* DNA polymerase
Water	X μl	enough to make 30 μl total volume

Add 30 μl of reaction mix to each well followed by 10 μl of each of the two mutagenic primers (2.5 μM) for a particular amino acid. Repeat this for each residue of the protein.

Cycling Protocol

An initial denaturation of the reaction components at 95° for 30 s is followed by 20 cycles of amplification; 95° for 45 s, 50° for 45 s, and 68° for 2 min for each kilobase of the template (vector plus insert). This is followed by a 5-min incubation at 68°.

Dpn*I* Digest

For each reaction, prepare the following digestion mix: 8.0 μl water, 1.0 μl 10× buffer, and 1.0 μl *Dpn*I. Add 10 μl of this mix to each reaction well and incubate at 37° for 4–5 h.

Amplification Quality Control

Run a small aliquot (5 μl) of each reaction on an agarose gel. A strong band of the appropriate size should be visible. If the product is not a single band, optimize the amplification conditions (see Notes).

Notes

1. This protocol may be run with different total reaction quantities. Adjust all reaction components proportionally.

2. Any nonstrand displacing thermostable DNA polymerase can be used in this protocol. *Taq* DNA polymerase and enzyme mixes containing *Taq* DNA polymerase are not suitable for this process as they will displace and cleave the mutagenic primers as they complete the synthesis of the full template.

3. The primers are resuspended or diluted in 5 mM Tris–HCl (pH 8.0), 0.1 mM EDTA (5T0.1E), or water to a concentration of 2.5 μM. Successful results have been obtained with primers purchased from a number of suppliers. A variety of purification protocols have also been found to be suitable. It is recommended that desalted primer preparations be utilized.

4. *Pfu Turbo* DNA polymerase[26] is generally slower than *Taq* and may require extension times of up to 2 min per kilobase of plasmid template.

5. The entire plasmid is the template; therefore, the template size is vector plus insert.

6. Do not reduce the template quantity. This protocol is a linear amplification protocol, not PCR, and therefore does not result in a logarithmic amplification of product.

7. As with any protocol of this type, it may be necessary to optimize times and temperatures for the amplification of template. When using templates that have a high percentage of G and C residues, it is sometimes beneficial to add dimethyl sulfoxide to the reaction. Successful amplification with the standard protocol is generally $\geq 95\%$.

Transformation

Protocol. Use 2.0 μl of each reaction product to transform 50 μl of highly competent *E. coli* cells ($\geq 10^9$ cfu/μg) according to the manufacturer's protocol.

Notes

1. These volumes have been useful using XL1 Blue[27] supercompetent *E. coli* cells (Stratagene) and may need to be adjusted when using other host cells. Perform a dilution series with a few of the reaction products. Using more reaction mix to transform the cells does not always result in more transformants.

2. Be certain to transform identical sets of the negative control. These reaction products should not yield any colonies. If significant numbers of colonies do result from the transformation of the negative control reactions, test the integrity of the *Dpn*I restriction digest. If the *Dpn*I digest is insufficient to digest all of the template DNA, any remaining supercoiled plasmid template DNA will transform the *E. coli* host cells with a much higher frequency than the staggered overlap circles that are the product of the amplification reaction.

Statistical Considerations

One important aspect of any mutagenesis technique is the bias introduced during the process. During the course of multiple GSSM experiments, a slight bias has been observed on occasion. When DNA sequences

[26] K. S. Lundberg, D. D. Shoemaker, M. W. Adams, J. M. Short, J. A. Sorge, and E. J. Mathur, *Gene* **108,** 1 (1991).
[27] W. O. Bullock, J. M. Fernandez, and J. M. Short, *Biotechniques* **5,** 376 (1987).

of random clones were determined, it was noticed that the annealing of primers with fewer mismatches from the template sequence was occasionally slightly favored. Primers with no mismatches with the template DNA were up to twice as likely to anneal and extend as primers that had three mismatches at the site of mutagenesis. Primers with one or two mismatches annealed and extended with intermediate efficiency. Screening more clones easily compensates for this level of nonuniformity.

It is important to recognize the statistical nature of the sampling of the different amino acid changes. The nature of the genetic code is such that the amino acids are represented by different numbers of triplet codons. The representation varies as much as sixfold. When using NNN at the site of randomization, the following representation exists

One codon	Met, Trp
Two codons	Cys, Asp, Asn, Glu, Gln, His, Phe, Lys, Tyr
Three codons	Ile, STOP
Four codons	Ala, Gly, Pro, Thr, Val
Six codons	Leu, Arg, Ser

When using NNK, the representation is

One codon	Cys, Asp, Asn, Glu, Gln, Phe, His, Ile, Lys, Met, Trp, Tyr, STOP
Two codons	Ala, Gly, Pro, Thr, Val
Three codons	Leu, Arg, Ser

Standard sampling statistics govern the probability of sampling a particular amino acid substitution at a particular position in the protein. Using the NNN strategy, sampling 200 variants will sample all possible amino acid substitutions with 90% confidence. Using the NNK strategy, sampling 150 clones will sample all possible amino acid substitutions with a 90% confidence.

Results

GSSM has been used to improve the thermostability of a haloalkane dehalogenase by 30,000-fold.[23] When the mutations were analyzed at the eight single sites that individually improved thermostability, three of the substitutions would likely never have been accessed by routine procedures such as error-prone PCR (EPP) independent of sampling scale (see Table I). Usually an EPP experiment is set up to introduce only a few nucleotide substitutions/gene; therefore, statistically, only a single nucleotide substitution per codon can be expected to occur. Single nucleotide substitutions result in a theoretical maximum of 5–6 amino acid changes per codon with only 2–3 amino acid substitutions generally accessed with standard protocols. In contrast, GSSM accesses all 20 amino acids at a given position.

TABLE I
OBSERVED AMINO ACID SUBSTITUTIONS IN A THERMOSTABLE DEHALOGENASE[23]

Amino acid substitution	Nucleotide substitution	EPP/GSSM[a]
D89G	GAT→GGT	EPP/GSSM
F91S	TTC→TCC	EPP/GSSM
T159L	ACC→CTC	GSSM only
G182Q	GGT→CAG	GSSM only
I220L	ATC→CTC	EPP/GSSM
N238T	AAC→ACC	EPP/GSSM
W251Y	TGG→TAT	GSSM only
P302A	CCC→GCC	EPP/GSSM

[a] EPP, error-prone PCR; GSSM, gene site saturation mutagenesis. For any given substitution it could have been found by one or both methods.

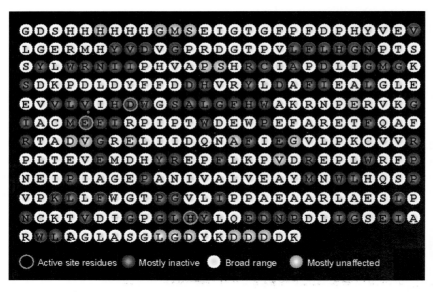

FIG. 3. Residue-specific analysis of the haloalkane dehalogenase from *Rhodococcus rhodochrous*. The mutant enzymes were tested for dehalogenase activity and compared to the wild-type enzyme. (See color insert.)

A further advantage of GSSM is that it enables a complete analysis of every position in a given gene. Substitutions at critical positions will have much larger effects (positively and negatively) than positions that are more flexible. An analysis of screening data allowed the construction of a map of the haloalkane dehalogenase (Fig. 3). These types of maps allow inferences

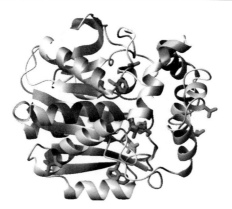

FIG. 4. Schematic representation of the crystal structure of the *R. rhodochrous* haloalkane dehalogenase. Highlighted in blue are residues that cannot be changed without severely affecting enzyme activity, in white are residues that can accommodate a broad spectrum of amino acids, and in gold are residues that can be replaced with any residue without affecting activity. The ball and stick figures in turquoise are catalytic residues, in red are wild-type residues, and in green are thermostable residues (PDB ref. 1BN7). (See color insert.)

to be made about structure–function relationships within a particular protein. If the crystal structure is known for the protein under investigation, the substitutions can be visualized in three-dimensional space (see Fig. 4) and hypotheses can be formulated as to the molecular basis for the improvement.

Acknowledgment

We thank Peter Luginbuhl for generating the schematic image shown in Fig. 4.

[2]   Single-Stranded DNA Family Shuffling

By OSAMU KAGAMI, MIHO KIKUCHI, and SHIGEAKI HARAYAMA

DNA shuffling is a technique that allows accelerated and directed protein evolution *in vitro*. This technique has been shown to be powerful in combining independently isolated mutations of a gene into a single progeny.[1–8] When DNA shuffling is done between closely related genes (i.e., orthologs) instead of a set of mutant genes derived from a single gene, this technique is called DNA family shuffling.[9] DNA family shuffling utilizes the naturally occurring genetic diversity of family genes as the driving force for *in vitro* evolution. Application of the DNA family shuffling strategy has been demonstrated in a number of publications.[10–14]

One difficulty in directed evolution with the original DNA family shuffling method[9] is a strong tendency for this method to reconstruct the unshuffled (i.e., parental) gene structures: a low frequency of mosaic structures in the pool of DNA family shuffling products has been reported.[15,16] We have assumed that the formation of parental gene structures by DNA family shuffling results from preferential annealing of the perfectly complementary strands of DNA fragments derived from the same gene (homoduplex formation; Fig. 1), whose probability may be higher than that of annealing between two homologous sequences (heteroduplex formation).[17,18] Several techniques have been developed to overcome this problem and to increase the crossover rate in DNA family shuffling.[17–20]

We have developed a DNA family shuffling method that uses single-stranded DNAs (ssDNA).[18] DNase I can cleave not only double-stranded DNAs (dsDNAs), but also single-stranded DNAs.[21] Accordingly, ssDNAs

[1] W. P. C. Stemmer, *Proc. Natl. Acad. Sci. USA* **91,** 10747 (1994).

[2] W. P. C. Stemmer, *Nature* **370,** 389 (1994).

[3] A. Crameri, E. A. Whitehorn, E. Tate, and W. P. C. Stemmer, *Nature Biotechnol.* **14,** 315 (1996).

[4] A. Crameri, G. Dawes, E. Rodriguez, Jr., S. Silver, and W. P. C. Stemmer, *Nature Biotechnol.* **15,** 436 (1997).

[5] J. C. Moore, H. M. Jin, O. Kuchner, and F. H. Arnold, *J. Mol. Biol.* **272,** 336 (1997).

[6] T. Yano, S. Oue, and H. Kagamiyama, *Proc. Natl. Acad. Sci. USA* **95,** 5511 (1998).

[7] J. H. Zhang, G. Dawes, and W. P. C. Stemmer, *Proc. Natl. Acad. Sci. USA* **94,** 4504 (1997).

[8] A. L. Kurtzman, S. Govindarajan, K. Vahle, J. T. Jones, V. Heinrichs, and P. A. Patten, *Curr. Opin. Biotechnol.* **12,** 381 (2001).

[9] A. Crameri, S. A. Raillard, E. Bermudez, and W. P. C. Stemmer, *Nature* **391,** 288 (1998).

[10] P. A. Patten, R. J. Howard, and W. P. C. Stemmer, *Curr. Opin. Biotechnol.* **8,** 724 (1997).

[11] S. Harayama, *Trends Biotechnol.* **16,** 76 (1998).

[12] T. Kumamaru, H. Suenaga, M. Mitsuoka, T. Watanabe, and K. Furukawa, *Nature Biotechnol.* **16,** 663 (1998).

[13] C.-C. Chang, T. T. Chen, B. W. Cox, G. N. Dawes, W. P. C. Stemmer, J. Punnonen, and P. A. Patten, *Nature Biotechnol.* **17,** 793 (1999).

[14] L. O. Hansson, R. B. Grob, T. Massoud, and B. Mannervik, *J. Mol. Biol.* **287,** 265 (1999).

[15] F. H. Arnold, *Nature Biotechnol.* **16,** 617 (1998).

[16] S. W. Michnick and F. H. Arnold, *Nature Biotechnol.* **17,** 1159 (1999).

[17] M. Kikuchi, K. Ohnishi, and S. Harayama, *Gene* **236,** 159 (1999).

[18] M. Kikuchi, K. Ohnishi, and S. Harayama, *Gene* **243,** 133 (2000).

[19] W. M. Coco, W. E. Levinson, M. J. Crist, H. J. Hektor, A. Darzins, P. T. Squires, D. Monticello, and J. Monticello, *Nature Biotechnol.* **19,** 354 (2001).

[20] V. Abecassis, C. Pompon, and G. Truan, *Nucleic Acids Res.* **28,** E88 (2000).

[21] E. Melgar and D. A. Goldthwait, *J. Biol. Chem.* **243,** 4409 (1968).

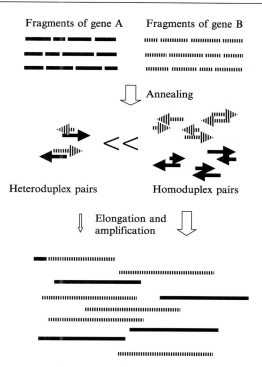

FIG. 1. Low recombination efficiency by DNA family shuffling. In the annealing process for PCR, homoduplex pairs may be formed at a probability much higher than that of heteroduplex pairs, if the homology between genes A and B is low. Thus, the homoduplex molecules may be selectively amplified by PCR in the DNA family shuffling process. The overall frequency of recombination in the shuffling products may also therefore be low.

can be used as materials for DNA family shuffling. By using the sense-strand ssDNA of one gene and the antisense-strand ssDNA of another gene as starting materials for DNA family shuffling, high recombination efficiency can be expected, as only heteroduplex molecules are formed in the first round of polymerase chain reaction (PCR) (Fig. 2). This article describes details of the ssDNA family shuffling method.

Principles and Requirements

The ssDNA family shuffling method can be used when conventional DNA shuffling does not yield recombinants at a high frequency due to a low degree of homology between two parental sequences (e.g., 70–95% identity). ssDNAs of two parental genes should be prepared for ssDNA

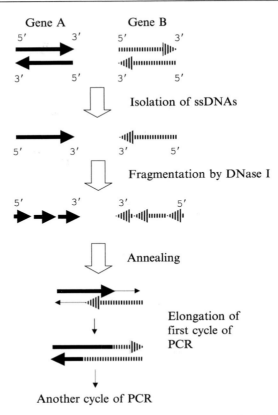

Fig. 2. Schematic diagram of the ssDNA family shuffling method. ssDNAs of genes A and B are prepared. The ssDNAs of these two genes are complementary to each other. These ssDNAs are used for family DNA shuffling: after fragmenting these ssDNAs by DNase I, the generated fragments are subjected to PCR without primers. Only heteroduplex molecules are formed in the first cycle of PCR.

family shuffling. The ssDNA of one gene is the sense (coding) strand, while that of the other gene is the antisense strand. In other words, the ssDNAs of the two parental genes should be complementary to each other. Although this method generally shuffles two parental sequences at once, more genes can be shuffled if the process is repeated.

Parental DNAs

The diversity of the recombinant library may be low if the two parental sequences are highly homologous, while productive recombination may only occur at a low frequency if the two parental sequences are strongly divergent. It is difficult to predict the minimum degree of homology between two parental sequences that yields crossover; we generally use two parental genes with 70–90% nucleotide sequence identity. It is preferable that two parental genes can be aligned without introducing gaps; if the lengths of the two parental genes are not equal, the probability of nonfunctional progenies may be increased in the recombinant library.

Single-Stranded DNAs

Single-stranded DNAs can be prepared by cloning into a phagemid vector such as pBluescript. The phagemid replicates as a double-stranded plasmid in an *Escherichia coli* host. After coinfecting with a helper phage derived from filamentous bacteriophage M13 (e.g., M13KO7 or VCSM13), ssDNA of the phagemid is synthesized and packaged into phage particles.[21] The ssDNA can be extracted easily from these phage particles.[21] To prepare two parental ssDNAs that are complementary to each other, one gene should be cloned into a phagemid vector in the same orientation as that of the origin of ssDNA replication (f1 origin) of the vector, while the other gene should be cloned in the opposite orientation with respect to the f1 origin. In the case of pBluescript, orientations of the multiple cloning site region in pBluescript SK(+) and that in pBluescript KS(+) are opposite to each other. Thus, one parental gene can be cloned into pBluescript SK(+), while the other parental gene can be cloned into pBluescript KS(+) using the same restriction sites. The recombinant plasmids thus constructed are then introduced into *E. coli*. A strain harboring the F or F' episome (e.g., JM109 or TOP10F') should be used for the transformation, as coinfection of the host *E. coli* with the helper phage requires F pili, which can serve as a receptor for the filamentous phage. After infecting the *E. coli* transformants with the helper phage, the single-stranded form of recombinant phagemid DNA is rescued through packaging into M13 phage particles.[22] Longer incubation after coinfecting with the helper phage will result in contamination of the M13 phage preparation with bacterial chromosomal DNA. The entire single-stranded DNA of recombinant phagemids (i.e., DNA of the vector plus cloned gene) can serve as the starting material for DNA shuffling. ssDNAs of the same strand are rescued from the vector sequences of pBluescript KS(+) and pBluescript SK(+) (i.e., ssDNAs derived from the vector sequences are not complementary to each other), and therefore the vector sequences may not be amplified by PCR.

[22] J. Vieira and J. Messing, *Methods Enzymol.* **153**, 3 (1987).

Single-Stranded DNA Fragmentation by DNase I

To achieve effective DNA shuffling, the length of the DNA fragments generated by DNase I digestion is an important factor. Results[17,18] indicate that 40–100 bases are suitable for shuffling two parental DNAs with about 80% nucleotide sequence identity. The degree of digestion by DNase I is determined by the amount of the enzyme, incubation time, temperature, etc. The reaction conditions should be optimized by monitoring the extent of digestion by polyacrylamide gel electrophoresis. After DNase I treatment and subsequent incubation at 90°, a brown precipitation may be produced. This can be removed by centrifugation, and the resulting supernatant is then used for PCR.

Recombination and Amplification by PCR

The annealing temperature, elongation time, and number of cycles are crucial in PCR, and factors affecting each step of PCR have already been described in many literature sources; therefore, these aspects are not addressed in this article. The PCR procedure used for ssDNA shuffling is basically carried out in two steps. The first PCR step is conducted without primers in order to reassemble the DNase I-digested fragments into longer pieces. However, the second PCR step is carried out to amplify the full-length gene sequences using 5' end and 3' end primers flanking both ends of the parental genes. PCR amplification, especially in the first step, is sensitive to the size of the target genes: the larger the size of the target genes, generally the lower the amplification efficiency. If the amount of products from the first PCR step is small, the number of cycles used in the first PCR procedure should be increased. The second PCR amplification step (using the primers) seems to be inhibited when the concentration of the reassembled PCR products (i.e., the products from the first PCR step without primers) is high. If the second PCR amplification is not efficient, the concentration of first-step PCR products to be used in the second PCR step can be reduced. An examination of the concentrations of first- and second-step PCR products by agarose gel electrophoresis is useful for troubleshooting.

Evaluation of Shuffling Efficiency

To optimize the DNA shuffling processes, a method for evaluating the shuffling efficiency is required. The percentage of recombinant molecules or the number and positions of crossovers may be used as an index to evaluate the recombination efficiency. Although this can be determined by sequencing many shuffling products, such an analysis is labor intensive and time-consuming. Joern et al.[23] have reported an effective method for analyzing a shuffled gene library by probe hybridization in a macroarray format.

Materials and Methods

The enzymes used are from New England BioLabs. pBluescript SK(+), pBluescript KS(+), helper phages, Pfu Turbo DNA polymerase, and dNTPs are all from Stratagene, E. coli JM109 is from Invitrogen, and the QIAquick gel extraction kit is from Qiagen.

Solutions and Media

2× YT medium: 16 g of bacto tryptone, 10 g of yeast extract, and 5 g of NaCl per liter

Helper phage stocks: the titer should generally be higher than 10^{10}/ml

Phage precipitation solution: 20% (w/v) polyethylene glycol 8000 (PEG8000) in 2.5 M NaCl; sterilize by filtration, stable at room temperature

Phenol/chloroform/isoamyl alcohol (25:24:1)

Chloroform/isoamyl alcohol (24:1)

7.5 M ammonium acetate (pH 5.2)

100 and 70% (v/v) ethanol

DNase I: store at $-20°$; the diluted enzyme solution must be prepared each day

10× DNase I digestion buffer: 500 mM Tris–HCl (pH 7.4) containing 100 mM MnCl$_2$

dNTP mix: 2 mM each of dNTP; store at $-20°$

10× Pfu Turbo DNA polymerase buffer: 200 mM Tris–HCl (pH 8.8), 20 mM MgSO$_4$, 100 mM KCl, 100 mM (NH$_4$)$_2$SO$_4$, 1% (w/v) Triton X-100, 1 mg/ml of nuclease-free bovine serum albumin (BSA); store at $-20°$

Forward and reverse primers: 20 μM stocks; store at $-20°$

Taq/Pfu Turbo DNA polymerase mixture: 1:1

Pfu Turbo DNA polymerase: store at $-20°$

Taq DNA polymerase: store at $-20°$

[23] J. M. Joern, P. Meinhold, and F. H. Arnold, J. Mol. Biol. 316, 643 (2002).

Procedure 1: Preparation of Single-Stranded DNAs

1. Clone two homologous genes separately in pBluescript SK(+) and pBluescript KS(+) between the same restriction sites on these phagemid vectors.
2. Introduce the recombinant phagemids thus constructed into "male" *E. coli* such as JM109 (infection with a helper phage requires F pili).
3. Inoculate a 1-ml culture of the 2× YT medium supplemented with an appropriate amount of antibiotics with a single JM109 transformant colony.
4. Grow the culture at 37° while shaking (e.g., rotary shaking at 250 rpm) to an A_{600} value of 2.0.
5. Inoculate 25 ml of 2× YT containing an appropriate amount of antibiotics in a 250-ml Erlenmeyer flask with 0.5 ml of the foregoing culture.
6. Incubate for 1 h at 37° while shaking.
7. Infect with a helper phage (M13KO7 or VCMS13) at a multiplicity of infection (moi) of 10–20.
8. Incubate for 2–6 h at 37° while shaking.
9. Harvest the cells by centrifuging at 12,000g at 4° for 15 min.
10. Transfer the supernatant (containing the phage particles) to a fresh tube. Do not introduce the centrifuged pellet into the tube.
11. Spin the supernatant again before transferring it to a fresh tube.
12. Add a 0.25 volume of the phage precipitation solution to the supernatant. Leave on ice for at least 1 h or overnight at 4°.
13. Centrifuge at 12,000g for 20 min at 4°. Phage particles precipitate in the presence of PEG.
14. Remove the supernatant and resuspend the pellet in 400 μl of the TE buffer and transfer the suspension to a 1.5-ml tube.
15. Add 1 volume of TE-saturated phenol:chloroform:isoamyl alcohol (25:24:1) to the sample, vortex for at least 1 min, and then centrifuge at 12,000g for 5 min.
16. Transfer the upper phase (containing phagemid DNA) to a fresh tube without disturbing the interface. Repeat the organic solvent extraction procedure until no visible material appears at the interface.
17. Add 0.5 volume (200 μl) of 7.5 M ammonium acetate plus 2 volumes (1.2 ml) of 100% ethanol. Mix and leave at −20° for 30 min to precipitate the phagemid DNA.
18. Centrifuge at 12,000g for 5 min, remove the supernatant, and carefully rinse the pellet with ice-cold 70% ethanol. If the pellet is

disturbed, centrifuge again for 2 min. Drain the tube and dry the pellet under vacuum.

19. Agarose gel electrophoresis usually produces two major bands corresponding to the helper phage DNA and ssDNA from the recombinant phagemid. A small amount of chromosomal DNA and RNA released by cell lysis may also be present. ssDNA migrates faster than dsDNA of the same length. The presence of the helper phage DNA does not interfere with the subsequent reactions.

Procedure 2: Fragmentation of Single-Stranded DNAs by DNase I

1. Dilute 2 μg (or 3 pmol) of each ssDNA in 45 μl of the TE buffer and add 5 μl of the 10$\times$ DNase I digestion buffer.
2. Incubate the solution at 15° for 5 min and then add 0.3 U of DNase I.
3. Incubate for a further 2 min and then increase the temperature to 90° and incubate for 10 min more to terminate the reaction.
4. Run a 2–3% low melting point agarose gel electrophoresis to purify the fragments of 40–100 bp (longer fragments could also be used) using, for example, a QIAquick gel extraction kit.
5. Resuspend the DNA fragments in 50 μl of the TE buffer.

Procedure 3: PCR Reassembly and Amplification

1. Mix 10 μl of the 10$\times$ Pfu Turbo DNA polymerase buffer, 10 μl of the dNTP mix, 10 μl of the purified ssDNA fragments, 2.5 U of Pfu Turbo DNA polymerase, and H$_2$O to a total volume of 100 μl.

2. Perform the first PCR step: 40 cycles of denaturation at 94° for 1 min, annealing for 1 min, and elongation at 72°. Select an annealing temperature 5° below the average T_m of the two parental DNAs. The elongation time in the first cycle is 1 min $\times$ (parental DNA length in kb) with increments of 5 s for each cycle; this gives an elongation time for the first cycle of 1.5 min with 1.5 kb DNA and an elongation time for the 40th cycle of 195 s longer than that for the first cycle.

3. Add to a new 0.5-ml tube 1–5 μl of the first PCR product, 10 μl of the 10$\times$ Pfu Turbo DNA polymerase buffer, 10 μl of the dNTP mix, 2 μl each of the forward and reverse primers, 2.5 units of the *Taq*/Pfu Turbo DNA polymerase mixture, and H$_2$O to a total volume of 100 μl.

4. Perform the second PCR step: 96° for 2 min followed by 25 cycles of denaturation at 94° for 30 s, annealing for 1 min, and elongating at 72°. The annealing temperature and elongation time are the same as those used in the first PCR step.

5. Separate the PCR products on an agarose gel and recover a band corresponding to the size of the full-length gene.

6. Purify the single band of the correct size from the agarose gel by using, for example, a QIAquick gel extraction kit and subclone after appropriate restriction endonuclease digestion into an appropriate vector.

7. Select the desired mutants from among the resulting transformants.

Example

Isolation of Thermostable Catechol 2,3-dioxygeneses (EC 1.3.11.2, C23Os)

Single-stranded DNA family shuffling has been applied to recombine the two C23O genes,[24] *xylE* and *nahH*, sharing 80% identity in their nucleotide sequences.[20] ssDNAs of *nahH* and *xylE* are prepared by cloning in the phagemid vectors pBluescript KS(+) and pBluescript SK(+), respectively. *nahH* is inserted between the *Not*I and *Eco*RI sites of pBluescript SK(+), whereas *xylE* is inserted between the *Not*I and *Eco*RI sites of pBluescript KS(+). *Escherichia coli* TOP10F′ cells are then transformed with these recombinant phagemids. To produce phage particles containing ssDNAs, each transformant is infected with the M13KO7 helper phage.[21] Each ssDNA is then fragmented randomly with DNase I, and fragments in the size range of 40–100 bases are mixed and subjected to ssDNA family shuffling. The reconstructed genes are then reintroduced in pBluescript SK(+). After transforming the *E. coli* TOP10F′ cells, about 60% of the colonies grown on the plates showed C23O activity. When 50 randomly selected clones exhibiting C23O activity were analyzed for the nucleotide sequences of their C23O genes, 7 of them (14%) were chimeric, with the others being either parental genes (40 clones) or their point mutants (3 clones). This chimera formation rate is much higher than that by the double-stranded method (i.e., the original DNA shuffling method), with which fewer than 1% of clones contained chimera genes.[20]

Isolation of Thermally Stable C23Os

Thermally stable clones were screened from 750 colonies that had been obtained by ssDNA family shuffling. After the treatment at 65° for 10 min, by which *XylE* and *NahH* had been inactivated, 10 colonies exhibited residual C23O activity, showing that they contained enzymes that were thermally more stable than those in the wild-type C23Os. The amino acid sequences of these thermally stable C23Os are shown in Fig. 3. Although

[24] P. Cerdan, M. Rekik, and S. Harayama, *Eur. J. Biochem.* **229**, 113 (1995).

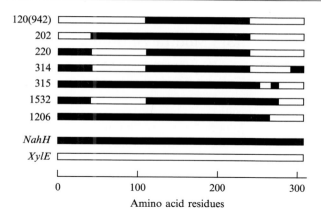

FIG. 3. Structures of thermally stable C23Os obtained by ssDNA family shuffling. Seven clones were thermally more stable than *NahH* and *XylE*. Filled and unfilled boxes, respectively, show the *NahH* and *XylE* sequences.

the nucleotide sequences of all 10 clones were different, the deduced amino acid sequences of clones 120 and 942, clones 202 and 450, and clones 315 and 1527 were identical.

Conclusion

The technology for recursive genetic recombination represented by DNA shuffling has exploded in recent years and has become one of the most powerful methods for enzyme evolution. This technology effectively mimics natural evolution and even offers some advantages. The recombination efficiency has been improved greatly by newly developed methods, including ssDNA family shuffling. It is important for the proficient production of improved enzymes to design a powerful screening method. An effective DNA shuffling method combined with a high-throughput screening method will provide the technical solutions for engineering proteins in basic and applied scientific applications.

Acknowledgment

This work was supported by the New Energy and Industrial Technology Development Organization (NEDO).

[3] Module Shuffling

By SATOSHI KANEKO

Introduction

The exon shuffling hypothesis asserts that nature constructs proteins by "shuffling" blocks of coding sequence (exons) mediated by recombination between noncoding regions of genes (introns), yielding rearranged genes with altered functions.[1,2] Because the average intron is much longer than the average exon and the recombination frequency is proportional to DNA length, the vast majority of crossovers occur in the noncoding sequences. In addition, introns contain large numbers of transposable elements and repetitive sequences, which promote the mismatching and recombination of nonhomologous genes. Consequently, it has been proposed that introns are hot spots for genetic recombination and that exon shuffling has been a major factor in protein evolution. Although exons generally lack correspondence to a single aspect of protein structure, there are many instances of the conservation of exon structure between homologous genes in different organisms.

A "module" is a contiguous polypeptide segment of a protein that has a compact conformation within a globular domain. A module is defined by the distance between $C\alpha$ atoms, and on average a module is about 15 amino acid residues long.[3,4] It should be noted that the module boundaries correlate to the border between exons and introns with considerable statistical significance.[3,5] This indicates that the location of the introns existing in eukaryotic genes is not random and supports the concept that introns play an important role in protein evolution as mediators of exon shuffling. Therefore, module shuffling *in vitro* mimics one of the natural mechanisms of protein evolution. Thus, we have chosen module shuffling as a tool for the elucidation of structure–function relationships and protein evolution of xylanases.[6–8] This article focuses on the methodology of construction and

[1] W. Gilbert, *Nature* **271,** 501 (1978).

[2] W. Gilbert, *Cold Spring Harb. Symp. Quant. Biol.* **52,** 901 (1987).

[3] M. Go, *Nature* **291,** 90 (1981).

[4] M. Go and M. Nosaka, *Cold Spring Harb. Symp. Quant. Biol.* **52,** 915 (1987).

[5] Y. Sato, Y. Niimura, K. Yura, and M. Go, *Gene* **238,** 93 (1999).

[6] S. Kaneko, A. Kuno, Z. Fujimoto, D. Shimizu, S. Machida, Y. Sato, K. Yura, M. Go, H. Mizuno, K. Taira, I. Kusakabe, and K. Hayashi, *FEBS Lett.* **460,** 61 (1999).

[7] S. Kaneko, S. Iwamatsu, A. Kuno, Z. Fujimoto, Y. Sato, K. Yura, M. Go, H. Mizuno, K. Taira, T. Hasegawa, I. Kusakabe, and K. Hayashi, *Protein Eng.* **13,** 873 (2000).

characterization of chimeric enzymes using our xylanase work as an example. The possibility of protein evolution by module shuffling is also discussed.

Module Shuffling of GH10 Xylanases

β-Xylanase (EC 3.2.1.8) is a glycoside hydrolase that randomly hydrolyzes β-1,4-glycosidic linkages within xylan, a major polysaccharide in plant cell walls. Xylanases have mostly been classified into two glycoside hydrolase families (GH), GH10 and 11, due to primary structural similarities.[9] On the basis of structural and mechanistic information, family GH10 is classified as part of clan GH-A, which is the major clan of glycoside hydrolases and also contains families 1, 2, 5, 17, 26, 30, 35, 39, 42, 51, 53, 59, 72, 79, and 86.[9] Enzymes belonging to clan GH-A are known to have a $(\beta/\alpha)_8$-barrel structure and their enzyme mechanism is known as "retaining" due to the double displacement mechanism, which is catalyzed by two glutamic acid residues acting as an acid/base pair and a nucleophile.[9] The three-dimensional structures of 10 GH10 xylanases have now been solved.[10–17] They all have very similar structures, comprising $(\beta/\alpha)_8$-barrels as well as additional helices and loops, which are arranged in a basic TIM-barrel structure forming the active site cleft.[14] The cleft forms deep grooves consistent with the *endo* mode of action and comprises a series of subsites, each one capable of binding a xylose moiety.[17] The subsites that bind the glycone and aglycone regions of the substrate are prefixed by $(-)$ and $(+)$, respectively, and their numbers are related to the proximity to the site of

[8] S. Kaneko, H. Ichinose, Z. Fujimoto, A. Kuno, K. Yura, M. Go, H. Mizuno, I. Kusakabe, and H. Kobayashi, *J. Biol. Chem.* (in press).

[9] P. M. Coutinho and B. Henrissat, Carbohydrate-Active Enzymes Server at URL: http://afmb.cnrs-mrs.fr/Cpedro/CAZY/db.html (1999).

[10] U. Derewenda, L. Swenson, R. Green, Y. Wei, R. Morosoli, F. Shareck, D. Kluepfel, and Z. S. Derewenda, *J. Biol. Chem.* **269**, 20811 (1994).

[11] G. W. Harris, J. A. Jenkins, I. Connerton, N. Cummings, L. Lo Leggio, M. Scott, G. P. Hazlewood, J. I. Laurie, H. J. Gilbert, and R. W. Pickersgill, *Structure* **2**, 1107 (1994).

[12] A. White, S. G. Withers, N. R. Gilkes, and D. R. Rose, *Biochemistry* **33**, 12546 (1994).

[13] R. Dominguez, H. Souchon, S. Spinelli, Z. Dauter, K. S. Wilson, S. Chauvaux, P. Beguin, and P. M. Alzari, *Nature Struct. Biol.* **2**, 569 (1995).

[14] A. Schmidt, A. Schlacher, W. Steiner, H. Schwab, and C. Kratky, *Protein Sci.* **7**, 2081 (1998).

[15] R. Natesh, P. Bhanumoorthy, P. J. Vithayathil, K. Sekar, S. Ramakumar, and M. A. Viswamitra, *J. Mol. Biol.* **288**, 999 (1999).

[16] Z. Fujimoto, A. Kuno, S. Kaneko, S. Yoshida, H. Kobayashi, I. Kusakabe, and H. Mizuno, *J. Mol. Biol.* **300**, 575 (2000).

[17] A. Canals, M. C. Vega, F. X. Gomis-Ruth, M. Diaz, R. R. I. Santamaria, and M. Coll, *Acta Crystallogr. D: Biol. Crystallogr.* **59**, 1447 (2003).

A

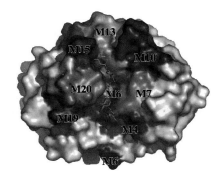

B

| SoXyn10A | M1 | M2 | M3 | M4 | M5 | M6 | M7 | M8 | M9 | M10 | M11 | M12 | M13 | M14 | M15 | M16 | M17 | M18 | M19 | M20 | M21 | M22 |

CfXyn10A

FCF-C4

FCF-C10

FCFCF-C4,10

FIG. 1. Surface model of the GH10 xylanase and bar diagram depictims of chimeric xylanases. (A) Modules M4, M6, M7, M10, M13, M15, M19, and M20 are close to the substrate binding pocket and contain important amino acids for the catalytic activity and substrate recognition. (B) Bar diagrams of chimeric xylanases including FCF-C4, FCF-C10, and FCFCF-C4, 10.

bond cleavage (i.e., the glycosidic bond between the xylose residues at the +1 and −1 subsites is cleaved by the enzyme[18]).

Based on the three-dimensional structure of a xylanase, CfXyn10A, from *Cellulomonas fimi* (formerly known as Cex), the catalytic domain of GH10 xylanase was divided into 22 modules.[5] The enzyme SoXyn10A from *Streptomyces olivaceoviridis* E-86 (formerly known as FXYN) was selected as one of the parent enzymes. The similarity of the catalytic domains of CfXyn10A and SoXyn10A was 49%.

The modules M4, M6, M7, M10, M13, M15, M19, and M20 contain important amino acids for catalytic activity and substrate binding (Fig. 1A). Therefore, we attempted to replace modules M4 and M10 as an example of single module shuffling. As shown in Fig. 2A, the catalytic domain of GH10 xylanases is subdivided into two parts, which consist of the N-terminal larger region and the C-terminal smaller region. The boundary between

[18] G. J. Davies, K. Wilson, and B. Henrissat, *Biochem. J.* **321,** 557 (1997).

A

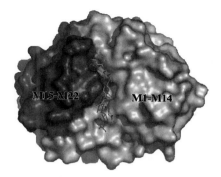

B

SoXyn10A	M1	M2	M3	M4	M5	M6	M7	M8	M9	M10	M11	M12	M13	M14	M15	M16	M17	M18	M19	M20	M21	M22

CfXyn10A																						

FC-14-15																						

FIG. 2. Surface model of GH10 xylanase and a bar diagram of a chimeric xylanase. (A) The boundary between module M14 and M15 along with the substrate binding cleft. (B) Bar diagram of a chimeric xylanase such as FC-14-15.

module M14 and M15 corresponds to the border between these two regions. Therefore, we selected the boundary between modules M14 and M15 for shuffling. This resulted in an interesting combination of the properties from the parent enzymes by module shuffling.

Construction of Chimeric Gene

Three polymerase chain reactions (PCRs) were employed for the construction of chimeric genes. The gene of actinomycetes contains rather high G + C contents (about 70%). However, we successfully performed PCR with modifications to only a few steps in the procedure.[19] By increasing the denaturing temperature to 98° and introducing a few silent mutations into the primers, we avoided making the tight structures (such as hairpins) in the gene and primers often caused by high G + C contents. In our experience, using LA-taq polymerase (Takara Bio Inc., Shiga, Japan), with relatively long primers (about 25 bp long) and employing shuttle steps of denaturing and extension, we achieved good results for amplifying high G + C content genes.[6–8,19] The overall procedure used to construct chimeric

[19] A. Kuno, D. Shimizu, S. Kaneko, Y. Koyama, S. Yoshida, H. Kobayashi, K. Hayashi, K. Taira, and I. Kusakabe, *J. Ferment. Bioeng.* **86**, 434 (1998).

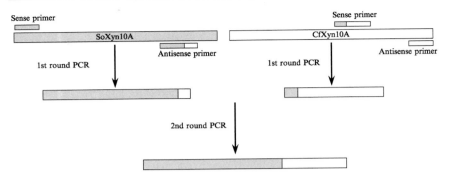

FIG. 3. Construction of module replaced chimeric enzymes. The chimeric xylanases were constructed using three rounds of PCR with overlapping primers.

xylanase genes is summarized in Fig. 3. The merit of this procedure is the lack of problems at the restriction enzyme site and the ease of combining DNA fragments at any desired location. Construction of the chimeric gene named FC-14-15 is provided as an example.[8]

The catalytic domains of SoXyn10A and CfXyn10A are subcloned separately into the pQE60 vector. Construction of the chimera is performed by PCR using overlapping primers at their respective module boundaries (Fig. 3). DNA fragments from SoXyn10A coding modules M1 to M14 and DNA fragments from CfXyn10A coding modules M15 to M22 are amplified using the following sets of primers (primer 1: M1–M14 sense; 5'-CCA TGG GCT CCT ACG CCC TTC CCA GAT CAG-3', primer 2: M1–M14 antisense; 5'-GCG ACT GGA AGC CAA CGC AGT CGA TTG GCA CGC C-3', primer 3: M15-M22 sense; 5'-CTG CGT CGG CTT CCA GTC GCA CCT CAT CGT CGG CC-3', primer 4: M15–M22 antisense; 5'-GGA TCC GAA GGC TTC CAT CAC GGC GGC GTA GG-3'). Each of the 25 amplification cycles consists of denaturation at $98°$ for 1 min and annealing and primer extension at $72°$ for 1 min. The 10-bp overlapping regions (underlined) of the primers are designed to be complementary to their respective module boundaries. The first round of PCR products is separated by agarose gel electrophoresis, followed by gel extraction, and then used for the second round PCR without primers. Each of the 20 amplification cycles consists of denaturation at $98°$ for 1 min, annealing at $60°$ for 25 min, and primer extension at $72°$ for 5 min. Strands having matching sequences at their respective module boundaries overlap and act as primers for each other. On the third round of PCR, the combined fragment is amplified with primers 1 and 4 using 25 cycles of shuttle PCR with denaturation at $98°$ for 1 min and annealing and primer extension at $72°$ for 1 min.

This is an example of combining two DNA fragments; however, we have also succeeded in combining three DNA fragments using this method.[6,7] It is likely that a larger number of fragments can be combined using this procedure. Synthetic single strand small fragments encoding a single module might also be employed during recombination.

Expression of Chimeric Genes

The expression of chimeric genes can sometimes be the rate-limiting step in this procedure, as the expression of a soluble protein can often fail due to the introduced mutations and their deletenous effect on protein folding and stability. We achieved a relatively high level of expression in *Escherichia coli* when the induction was performed at low temperatures, such as at 25°.

For expression in *E. coli* and purification of the SoXyn10A, CfXyn10A, and FC-14-15 genes, the pET expression system (Novagen, Madison, WI) is employed. Thus, each gene is inserted individually into the pET28 vector. The enzymes are expressed as fusion proteins that consist of each enzyme plus a carboxyl-terminal tag with six histidine residues and are secreted into the medium with the aid of a signal sequence derived from *Streptomyces*. The plasmids are used to transform *E. coli* BL21 (DE3), and transformants are cultivated at 25° in LB medium (1 liter) that contains kanamycin (20 μg/ml) until the optical density at 600 nm reaches 0.4. After the addition of isopropyl-1-thio-β-D-galactoside (IPTG) to a final concentration of 1 mM, the culture is incubated at 25° for 24 h. After the *E. coli* cells are removed from the culture by centrifugation (6000g, 10 min), ammonium sulfate is added to a 70% saturation level and the resulting mixture is kept at 4° for 16 h. The precipitate is collected by centrifugation (10,000g, 20 min) and dissolved in a small amount of distilled water followed by dialysis against deionized water. After the removal of insoluble material by centrifugation (12,000g, 30 min), the obtained solution is then loaded onto a HisTrap chelating column (Amersham Bioscience, Piscataway, NJ). The bound enzyme is eluted with a 50 mM phosphate buffer (pH 7.0) containing 250 mM imidazole.

In some cases, chimeric proteins do not fold in the same way as the parental proteins. We constructed almost 20 different types of chimeras; however, some chimeric enzymes were not active enzymes. In most cases, replacement of a few central modules resulted in an active chimeric enzyme. However, when two large fragments were combined at the middle of a sequence, this almost always resulted in an inactive chimeric enzyme. As shown in Fig. 4A, interaction of the N-terminal and C-terminal regions in SoXyn10A is observed in the structure. We found this interaction to be

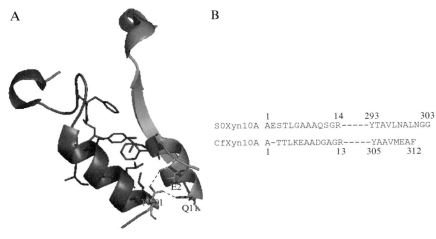

```
            1         14    293         303
SoXyn10A AESTLGAAAQSGR-----YTAVLNALNGG

CfXyn10A A-TTLKEAADGAGR-----YAAVMEAF
         1           13    305       312
```

FIG. 4. Interaction of N-terminal and C-terminal regions in SoXyn10A. (A) Structure of N-terminal and C-terminal regions in SoXyn10A. SoXyn10A consists of nine α helices (α0–8) and eight β sheets (β1–8). Interactions were observed between the α helices in the N-terminal (α0) and C-terminal (α8) regions of the catalytic domain of SoXyn10A. Specifically, the α8 helix (including L300) was observed to form a hydrophobic core in association with α0, stabilizing the structure of SoXyn10A. The hydrogen bond between N301 in α8 and E2 and Q11 in α0 surrounds the hydrophobic core, limiting the exposure of the hydrophobic amino acids on the outer surface of the molecule. This interaction appears to stabilize the structure of SoXyn10A. (B) Comparison of the amino acid sequences in the N-terminal and C-terminal regions of SoXyn10A and CfXyn10A.

quite important for the stability of the enzyme.[20] When the structure of SoXyn10A is compared with CfXyn10A, differences in the structures of these regions are observed. The catalytic domain of SoXyn10A consists of nine α helices (α0–8) and eight β sheets (β1–8). Interactions were observed in the α helices of the N-terminal (α0) and C-terminal (α8) regions of the catalytic domain of SoXyn10A consisting of four hydrogen bonds and eight hydrophobic interactions. The hydrogen bond between N301 in α8 and E2 and Q11 in α0 was observed to surround the hydrophobic core, limiting the exposure of the hydrophobic amino acids on the outer surface of the molecule. This interaction appears to stabilize SoXyn10A. In contrast, CfXyn10A does not contain the amino acids corresponding to E2, Q11, and N301 (Fig. 4B), and so CfXyn10A does not contain the hydrogen bond at the N- and C-terminal ends. When three amino acids such as NGG

[20] S. Kaneko, A. Kuno, Z. Fujimoto, S. Ito, S. Iwamatsu, H. Mizuno, T. Hasegawa, and H. Kobayashi (in press).

were added to the C-terminal end of CfXyn10A, some of the resultant chimeric enzymes became active. Therefore, care must be taken regarding the environment around the N- and C-terminal when constructing chimeric proteins.

Characterization of Chimeric Enzymes

Chimeras named FCF-C4, FCF-C10, and FCFCF-C4, 10 are the first discussed.[6,7] Bar diagrams of the constructed chimeras are given in Fig. 1B. The expressed chimeric enzymes are purified and their folded state analyzed using CD spectroscopy. Kinetic parameters for the hydrolysis of p-nitrophenyl β-D-xylobioside (PNP-X$_2$) and p-nitrophenyl β-D-xylobioside (PNP-G$_2$) are determined to investigate the effect of the replacement of modules an enzymatic activity. The effect of replacing module M10 in SoXyn10A with that of CfXyn10A is to delete one amino acid residue and to change three other amino acids in the enzyme.[6] The kinetic parameters for PNP-G$_2$ are not very different from SoXyn10A. In contrast, K_m and k_{cat} values of the module M10 replaced chimera, FCF-C10, for PNP-X$_2$ (K_m: 0.2 mM and k_{cat}: 1.9 s^{-1}) are decreased 10-fold from those of SoXyn10A (K_m: 2.0 mM and k_{cat}: 20 s^{-1}).[6] However, the effectiveness of the enzyme is not reduced at low substrate concentration.[6] Module M10 includes catalytic and substrate binding amino acid residues. E128 of SoXyn10A is an acid/base catalyst of the double displacement mechanism and N127 surrounds the -1 site in the substrate-binding cleft.[17] Because these amino acid residues are conserved in all GH10 xylanases, replacement of module M10 did not change these amino acids. As shown in Fig. 5, there are differences between SoXyn10A and CfXyn10A in the environment of module M10. The distance from the side chains of D132 of SoXyn10A to those of Y172 and N173 are 2.6 and 4.1 Å, indicating that D132 hydrogen bonds with Y172 and that Y172 also hydrogen bonds with R139, which exists in module M10.

In contrast, in CfXyn10A, the distance from D131 to Y171 is 3.6 Å, which indicates that D131 cannot form a hydrogen bond with Y171. Instead, D131 hydrogen bonds with N172, and R136 forms hydrogen bonds with Y171 and D138. In SoXyn10A, Y172 forms the (+) side of the substrate binding cleft together with N173 (Fig. 5).[17] Y172 seems to be an important residue for recognizing the phenolic rings of PNP-X$_2$ and PNP-G$_2$. The replacement of amino acids in module M10 must change the position of D132, which in turn will affect the position of Y172. Due to the change in position or orientation of Y172, it appears that the enzyme is able to form stronger interactions with phenolic rings of PNP-X$_2$ and PNP-G$_2$. Further changes in kinetic parameters occurred with the addition of

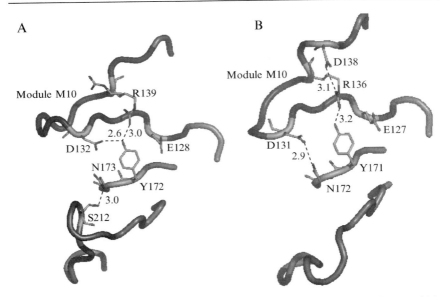

Fig. 5. The environment of module M10 in SoXyn10A and CfXyn10A. (A) SoXyn10A. (B) CfXyn10A. Hydrogen bonds are shown as dashed lines. Distances between the atoms are shown in angstroms.

Q140 and may be explained by an elimination of hydrogen bonds between R139 and Y172, which would result in a more CfXyn10A-like environment around module M10. Therefore, the kinetic properties of FCF-C10 changed from SoXyn10A to CfXyn10A-like.

When the module M4 of SoXyn10A was replaced by that of CfXyn10A, the situation was almost identical to the replacement of module M10.[7] Actually, despite only two amino acid changes occurring with the replacement of module M4, the environment around module M4 was altered, and the position of functionally important amino acids were shifted. This resulted in changes to the kinetic parameters for PNP-G_2 and PNP-X_2.[7] The K_m values of the module M4 replaced chimera, FCF-C4, for PNP-G_2 was 112 mM, almost the same value as that of SoXyn10A (97 mM).[7] The k_{cat} value of FCF-C4 for PNP-G_2 was significantly higher (at 7.0 s^{-1}) than the value of SoXyn10A (2.2 s^{-1}). The K_m and k_{cat} values of FCF-C4 for PNP-X_2 were 0.88 mM and 21 s^{-1}.[7] The k_{cat}/K_m value of FCF-C4 was unchanged relative to that of SoXyn10A.[7]

The synergistic effects of the module replacements were observed when module M10 of FCF-C4 was replaced by the corresponding module of CfXyn10A.[7] The constructed chimera, FCFCF-C4,10, displays a k_{cat}/K_m

value some 2.5-fold lower for PNP-G_2 and 2.8-fold lower for PNP-X_2 relative to the FCF-C4 chimera.[7] The kinetic parameters of the FCFCF-C4,10 chimeric enzyme were closer to those of FCF-C10 than to those obtained with FCF-C4. The K_m value for FCFCF-C4,10 toward PNP-G_2 was 46 mM, which was closer to that of FCF-C10 (64 mM) than that observed for FCF-C4 (112 mM).[7] The k_{cat} value of FCFCF-C4,10 toward PNP-G_2 (1.2 s^{-1}) was also similar to the value observed for FCF-C10 (1.8 s^{-1}).[7] The K_m value of FCFCF-C4,10 toward PNP-X_2 was 0.14 mM, close to that observed for FCF-C10 (0.24 mM) and 6.3 times less than that seen for FCF-C4 (0.88 mM).[7] The k_{cat} value obtained for FCFCF-C4,10 toward the substrate PNP-X_2 (1.2 s^{-1}) was also reduced to a level 17.5 times less than that seen with FCF-C4 (21 s^{-1}).[7] When module M10 of FCF-C4 was replaced by that of CfXyn10A, the K_m value for PNP-X_2 was also decreased from 0.88 m to 0.14 mM together with the k_{cat} value from 21 to 1.2 s^{-1}.[7] These are quite similar situations as that observed when module M10 of SoXyn10A was replaced by that of CfXyn10A. Therefore, the effect of module M10 was nearly the same for SoXyn10A and FCF-C4 so that in FCFCF-C4,10, the effects of replacement of modules M4 and M10 were almost additive.[7]

Next, the chimera named FC-14-15 will be introduced.[8] A bar diagram of the constructed chimera is shown in Fig. 2B. The border of module M14 and M15 lies along the substrate binding cleft. The substrate binding clefts of SoXyn10A and CfXyn10A are shown in Fig. 6. It is apparent that the structure of the glycon side of the subsite in GH10 xylanases is highly conserved. However, the structure of the aglycon side is very different. In SoXyn10A, the (+) side of the catalytic cleft displays variation in the upper left position of the cleft in Fig. 6, especially in the vicinity of the loop region Asn209-Pro213, which protrudes out to the end of the (+) side of the catalytic cleft compared to the other xylanases. The catalytic cleft of FC-14-15 retained the salient features of both parent enzymes. In FC-14-15, the left half of SoXyn10A in Fig. 6A was replaced by the left half of CfXyn10A in Fig. 6B.

Kinetic data were collected for FC-14-15 along with parental SoXyn10A and CfXyn10A using PNP-G_2 and PNP-X_2 as the substrates. The K_m and k_{cat} values of FC-14-15 for PNP-X_2 and PNP-G_2 were 0.22 mM and 74 s^{-1} and 13 mM and 3.4 s^{-1}, respectively.[8] These values were halfway between those of parental SoXyn10A and CfXyn10A, suggesting that the chimera inherits half its characteristics from each parental enzyme.

The topology of the substrate binding cleft of FC-14-15 was investigated.[8] The activity of the chimeric xylanase, along with SoXyn10A and CfXyn10A against different degrees of polymerization of xylooligosaccharides, is shown in Fig. 7A. All of the enzymes showed a similar trend

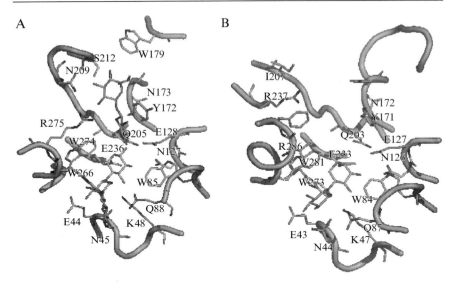

FIG. 6. Substrate binding cleft of SoXyn10A and CfXyn10A. (A) SoXyn10A. (B) CfXyn10A. The major differences between the structures are centered around the upper left region of the tertiary structures as shown.

toward xylooligosaccharide length. The hydrolysis rate for xylotriose of FC-14-15 was similar to that of SoXyn10A; however, the rates for xylooligosaccharides longer than xylotetraose were the same as those of CfXyn10A, indicating that these properties had been determined in accord to the structure of the aglycone side of the substrate binding cleft.

Bond cleavage frequencies of FC-14-15, SoXyn10A, and CfXyn10A were investigated using the substrate series methyl xylotrioside to methyl xylohexaoside (Fig. 7B).[8] The amino acids that comprise subsites −1, −2, −3, and +1 of the structure are almost completely conserved. In the case of the hydrolysis of xylotrisaccharide, the subsites −1, −2, and +1 are used to produce the major hydrolysis product. Therefore, there was not a great difference between both parental and chimeric enzymes regarding bond cleavage frequency. In contrast, there were differences in cases longer than xylotriose due to the influence of the effect of subsite +2 on the properties of the parental enzymes. In the case of the hydrolysis of xylotetraose, SoXyn10A cleaved the first and second linkages from the reducing end, whereas CfXyn10A only hydrolyzed the second linkage. The chimeric enzyme had BCF for xylotetraose more similar to that of CfXyn10A than to that of SoXyn10A. The BCFs of chimera for xylopentaose again showed similar properties to CfXyn10A.

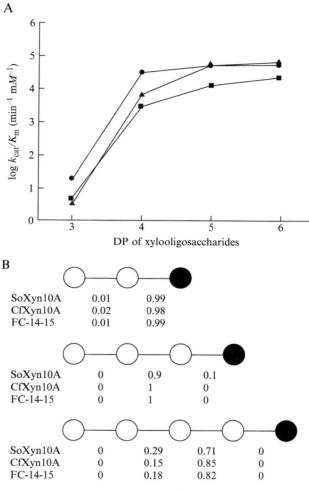

FIG. 7. Rate of hydrolysis and bond cleavage frequencies of xylooligosaccharides. (A) Rate of xylooligosaccharide hydrolysis catalyzed by CfXyn10A, SoXyn10A, and FC-14-15. (B) Bond cleavage frequencies of xylooligosaccharides catalyzed by CfXyn10A, SoXyn10A, and FC-14-15. SoXyn10A (■), CfXyn10A (●), and FC-14-15 (▲) were incubated with xylooligosaccharides of different lengths, and the rate of substrate hydrolysis was used to calculate k_{cat}/K_m in A. Xylose is indicated as open circles and methyl xyloside is indicated as closed circles in B.

Next, activities for the hydrolysis of xylan by SoXyn10A, CfXyn10A, and FC-14-15 and their hydrolysis products were examined.[8] The enzyme reaction mixtures, which produced the same levels of reducing power, were subjected to HPLC. The hydrolysis products of FC-14-15 were similar to those of CfXyn10A. However, no xylose peak was detected from the hydrolysis product of FC-14-15, whereas parental SoXyn10A and CfXyn10A both produced xylose.[8] FC-14-15 inherits half the characteristics of each parent, SoXyn10A and CfXyn10A, thereby retaining interesting combinations of properties of the parent enzymes, resulting in a low level production of xylose.

Conclusion

Using the example of xylanases, the potential of protein evolution by module shuffling is demonstrated. Our results indicate that in xylanases, modules M4 and M10, including catalytic residues and substrate binding residues, are not only related directly to enzyme activity, but also interact with adjacent modules, which are involved in forming the substrate binding cleft. The double replacement of modules M4 and M10 displayed additive effects in K_m and k_{cat} compared to replacing modules M4 and M10 individually. FC-14-15 inherited approximately half the characteristics of each parent, SoXyn10A and CfXyn10A, resulting in an interesting combination of catalytic properties.

In a family of enzymes (proteins), the important amino acids for function are generally conserved. Therefore, the important amino acids are not replaced by module shuffling. However, the replacement of modules affects the position of the important amino acids by changing the environment of adjacent modules. This results in a change of enzyme character. These aspects are often observed in well-evolved enzymes by random mutagenesis that being mutations often occurred in places not related directly to activity. Therefore, random module shuffling seems to have almost the same potential as random mutagenesis for evolving protein function. In the case of the GH10 xylanase, the catalytic domain is divided into 22 modules, therefore if only two parents are used, the number of possible combinations of modules is 2^{22}, which is the sufficient number for the gene library to select desirable properties. Thus, module shuffling can be expected to be a useful process for the evolution of proteins.

[4] SCHEMA-Guided Protein Recombination

By Jonathan J. Silberg, Jeffrey B. Endelman, and
Frances H. Arnold

Introduction

SCHEMA is a scoring function that predicts which elements in homologous proteins can be swapped without disturbing the integrity of the structure.[1] Using the structural coordinates of the parent proteins, SCHEMA identifies pairs of residues that are interacting and determines the number of interactions, E, that are broken when a chimeric protein inherits portions of its sequence from different parents. E appears to be a good metric for anticipating structural conservation when homologous proteins are recombined. Analysis of well-defined libraries of β-lactamase chimeras revealed that chimeras with low E retained function with higher probability than chimeras with the same effective level of mutation but higher E or chosen at random.[2] Another study also showed that E is a useful measure for anticipating disruption in chimeras of a larger, cofactor-containing protein, cytochrome P450.[3]

Using SCHEMA, libraries of chimeras can be compared *in silico* to determine which one is expected to contain the highest fraction of folded (and potentially interesting) sequences for laboratory evolution studies.[2–4] These libraries can be synthesized *in vitro* using site-directed recombination methods (see Fig. 1), which allow for the simultaneous recombination of two or more parents at specified locations.[2,5] This approach can be used to make chimeric libraries from any parent sequences. In addition, the sequence diversity of folded and functional chimeras encoded in the library can be controlled, i.e., the number of possible unique sequences and the average level of mutation of chimeras predicted to retain structure, can be

[1] C. A. Voigt, C. Martinez, Z. G. Wang, S. L. Mayo, and F. H. Arnold, *Nature Struct. Biol.* **9,** 553 (2002).

[2] M. M. Meyer, J. J. Silberg, C. A. Voigt, J. B. Endelman, S. L. Mayo, Z. G. Wang, and F. H. Arnold, *Protein Sci.* **12,** 1686 (2003).

[3] C. R. Otey, J. J. Silberg, C. A. Voigt, J. B. Endelman, G. Bandara, and F. H. Arnold, *Chem. Biol.* **11,** 309 (2004).

[4] D. A. Drummond, J. J. Silberg, J. B. Endelman, C. A. Wilke, and F. H. Arnold, submitted for publication.

[5] K. Hiraga and F. H. Arnold, *J. Mol. Biol.* **330,** 287 (2003).

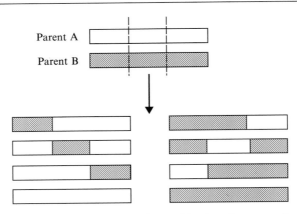

FIG. 1. Library synthesis by site-directed recombination. Sequence elements encoding structurally related polypeptides are swapped at defined locations (dashed lines) in two or more homologous proteins. This yields a library containing $h^y - h$ unique chimeras, where h is the number of parents recombined and y is the number of sequence elements that are exchanged.

used to guide the selection of crossover locations and crossover number. In contrast, annealing-based recombination or "DNA shuffling" techniques, such as Stemmer shuffling,[6,7] StEP,[8] and *in vivo* methods,[9] generate crossovers only in regions of sequence identity and therefore can not generate diverse libraries from more distant parent sequences. The sequence-independent random recombination methods now available (SHIPREC,[10] ITCHY,[11] or SCRATCHY[12]) do not make multiple crossovers efficiently and therefore create libraries of very limited diversity.

This article outlines the procedure used for calculating E for a chimera and discusses ideas for optimizing the design of combinatorial libraries for directed evolution.

[6] W. P. Stemmer, *Proc. Natl. Acad. Sci. USA* **91,** 10747 (1994).

[7] W. P. Stemmer, *Nature* **370,** 389 (1994).

[8] H. Zhao, L. Giver, Z. Shao, J. A. Affholter, and F. H. Arnold, *Nature Biotechnol.* **16,** 258 (1998).

[9] A. A. Volkov, Z. Shao, and F. H. Arnold, *Nucleic Acids Res.* **27,** e18 (1999).

[10] V. Sieber, A. Pluckthun, and F. X. Schmid, *Nature Biotechnol.* **16,** 955 (1998).

[11] M. Ostermeier, A. E. Nixon, and S. J. Benkovic, *Bioorg. Med. Chem.* **7,** 2139 (1999).

[12] S. Lutz, M. Ostermeier, G. L. Moore, C. D. Maranas, and S. J. Benkovic, *Proc. Natl. Acad. Sci. USA* **98,** 11248 (2001).

Methods

Calculating SCHEMA Disruption

Based on the structure of the parent proteins, SCHEMA determines which residues are interacting, defined as those residues within a cutoff distance, and generates a contact matrix.[1] When recombining two parents, the contacts are scaled by the sequence identity of the parents being recombined, i.e., all contacts that cannot be broken by recombination are removed from the matrix. E is determined by counting the number of contacts broken when a chimeric protein inherits portions of its sequence from different parents.

The SCHEMA disruption E of a chimeric sequence s, made by recombining sequence elements from h homologous proteins, is given by

$$E = \sum_{i=1}^{N} \sum_{j=i+1}^{N} C_{ij} P(i, j, s_i, s_j), \tag{1}$$

where N is the number of residues that have defined coordinates in the parental structure, $C_{ij} = 1$ if residues i and j are within the cutoff distance d_c (otherwise $C_{ij} = 0$), and s_i designates the parent incorporated at position i in the chimera (e.g., $s_i = 1$ if the sequence is derived from parent #1, $s_i = 2$ if derived from parent #2, etc.). $A(s_i, k)$ is the identity of the residue in parent s_i at position k within the parental amino acid sequence, and $P(i, j, s_i, s_j) = 0$ if any parent has residue $A(s_i, i)$ at position i and residue $A(s_j, j)$ at position j [otherwise $P(i, j, s_i, s_j) = 1$]. It is essential that structurally related residues in each parent are numbered identically, e.g., $A(1, k)$ and $A(2, k)$ should represent structurally related residues in each parent, to ensure that sequence identities among the parents are properly accounted for when calculating E.

Typically, we use $d_c = 4.5$ Å, and hydrogen, backbone nitrogen, and backbone oxygen atoms are excluded from the calculation of E. Small deviations from this value of d_c or the use of all atoms to calculate C_{ij} does not significantly affect the relative E of chimeras being compared, although the magnitude of E changes. When cofactor-containing proteins are recombined, contacts between the cofactor and residues in the proteins are also excluded from calculation of E.[3] In this simple model, contacts between the cofactor and the protein cannot be broken by the recombination of related proteins.

Ideally, PDB coordinates for all the parent sequences are available, and a structure-based alignment is performed. For parents whose sequences differ in length, this ensures that structurally related residues in each parent are numbered identically. This can be done using free software packages

Fig. 2. Treatment of gaps in sequence alignments. A hypothetical sequence alignment used in SCHEMA calculations is shown. PDB coordinates corresponding to the top sequence are being used by SCHEMA to calculate C_{ij}, and the bottom sequence represents a homologous protein for which no structural information is available. At position 1, the atomic coordinates of glycine are defined in the PDB file, so the gap in the second parent is treated as a mutation relative to G when computing E and m. Because there are no coordinates for position 2, it is ignored when computing E and m.

such as SwissProt or the combinatorial-extension algorithm.[13,14] If structural coordinates are available for different conformational states of the parents being recombined, it is best to assess E using the coordinates for each conformation to ensure that both states of the chimeras are likely to exhibit similar low disruption. When the structure of only one parent is available, sequence alignments can be performed using the BLAST algorithm.[15]

Often alignment of the parents requires the insertion of gaps within the primary amino acid sequence of one or more of the parents (see Fig. 2). When gaps are introduced into the parent whose structural coordinates are being used to generate the contact matrix C_{ij}, the residues found in the other parents are ignored when calculating E because there is no corresponding structural information. In contrast, when gaps occur in any parent other than the one used for structural information, they are treated like real residues that differ in identity from the residues in the other parents.

From Disruption to Probabilities

The fraction of chimeras retaining function has been found to decrease exponentially with E.[2] If we posit that any disrupted contact has a probability f_d of yielding a nonfunctional chimera and each contact acts statistically independently of the others, the fraction of chimeras at each E predicted to retain function is given by $P_f = (1 - f_d\ E/N)^N$. In this case, N equals the total number of contacts that could be broken by recombination. When N becomes large, as with proteins, this equation can be approximated by a simple exponential, $P_f = e^{-f_d E}$. This relationship between E and P_f is likely to hold for the recombination of any homologous proteins. However, functional data from different chimeric libraries may yield a range of f_d values. We have found that the sensitivity of the

[13] N. Guex and M. C. Peitsch, *Electrophoresis* **18**, 2714 (1997).
[14] I. N. Shindyalov and P. E. Bourne, *Protein Eng.* **11,** 739 (1998).
[15] S. Henikoff and J. G. Henikoff, *Proc. Natl. Acad. Sci. USA* **89,** 10915 (1992).

functional assay alone can significantly affect the P_f at each E. A more sensitive functional screen for the conservation of lactamase function, for example, identified more functional chimeras and yielded a higher value of P_f at each E (and lower f_d) than a functional selection.[2,5] We also expect f_d to depend on the protein scaffold.

The value of f_d can be calibrated rapidly for any system by analyzing folding or function of a small population of chimeras with known E.[3] From approximately 10 to 20 chimeras that encompass a broad range of disruption, we have observed that those exhibiting the highest levels of E are mostly nonfunctional and those with the lowest E are almost all functional. From such a population of chimeras, the E where chimeras exhibit a $P_f = 0.5$ (designated $E_{1/2}$) can be estimated, and the f_d can be calculated, $f_d = \ln(2)/E_{1/2}$.

SCHEMA-Guided Library Design

Because screening and selection strategies can evaluate a limited number of protein variants for altered functions, we would like to make libraries that are enriched in folded chimeras. SCHEMA can help identify such libraries by computing the fraction of chimeras F expected to retain the parental function (and fold) in different libraries arising from recombination of the same parents at different crossover locations. The fraction of folded chimeras in a library is given by

$$F = \frac{1}{n}\sum_{i=1}^{n} e^{-f_d E_i}, \tag{2}$$

where n is the number of unique chimeras present.

Libraries with the highest possible F are not necessarily preferred for directed evolution. An additional issue to consider when choosing a library is the sequence diversity D of the functional chimeras in that library. The D of a library describes the average mutation level m of folded chimeras in that library, where m is the amino acid Hamming distance of each chimera to its closest parent. D is calculated from

$$D = \frac{1}{F}\frac{1}{n}\sum_{i=1}^{n} m_i e^{-f_d E_i}. \tag{3}$$

Unfortunately, we know little about the effect of m on the evolution of function. Studies examining the effect of m on the acquisition of novel functional properties in laboratory evolution studies suggest that variants with higher m are more likely to exhibit altered functional properties, whereas those with lower m tend to be more similar to the parents.[3,16,17]

In the case of cytochrome P450 chimeras, chimeras exhibiting altered substrate specificity had an average m of 34, and chimeras that displayed a parent-like substrate specificity had an average m of 22.[3] Because E also depends on m, there is a trade-off between F and D. Chimeras with low E on average also have low m; therefore, libraries that simply maximize F have limited diversity.

If one can enumerate all possible libraries, then the library design problem is easy: simply choose the one with the desired combination of D and F. Exhaustive enumeration, however, is not feasible with today's computers for average-sized proteins and the library sizes appropriate for directed evolution (thousands or more chimeras). Consider the design of a 10 crossover library between the β-lactamases PSE-4 and TEM-1, which have similar overall structure, similar length ($\sim$265 residues), and $\sim$40% sequence identity. There are 10^{15} possible libraries, each with $2^{11} = 2048$ unique chimeric sequences—too many for exhaustive enumeration. In this case, one can make progress by evaluating D and F for thousands, or even millions of random libraries. Figure 3 shows the F and D for 50,000 randomly chosen libraries. Here we observe libraries with diversities ranging from 7 to 41 (mutations per folded sequence) and F values from 0.5 to 7%. With SCHEMA we can identify the library with the highest F at a particular level of D. For the random sample in Fig. 3, this increases the number of folded chimeras up to fivefold compared to the average for libraries with the same diversity.

The random enumeration in Fig. 3 sampled a limited region of the (F, D) plane. Alternate enumeration schemes, e.g., enforcing minimum and maximum fragment sizes, can rapidly explore different combinations of F and D. Choosing different parents will also affect which regions of the (F, D) plane are accessible, even if the parents exhibit similar levels of sequence identities. Decreasing the sequence identity of the parents will, in general, lead to libraries with higher D but lower F values. The same trend is expected if one increases the number of parents recombined.

Practical Considerations in Library Design

The best methods available for creating the libraries identified by SCHEMA as enriched in folded chimeras are sequence-independent site-directed chimeragenesis and chemical synthesis.[2,5] These techniques can recombine parents with any level of sequence identity at multiple sites and easily create libraries encoding thousands of chimeras. However, these

[16] M. Zaccolo and E. Gherardi, *J. Mol. Biol.* **285,** 775 (1999).
[17] P. S. Daugherty, G. Chen, B. L. Iverson, and G. Georgiou, *Proc. Natl. Acad. Sci. USA* **97,** 2029 (2000).

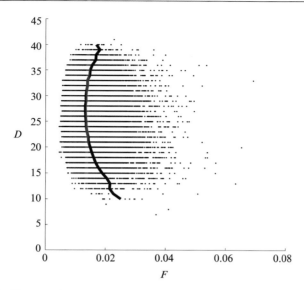

Fig. 3. SCHEMA analysis of computed libraries. Randomly chosen libraries made by allowing 10 crossovers between β-lactamases TEM-1 and PSE-4 were compared. The crossover locations in each library were enumerated by choosing 10 distinct, random integers between 1 and 153 (the length of the structural alignment minus the number of conserved residues minus 1). Each integer k represents a crossover between k^{th} and $(k + 1)^{th}$ nonidentical residues. This was repeated until 50,000 ten-integer sequences were generated. The structural coordinates of PSE-4 (PDB = 1G68)[18] were used to compute E for every chimera present in each library. F and D (rounded to the nearest integer) were computed for each library according to Eqs. (2) and (3) with $f_d = 0.095$.[2] The average F at each value of D is shown as a solid line.

methods are limited in where they can recombine distantly related proteins without introducing amino acid sequence changes not found in either parent. This happens because both techniques generate chimeras by ligating double-stranded gene modules together (encoding each swapped polypeptide), which requires that parents exhibit 2 to 4 bp of identity at the crossover boundaries. When the parents do not exhibit sufficient identity at the desired crossover locations, synonymous mutations can often be introduced to allow recombination at that site. If synonymous mutations are not sufficient, then other crossover positions should be chosen for the library. SCHEMA only calculates the disruption arising from recombination, not from mutation.

[18] D. Lim, F. Sanschagrin, L. Passmore, L. De Castro, R. C. Levesque, and N. C. Strynadka, *Biochemistry* **40,** 395 (2001).

Program Availability

Software for running SCHEMA calculations is available at the Arnold group web site at the California Institute of Technology (http://www.che.caltech.edu/groups/fha/).

[5]   Staggered Extension Process *In Vitro* DNA Recombination

By Huimin Zhao

Introduction

In vitro DNA recombination is an extremely powerful approach for the directed evolution of proteins and nucleic acids. Unlike random mutagenesis methods in which point mutations are introduced randomly into a single parent sequence to produce a library of progeny sequences, DNA recombination methods entail the block-wise exchange of genetic variations among multiple parent sequences created in the laboratory or existing in nature to produce a library of chimeric progeny sequences. The key advantage of DNA recombination is its ability to accumulate beneficial mutations while simultaneously removing deleterious mutations, which may greatly accelerate the evolution of a protein or nucleic acid molecule of interest toward a specific function. It was demonstrated in computational simulation studies that DNA recombination plays a critical role in the evolution of biological systems.[1] In the past decade, *in vitro* DNA recombination has been used successfully to alter and engineer many types of protein function, such as stability, activity, affinity, selectivity, substrate specificity, and protein folding/solubility.[2,3]

The first described *in vitro* DNA recombination method, or "DNA shuffling," was developed by Stemmer in 1994,[4,5] in which DNA fragments generated by the random digestion of parent genes with DNase I are combined and reassembled into full-length chimeric progeny genes in a polymerase chain reaction (PCR)-like process. Since then, a number of *in vitro* DNA recombination methods have been described,[6] such as

[1] S. Forrest, *Science* **261,** 872 (1993).
[2] O. Kuchner and F. H. Arnold, *Trends Biotechnol.* **15,** 523 (1997).
[3] C. Schmidt-Dannert, *Biochemistry* **40,** 13125 (2001).
[4] W. P. Stemmer, *Proc. Natl. Acad. Sci. USA* **91,** 10747 (1994).
[5] W. P. Stemmer, *Nature* **370,** 389 (1994).

staggered extension process (StEP) recombination,[7] random-priming recombination (RPR),[8] random chimeragenesis on transient templates (RACHITT),[9] degenerate homoduplex recombination (DHR),[10] and synthetic shuffling.[11] This article describes the method, protocol, and applications of StEP recombination. For additional technical discussions on the same topic, interested readers are referred elsewhere.[12,13]

Principle of the StEP Method

StEP recombination is based on template switching during polymerase-catalyzed primer extension.[7] As illustrated in Fig. 1, the StEP method uses full-length genes as templates for the synthesis of chimeric progeny genes. It consists of priming denatured templates, followed by repeated cycles of denaturation and extremely short annealing/extension steps. Recombinogenic events occur when the partially extended primers anneal randomly to different templates (template-switching events) based on sequence complementarity and extend further. StEP is continued until full-length genes are formed. If the product yield is low, the full-length chimeric genes can be amplified in a standard PCR. Compared to DNA shuffling and other polymerase-based recombination methods that require fragmentation or chemical synthesis of fragments, the StEP method is much simpler and less labor intensive and can be performed using a pair of flanking primers in a single PCR tube. It is noteworthy that the StEP method is somewhat similar to the process that retroviruses, including HIV, use to evolve their genomes.[14]

The recombination efficiency of the StEP method was compared to the most widely used DNA shuffling method using a green fluorescent protein (GFP)-based recombination test system.[15] A series of truncated GFP

[6] H. Zhao and W. Zha, *in* "Enzyme Functionality: Design, Engineering and Screening" (A. Svendsen, ed.), p. 353. Dekker, New York, 2003.

[7] H. Zhao, L. Giver, Z. Shao, J. A. Affholter, and F. H. Arnold, *Nature Biotechnol.* **16,** 258 (1998).

[8] Z. Shao, H. Zhao, L. Giver, and F. H. Arnold, *Nucleic Acids Res.* **26,** 681 (1998).

[9] W. M. Coco, W. E. Levinson, M. J. Crist, H. J. Hektor, A. Darzins, P. T. Pienkos, C. H. Squires, and D. J. Monticello, *Nature Biotechnol.* **19,** 354 (2001).

[10] W. M. Coco, L. P. Encell, W. E. Levinson, M. J. Crist, A. K. Loomis, L. L. Licato, J. J. Arensdorf, N. Sica, P. T. Pienkos, and D. J. Monticello, *Nature Biotechnol.* **20,** 1246 (2002).

[11] J. E. Ness, S. Kim, A. Gottman, R. Pak, A. Krebber, T. V. Borchert, S. Govindarajan, E. C. Mundorff, and J. Minshull, *Nature Biotechnol.* **20,** 1251 (2002).

[12] A. M. Aguinaldo and F. H. Arnold, *Methods Mol. Biol.* **192,** 235 (2002).

[13] A. A. Volkov and F. H. Arnold, *Methods Enzymol.* **328,** 447 (2000).

[14] W. S. Hu, E. H. Bowman, K. A. Delviks, and V. K. Pathak, *J. Virol.* **71,** 6028 (1997).

[15] A. A. Volkov, Z. Shao, and F. H. Arnold, *Nucleic Acids Res.* **27,** e18 (1999).

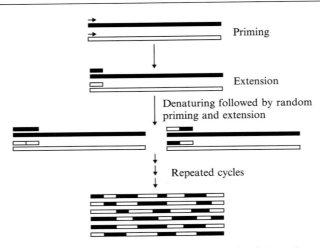

Fɪɢ. 1. Principle of the StEP recombination method. For simplicity, only one primer and single strands from two DNA templates (open and solid blocks) are shown. During priming, oligonucleotide primers anneal to the denatured templates. Short fragments are produced by brief polymerase-catalyzed primer extension. During subsequent random annealing-abbreviated extension cycles, fragments randomly prime the template (template switching) and extend further. The process is repeated for many cycles until the full-length genes are produced. The full-length chimeric genes can be amplified in a standard PCR (optional).

variants containing stop codon mutations that are nonfluorescent were created by site-directed mutagenesis at selected positions along the GFP gene. Recombination between truncated GFP variants generates the full-length wild-type gene and restores fluorescence. The percentage of fluorescent host *Escherichia coli* colonies indicates the recombination frequency or efficiency between two stop codon mutations of a given distance. As summarized in Table I, the StEP method and the DNA shuffling method are equally efficient. With DNase I fragmentation, using small fragments (<100 bp) yields a slightly higher efficiency than large fragments (100–200 bp).

Materials

DNA templates containing the target sequences to be recombined
Oligonucleotide primers
Taq DNA polymerase and its 10× reaction buffer: 500 mM KCl, 100 mM Tris–HCl, pH 8.3 (Promega, Madison, WI)
25 mM MgCl$_2$

TABLE I
COMPARISON OF StEP RECOMBINATION AND DNA SHUFFLING[15]

Distance between mutations (bp)	Fraction of fluorescent colonies (%)[a]		
	StEP	DNA shuffling (<100-bp fragments)	DNA shuffling (100- to 200-bp fragments)
423	18.5	20.5	19.2
315	13.1	14.5	9.7
207	9.8	11.5	8.3
99	8.2	9.6	8.4
24	4.8	5.8	5.1

[a] Percentage of fluorescent *E. coli* colonies obtained by recombining two green fluorescent protein templates containing stop codon mutations.

10× dNTP mix: 2 mM of each dNTP (Roche Diagnostics, Indianapolis, IN)

Agarose gel electrophoresis supplies and equipment

MJ PTC-200 thermocycler (MJ Research Inc., Watertown, MA)

QIAEX II gel extraction kit (QIAgen, Valencia, CA)

*Dpn*I restriction endonuclease (20 U/μl) and 10× supplied reaction buffer (New England Biolabs, Beverly, MA)

Experimental Approach

1. Prepare DNA template. Appropriate templates include plasmids carrying target sequences, cDNA or genomic DNA carrying the target sequences, sequences excised by restriction endonucleases, and PCR-amplified sequences.

2. Combine 5 μl of 10× *Taq* buffer, 5 μl of 10× dNTP mix (2 mM of each dNTP), 1.5 mM MgCl$_2$, 1–20 ng total template DNA, 30–50 pmol of each primer, sterile dH$_2$O, and 2.5 U *Taq* DNA polymerase in a total volume of 50 μl.

3. Run 80–100 extension cycles using the following program: 94° for 30 s (denaturation) and 55° for 5–15 s (annealing/extension).

4. Run a small aliquot (5–10 μl) of the reaction on an agarose gel. Possible reaction products are full-length amplified sequences, a smear, or a combination of both. If a discrete band with sufficient yield for subsequent cloning is obtained after the StEP reaction, no additional amplification step is needed. Proceed to step 8.

5. (Optional) If parent templates were isolated from a *dam* methylation-positive *E. coli* strain (e.g., DH5α, XL1-Blue), the products from extension reactions can be incubated with *Dpn*I endonuclease to remove parent DNA so as to reduce the background of nonchimeric genes. Combine 2 μl of the StEP reaction, 1× *Dpn*I reaction buffer, 6 μl of sterile dH₂O, and 1 μl of *Dpn*I restriction endonuclease. Incubate at 37° for 1 h.

6. Amplify the target sequence in a standard PCR. Combine 1 μl of the StEP reaction, 0.3–1.0 μ*M* of each primer, 10 μl of 10× *Taq* buffer, 1.5 m*M* MgCl₂, 10 μl of 10× dNTP mix (2 m*M* of each dNTP), and 2.5 U *Taq* DNA polymerase in a total volume of 100 μl. Run the PCR reaction using the following program: 96° for 2 min, 25 cycles of 30 s at 94°, 30 s at 55°, and 60 s at 72° for each 1 kb in length. The final step of elongation is at 72° for 7 min.

7. Run a small aliquot of the reaction mixture (5–10 μl) on an agarose gel. In most cases, a clear, discrete band of the correct size among a smear should be obtained.

8. Purify the product of correct size using the QIAEX II gel purification kit. Digest the product with the appropriate restriction endonucleases and ligate into the desired cloning vector.

Notes

1. *Primer design.* Primer design should follow standard criteria, including similar melting temperatures and elimination of self-complementarity or complementarity of primers to each other. Free computer programs such as Primer3 at Biology Workbench (http://workbench. sdsc.edu) can be used to design primers. Typically, primers should also include unique restriction sites for subsequent directional subcloning.

2. *Choice of a DNA polymerase.* The key to successful recombination by StEP is to tightly control the polymerase-catalyzed DNA extension. Too much extension during each StEP cycle will severely limit recombination events. Thermostable DNA polymerases currently used in DNA amplification are often very fast. Even very brief cycles of denaturation and annealing provide time for these enzymes to extend primers for hundreds of nucleotides. For example, extension rates of *Taq* DNA polymerase at various temperatures are: 70°, >60 nucleotides/s; 55°, ~24 nucleotides/s; 37°, ~1.5 nucleotides/s; 22°, ~0.25 nucleotides/s.[16] Thus, it is

[16] M. A. Innis, K. B. Myambo, D. H. Gelfand, and M. A. Brow, *Proc. Natl. Acad. Sci. USA* **85**, 9436 (1988).

not unusual for the full-length gene product to appear after only 10–15 cycles. Unfortunately, the faster the full-length gene product appears in the extension reaction, the lower the recombination frequency due to the fewer number of the template switching events. To increase the recombination frequency, various measures should be taken to minimize the time spent in each StEP cycle, including selecting a faster thermocycler, reducing the reaction volume, and using smaller PCR tubes with thin walls.

Alternatively, thermostable DNA polymerases with proofreading activity can be used. It was reported that the proofreading activity of high-fidelity DNA polymerases can significantly slow down their extension rates.[17] For example, *Vent* DNA polymerase has an extension rate of 1000 nucleotides/min and a processivity of 7 nucleotides/(initiation event) as compared to >4000 nucleotides/min and 40 nucleotides/(initiation event) for *Taq* DNA polymerase at a certain extension temperature. In addition, use of these alternative polymerases is highly recommended during DNA amplification to minimize the mutagenic rate of point mutations. Commercially available thermostable DNA polymerases with proofreading activity include *Pfu* DNA polymerase (Stratagene, La Jolla, CA), *Vent* DNA polymerase (New England Biolabs, Beverly, MA), and *Pfx* DNA polymerase (Invitrogen Life Technolgies, Carlsbad, CA). When setting up reactions with these polymerases, it is very important to add the polymerase last, as in the absence of dNTPs, the 3′ to 5′ exonuclease activity of the polymerase can degrade DNAs.

3. *Choice of annealing/extension temperatures and times.* As a general rule, the annealing temperature should be a few degrees lower than the melting temperature of the primers. The annealing temperature should be decreased when higher recombination frequency is required or when templates have low GC content. However, it should not be reduced too much in order to minimize nonspecific annealing events. The annealing/extension times are chosen based on the desired recombination frequency. Both shorter extension times and lower annealing temperatures will increase the recombination frequency. The number of the annealing/extension cycles is determined by the size of the full-length gene product.

4. *Extension products.* The progress of the StEP reaction can be monitored by taking aliquots of the reaction mixture at various time points and separating the DNA fragments by agarose gel electrophoresis. The appearance of the extension products may depend on the specific

[17] M. S. Judo, A. B. Wedel, and C. Wilson, *Nucleic Acids Res.* **26,** 1819 (1998).

sequences recombined or the type of templates used. Small templates will likely show gradual accumulation of the full-length gene products with an increasing number of cycles. For example, during StEP recombination of two subtilisin E genes ($\sim$1 kb), the average size of the extension products increases gradually with increasing cycle number: 100 bp after 20 cycles, 400 bp after 40 cycles, 800 bp after 60 cycles, and a clear discrete band around 1 kb (the desired size) after 80 cycles.[7] However, using large templates such as whole plasmids and long genes may result in nonspecific annealing of primers and their extension products throughout the templates. Although it may appear as a smear on the agarose gel, the increase of the size of their extension products may not be so obvious.

5. *PCR amplification.* If the PCR amplification reaction is not successful, i.e., no discrete band with sufficient yield is produced, repeat amplification using serial dilutions of the StEP reaction mixture: 1:10 dilution, 1:20 dilution, and 1:50 dilution. Run small aliquots of the amplified products on an agarose gel to determine the yield and quality of amplification. Select the reaction with a higher yield and lower amount of nonspecific products for subsequent cloning. An alternative solution is to use nested internal primers separated by 50–100 bp from the original primers to amplify the target sequences.

Applications of the StEP Method

The StEP method has been used successfully to recombine gene variants created from random mutagenesis and naturally occurring homologous genes that are approximately 80% identical. For example, the StEP method was used to increase the temperature optimum of subtilisin E by 18° over that of the wild-type enzyme, essentially converting a mesophilic enzyme into its thermophilic counterpart.[18] The substrate specificity of biphenyl dioxygenase was broadened by StEP recombination of two homologous *bphA* genes encoding *Burkholderia cepacia* LB400 biphenyl dioxygenase and *Pseudomonas pseudoalcaligenes* KF707 biphenyl dioxygenase, respectively.[19] Unlike the two parental dioxygenases, which preferentially recognize either *ortho*-(LB400) or *para*-(KF707) substituted polychlorinated biphenyls, the evolved variants can degrade both congeners to the same extent.

[18] H. Zhao and F. H. Arnold, *Protein Eng.* **12,** 47 (1999).
[19] F. Bruhlmann and W. Chen, *Biotechnol. Bioeng.* **63,** 544 (1999).

By combining StEP recombination and DNA shuffling based directed evolution and structure-based rational design, Altamirano and co-workers[20] engineered a novel function in an α/β barrel enzyme by completely converting the activity of indole-3-glycerol-phosphate synthase (IGPS) to that of phosphoribosylanthranilate isomerase (PRAI). A structure-based design was used to modify the IGPS α/β barrel by incorporation of the basic design of the loop system of PRAI, yielding a chimeric variant with very low PRAI activity. DNA shuffling, StEP recombination, and genetic selection were then used to increase the PRAI activity, which led to the creation of a variant exhibiting sixfold higher activity than wild-type PRAI and no IGPS activity. Other applications of StEP recombination include the alteration of the regioselectivity of a *Bacillus* α-galactosidase,[21] the thermostabilization of cellulosomal endoglucanase EngB by recombining its gene with a homologous gene encoding the noncellulosomal endoglucanase EngD,[22] and the improvement of the protein expression level as well as the enzyme activity of a fugal laccase in *Saccharomyces cerevisiae*.[23]

Concluding Remarks

The recombination efficiency of the StEP method is similar to that of the most widely used *in vitro* gene recombination method, DNA shuffling. However, the StEP method does not require DNA fragmentation and can be carried out in a single tube. Thus, the simple and efficient StEP recombination method represents a powerful tool that can be applied to the directed evolution of genes, operons, and metabolic pathways for specific applications.

[20] M. M. Altamirano, J. M. Blackburn, C. Aguayo, and A. R. Fersht, *Nature* **403,** 617 (2000).
[21] M. Dion, A. Nisole, P. Spangenberg, C. Andre, A. Glottin-Fleury, R. Mattes, C. Tellier, and C. Rabiller, *Glycoconj J.* **18,** 215 (2001).
[22] K. Murashima, A. Kosugi, and R. H. Doi, *Mol. Microbiol.* **45,** 617 (2002).
[23] T. Bulter, M. Alcalde, V. Sieber, P. Meinhold, C. Schlachtbauer, and F. H. Arnold, *Appl. Environ. Microbiol.* **69,** 987 (2003).

[6] Using Incremental Truncation to Create Libraries of Hybrid Enzymes

By Alexander R. Horswill, Todd A. Naumann, and Stephen J. Benkovic

The creation of genetic diversity through mutagenesis and selection is essential for organisms to adapt to changing environments. In the laboratory, these processes can be enhanced and accelerated to evolve proteins for specific functions. The most commonly used method of protein engineering is DNA shuffling and the numerous variations of this technique.[1-4] However, all of these techniques require homology or DNA recombination that biases the crossovers to regions of high homology, limiting the sequence space that can be explored.[5] In our laboratory, an alternative method was developed to bypass the requirement for homologous recombination. This methodology is based on incremental gene truncation, where libraries of random length gene fragments are created using a combinatorial approach.

We have successfully used incremental truncation to convert monomeric enzymes into heterodimers and to create libraries of hybrid enzymes.[6,7] These methods are generally referred to as incremental truncation for the creation of hybrid enzymes (ITCHY). The creation of single-crossover hybrid enzymes using ITCHY is the focus of this article. A variant of the ITCHY method, termed THIO-ITCHY, is also described.

Time-Dependent ITCHY

Overview

This article presents the simplest scenario, where two genes of interest are selected and cloned into an expression vector in tandem, with a unique restriction site (R1) situated between them (Fig. 1). The goal is to generate a library of plasmids containing crossovers at every position between the

[1] A. Crameri *et al.*, *Nature* **391**, 288 (1998).
[2] J. E. Ness *et al.*, *Nature Biotechnol.* **20**, 1251 (2002).
[3] W. P. Stemmer, *Proc. Natl. Acad. Sci. USA* **91**, 10747 (1994).
[4] W. P. Stemmer, *Nature* **370**, 389 (1994).
[5] S. Lutz and S. J. Benkovic, *Curr. Opin. Biotechnol.* **11**, 319 (2000).
[6] M. Ostermeier *et al.*, *Proc. Natl. Acad. Sci. USA* **96**, 3562 (1999).
[7] M. Ostermeier *et al.*, *Nature Biotechnol.* **17**, 1205 (1999).

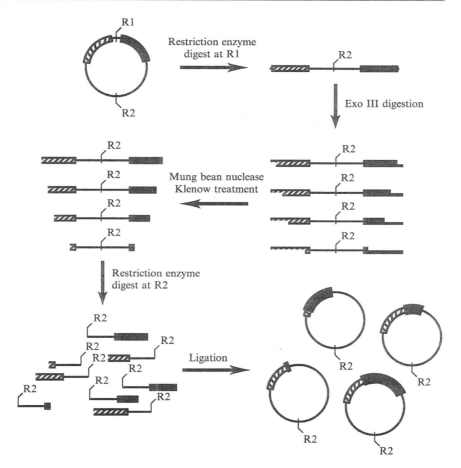

FIG. 1. Schematic of time-dependent ITCHY. Two gene fragments are cloned in tandem into an expression vector, separated by a unique restriction site (R1). Plasmid DNA is linearized at the R1 site, and the ITCHY method is performed as described in the text. The Exo III digestions are desynchronized by digestion with restriction enzyme R2, and the DNA fragments are ligated to generate a library of hybrid enzymes.

two genes. When constructing the plasmid, the R1 restriction site must generate an exonuclease III (Exo III)-sensitive 5′ overhang or blunt end, as 3′ overhangs are resistant to digestion.[8]

[8] M. Ostermeier et al., in "Protein-Protein Interactions: A Molecular Cloning Manual" (E. A. Golemis, ed.), p. 507. Cold Spring Harbor Laboratory Press, Plainview, NY, 2002.

To start the ITCHY method, the plasmid is linearized with restriction enzyme R1 (Fig. 1) and truncated incrementally using Exo III. The rate of digestion of DNA with Exo III at 37° is ~500 bp/min,[7] which is too rapid for the construction of complete libraries. To gain kinetic control, the rate of Exo III digestion is modulated by the addition of NaCl and lowered temperatures,[9] which allows aliquots to be removed and quenched at defined time points. At 22°, we have found that the rate of digestion (bp/min) is equal to $47.9 \times 10^{(-0.00644 \times N)}$, where N is the millimolar concentration of NaCl in the range of 0–150 mM.[8] Using this empirical relationship, Exo III digests DNA at a rate of ~10 bp/min in 100 mM NaCl. For each experiment, the amount of NaCl can be adjusted, depending on the size of the DNA fragments being truncated. After Exo III digestion, the single-stranded DNA ends are removed with mung bean nuclease and Klenow fragment, and the DNA is digested with a second restriction enzyme (R2) to desynchronize the Exo III deletions. Finally, ligation of the resultant gene fragments is performed and the ligated pool is transformed into a suitable *Escherichia coli* host. Transformants are screened by colony polymerase chain reaction (PCR) or directly selected as appropriate for the particular enzyme library.

Materials

Stock Solutions and Reagents

> Restriction enzyme reagents: 10× buffers and 10 mg/ml bovine serum albumin (BSA)
>
> Exonuclease buffer (10× stock solution): 660 mM Tris–HCl, pH 8.0, 6.6 mM MgCl$_2$ NaCl: 100 mM sodium chloride solution
>
> Mung bean nuclease buffer (10× stock solution): 500 mM sodium acetate, pH 5.0, 300 mM NaCl, 10.0 mM ZnSO$_4$
>
> Klenow buffer (10× stock solution): 100 mM Tris–HCl, pH 7.5, 50 mM MgCl$_2$
>
> Deoxynucleotide triphosphates (dNTPs): 2.0 mM stock solution
>
> Ammonium acetate solution: 7.5 M solution
>
> Ethanol solutions: 100 and 70% solutions
>
> Ligase buffer (10× stock solution): 300 mM Tris–HCl, pH 7.8, 100 mM MgCl$_2$, 100 mM dithiothreitol, 10 mM ATP
>
> TAE buffer (1× stock solution): 40 mM Tris–acetate, pH 8.0, 1 mM ethylenediaminetetraacetic acid (EDTA)
>
> Agarose gels in 1× TAE buffer [SeaKem GTG agarose from Bio Whittaker (Rockland, ME) with 0.5 μg/ml ethidium bromide]

[9] J. F. Tomb and G. J. Barcak, *Biotechniques* **7**, 932 (1989).

Taq DNA polymerase buffer (10× stock solution): 100 mM Tris–HCl,
pH 9.0, 500 mM KCl, 12.5 mM MgCl$_2$, 1% Triton X-100
Oligonucleotide primer: 10 μM stock solution
α-Phosphothioate (αS)-dNTPs stock: 100 mM stock solution from
Promega (Madison, WI).

Growth Media

Luria Bertani (LB) broth: Mix 10 g tryptone, 5 g yeast extract, 5 g
NaCl per liter, pH to 7.0 and autoclave
LB agar: Mix LB broth components with 15 g agar per liter and
autoclave
SOC broth: Mix 20 g tryptone, 5 g yeast extract, 0.5 g NaCl, 0.19 g
KCl per liter, pH to 7.0 and autoclave. After cooling, add MgCl$_2$ to
10 mM, MgSO$_4$ to 10 mM, and glucose to 0.5%
Antibiotic stock solutions.

Enzymes

Restriction enzymes, mung bean nuclease, and DNA polymerase I
(Klenow fragment) are all from New England BioLabs (Beverly, MA).
Exonuclease III, T4 DNA ligase, and Taq DNA polymerase are from
Promega.

Kits and Equipment

DNA purification kits are all from Qiagen (Valencia, CA). For plasmid
preps, the Spin Miniprep or Midi Prep kits are used. For agarose gel
extraction, the QIAquick gel extraction kit is used, and for routine DNA
purification, the QIAquick PCR purification kit is used. Standard electrophoresis equipment is employed for running agarose gels. A spectrophotometer, electroporation device, and PCR machine are also necessary for
the procedure.

Protocol

Exonuclease Digestion

1. Purify plasmid DNA with a Qiagen Midi Prep kit.
2. Linearize 10 μg of plasmid at the unique restriction site (R1) and
purify the DNA using the Qiagen QIAquick PCR purification kit.
Calculate the DNA concentration using a spectrophotometer.[10]

[10] J. Sambrook and D. W. Russell, "Molecular Cloning: A Laboratory Manual." Cold Spring
Harbor Laboratory Press, Cold Spring Harbor, NY, 2001.

3. To a microcentrifuge tube (tube A) add 2 μg of DNA, 6 μl of 10× Exo III buffer, NaCl, and water to 60 μl. The amount of NaCl to add should be calculated using the relationship described previously. Equilibrate this mixture at 22°.

4. Place 300 μl of PB buffer (from Qiagen QIAquick PCR purification kit) in a separate microcentrifuge tube (tube B) and keep on ice.

5. At time zero, add 200 units of Exo III to tube A and mix immediately by pipetting. Beginning at time = 30 s, remove 1-μl samples every 30 s, add samples to tube B, and mix well after each addition.

6. When all the samples have been taken, purify the Exo III digest using the Qiagen QIAquick PCR purification kit. Elute the DNA with 80 μl of EB buffer.

7. To the Exo III-digested product, add 10 μl of 10× mung bean nuclease buffer, mung bean nuclease (5 units per μg DNA), and water to 100 μl. Mix thoroughly and incubate at 30° for 30 min. Stop the reaction and purify the DNA as described earlier.

8. To the eluted DNA add 10 μl of 10× Klenow buffer, Klenow fragment (1 unit per μg DNA), and water to 95 μl. Mix thoroughly and incubate at 37° for 5 min. Then add 5 μl of dNTPs (2 mM stock, 100 μM final concentration), mix thoroughly, and incubate at 37° for 10 min. Stop the reaction and purify the DNA as described earlier.

9. Finally, digest the DNA with restriction enzyme R2 to desynchronize the Exo III deletions.

Ligation and Transformation

1. Concentrate DNA samples by ethanol precipitation by adding one-half volume ammonium acetate and two volumes ice-cold 100% ethanol. Incubate on ice or −20° for 30 min and pellet at 12,000g for 10 min. Wash the pellet with 70% ethanol and pellet at 12,000g for 5 min.

2. Prepare a 20-μl ligation: 2 μl 10× T4 ligase buffer, DNA product, 6 Weiss units T4 DNA ligase. Incubate at 16° for >12 h.

3. Remove salt from ligation mixtures by ethanol precipitation.

4. Thaw 50 μl of *E. coli*-competent cells on ice and add 20 ng of ligated DNA. Add the mixture to a chilled cuvette and perform the electroporation according to the manufacturer's instructions.

5. Recover cells with 1 ml of SOC medium and grow for 1 h at 37°.

6. Remove 10 μl of library transformation and dilution plate onto LB agar to calculate the library size.

7. Plate the rest of the cells on a LB agar library plate (245 × 245 mm) with appropriate antibiotics and incubate at 37° overnight.

Determining Library Quality

1. Colony PCR is usually performed at this stage to determine library quality. Prepare a master mix for the PCR at the following concentrations: $1\times$ *Taq* DNA polymerase buffer, 200 μM dNTPs, 1 μM primer A, 1 μM primer B, and *Taq* (0.1 units/μl). Usually 25-μl reactions are prepared, which yield enough DNA for agarose gel electrophoresis and DNA sequencing. Pick colonies with sterile toothpicks and mix into individual reactions. Typical PCR machine conditions: 30 cycles at 94° denaturation, 53° annealing, and 72° extension (1 min per kb).

2. Estimate library quality by agarose gel electrophoresis of PCR reactions.

3. Determine crossover distribution by sequencing the DNA of purified PCR samples.

4. If the library size and quality are sufficient, perform the desired screen or selection for enzyme activity as appropriate.

Considerations

Complete restriction digestion of the starting plasmid is important. If insufficient restriction enzyme is used, a large portion of the library will remain untruncated. However, excess restriction enzyme should also be avoided due to the potential for star activity.

The mung bean nuclease step can require adjustments. A simple approach to check this step is to try a control experiment with varying enzyme levels (0 to 20 units). With insufficient mung bean nuclease, DNA bands will appear undigested due to the single-strand overhangs left by Exo III, and overdigestion will result in degradation of even double-stranded DNA.

Escherichia coli DH5α-E cells have typically been used for the electroporation and storage of libraries. This *E. coli* strain is an efficient and stable cloning host and has provided consistently good results with this protocol.

For an alternative to Qiagen kits, each of the DNA purification steps can be performed using other methods. A number of techniques are available for the purification of plasmid DNA.[10] For the Exo III, mung bean nuclease, Klenow reactions, each of these enzymes can be stopped by heat inactivation (65° for 20 min), and the DNA products are purified by dialysis on 0.05-μm VM nitrocellulose membranes (Millipore, Bedford, MA).[11] The sample volumes will change during dialysis, which must be accounted for in any subsequent steps. Dialysis can also be used to desalt the ligation mixture prior to electroporation.

---

[11] I. Y. Goryshin *et al.*, *Nature Biotechnol.* **18**, 97 (2000).

THIO-ITCHY

Overview

One drawback to ITCHY is the need to take numerous aliquots in a defined time frame. To bypass this problem, another method for creating hybrid enzymes was developed, named THIO-ITCHY. Once again, two genes are cloned in tandem with a unique restriction site in between them (Fig. 2). The plasmid is linearized with a unique restriction enzyme (R1) and gel is extracted to remove any residual contaminants. Next, the linearized plasmid is PCR amplified with dNTPs that have been spiked with αS-dNTPs. DNA polymerases, such as *Taq*, can efficiently incorporate αS-dNTPs into new templates,[12] but Exo III cannot digest DNA containing αS-dNMPs, which allows for deletions of random lengths to be generated. By varying the ratio of dNTPs to αS-dNTPs in the PCR reaction, the extent of Exo III digestion can be controlled carefully so as to provide a uniform set of incrementally smaller deletions. In the PCR reaction, the primer binding sites (A and B) should be positioned close to restriction site R1, as shown in Fig. 2, in order to expose the correct DNA ends of the plasmid for Exo III digestion. After Exo III treatment, the rest of the procedure is identical to time-dependent ITCHY (see earlier discussion).

Method

Template Preparation and Exo III Digestion

1. Purify the plasmid DNA using the Qiagen Spin Miniprep kit. One microgram is digested with restriction enzyme R1 (see Fig. 2), and the digested DNA is loaded onto an agarose gel. The DNA is separated by gel electrophoresis, cut out of the gel, and extracted using the QIAquick gel extraction kit.

2. Prepare three 50-μl PCR reactions with the purified DNA sample as the template. Final concentrations in the reactions are as follows: $1\times$ *Taq* buffer, 1 μM primer A, 1 μM primer B, 10 ng of purified DNA template, dNTPs, and the appropriate amount of water. To reaction #1 add dNTPs to 200 μM, to reaction #2 add dNTPs to 175 μM and αS-dNTPs to 25 μM (8:1 ratio), and to reaction #3 add dNTPs to 180 μM and αS-dNTPs to 20 μM (10:1 ratio). Five units of *Taq* are added to start the reaction. Hot starts are usually performed to increase product yield and to minimize contaminating bands. Typical instrument settings are 30 cycles at 94° denaturation, 53° annealing, and 72° extension (1 min per kb).

---

[12] S. Lutz *et al.*, *Nucleic Acids Res.* **29,** E16 (2001).

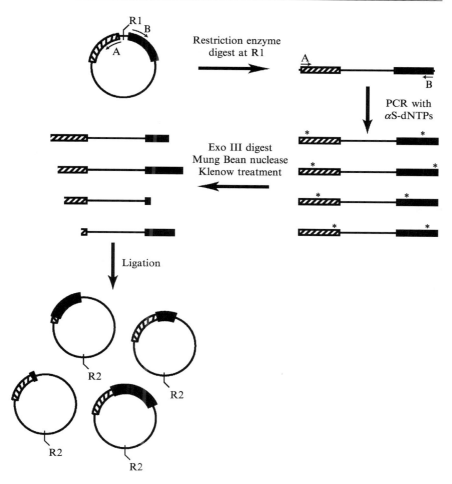

FIG. 2. Schematic of THIO-ITCHY. Plasmid DNA is linearized by digestion with a unique restriction enzyme, R1. The linearized plasmid is amplified by PCR to incorporate the αS-dNTPs using primers A and B. The THIO-ITCHY method is carried out as described in the text, and an intramolecular blunt ligation is prepared to generate a library of hybrid enzymes.

3. Verify the quality and yield of the PCR reactions by running 2 μl of the resultant PCR mixture on an agarose gel. Purify DNA using the QIAquick PCR purification kit and elute with 80 μl EB buffer. Determine the DNA concentration using a spectrophotometer.

4. Mix the following in a 100-μl reaction: the entire DNA sample, 10 μl 10× Exo III buffer, and water to 100 μl. Start the reaction with Exo III (120 units per μg DNA) and incubate at 37° for 30 min.

5. Stop the reaction with 500 μl Qiagen PB buffer and purify using the QIAquick PCR purification kit. Elute the DNA with 80 μl of EB buffer.

6. Perform the mung bean nuclease step and purify the DNA as described for time-dependent ITCHY (see previous discussion).

7. Remove 5 μl of DNA and run on an agarose gel to check the nuclease digestions. The sample without αS-dNTPs should be digested completely and not visible on the gel. The other samples containing 8:1 or 10:1 ratios of dNTPs to αS-dNTPs should appear as a smear on the gel, running from the size of the complete plasmid to the bottom of the gel. Incorrectly digested samples will have an overweighting of smaller bands or larger bands due to improper ratios of αS-dNTPs to dNTPs in the PCR reaction. The 8:1 and 10:1 ratios work well for most genes, but may need to be adjusted for large templates. If both ratios provide appropriate mixtures, the samples can be carried through the rest of the procedure and combined in the ligation step.

8. Perform the Klenow step and purify the DNA as described for time-dependent ITCHY.

9. Prepare a 400-μl intramolecular ligation mixture: 40 μl 10× T4 ligase buffer, purified DNA sample, and 5 Weiss units T4 DNA ligase (per μg DNA). Incubate at 16° for >12 h. Polyethylene glycol can be added to improve ligation efficiency, typically to 5–7.5% final concentration.[13]

Considerations for THIO-ITCHY

Note the considerations described earlier for time-dependent ITCHY. For THIO-ITCHY, gel purification of the starting vector is important. This step minimizes the potential for contamination in later steps of the procedure.

This protocol has been optimized for *Taq* DNA polymerase. Proof-reading polymerases, such as Pfu, should not be substituted for *Taq* in this procedure. The exonuclease activity of these enzymes allows them to continually remove and synthesize new 3′ DNA ends in a process known as idling. This process leads to an enrichment of α-phosphothioate nucleotides at the DNA ends, biasing the libraries toward untruncated products.

After adding Exo III, complete mixing of the reaction is essential. Pipetting up and down for 15–20 s is usually sufficient to get good results.

---

[13] B. H. Pheiffer and S. B. Zimmerman, *Nucleic Acids Res.* **11,** 7853 (1983).

Other Considerations

Vector Design

Careful planning can greatly expedite experiments with isolated hybrid enzymes. For example, the background expression from pET vectors (Novagen, Madison, WI) is often adequate to complement *E. coli* mutants for *in vivo* selections of active hybrid enzymes. Further, the pET vector could be used for the overexpression and purification of the isolated hybrid. Alternatively, the starting expression vector could be designed with restriction sites compatible with other vectors that might be required in later experiments.

When designing a plasmid for *in vivo* selections, it is necessary to use inactive gene fragments for the construction of incremental truncation libraries, as the wild-type enzyme will overwhelm the genetic selection and hybrid enzymes will not be isolated. Also, it is strongly recommended that wild-type enzymes be tested for complementation in the chosen vector prior to the generation of incremental truncation libraries. High expression levels of some enzymes can create deleterious growth effects, especially on minimal media. Examples of vector design are described elsewhere for time-dependent ITCHY and THIO-ITCHY.[7,8,12]

Library Quality

A quick estimation of minimal library size can be calculated from the product of the total number of bases truncated. For example, in the isolation of all crossovers between two 600-bp genes, the minimal library size would be $600 \times 600 = 360,000$ members. As a general rule, 10 times the library size is sufficient to ensure that all crossovers are isolated, indicating that a library size of 3.6 million members is needed. Because crossovers occur at every base pair, it is important to remember that two-thirds of all library members will be out of frame.[14]

For each ITCHY library, checking crossover distribution is also important. Usually, 20 isolates are sequenced from each library, and the crossover positions are plotted to check for adequate distribution. Libraries can become biased from insufficient or excessive Exo III digestion and need to be reconstructed. For further information on theoretical distribution crossovers, the reader is referred to Ostermeier.[15] To improve library quality, a size selection can be performed to focus the library. In this approach, the DNA is subjected to high-resolution gel electrophoresis following the

[14] M. Ostermeier *et al.*, *Bioorg. Med. Chem.* **7**, 2139 (1999).
[15] M. Ostermeier, *Biotechnol. Bioeng.* **82**, 564 (2003).

Klenow step, and only fragments of the correct size are extracted and ligated.[16]

Other Applications for ITCHY

Monomeric enzymes have been converted into functional heterodimers with ITCHY.[6] The reader is referred to an earlier review for more information on this approach.[8] Perhaps the most generally useful application of incremental truncation is the identification of protein domains or epitopes, such as regions involved in protein–protein interactions. Again, careful vector design is critical for this type of experiment.

Variants of the ITCHY Method

Since the original development of the ITCHY method, a number of related techniques have been described, such as the creation of hybrid enzymes through circular permutation, termed CP-ITCHY.[17] For constructing hybrid enzyme libraries without Exo III, methods have been developed that use DNase I,[18] BAL-31 exonuclease,[19] or Tn5 transposase.[20]

Acknowledgments

A.R.H. is a Damon Runyon Fellow supported by the Damon Runyon Cancer Research Foundation (DRG-#1729-02). T.A.N. was supported by a National Institute of Health Grant (GM068267-01). This work was supported in part by a National Institute of Health Grant (GM24129) to S.J.B.

[16] S. Lutz et al., Proc. Natl. Acad. Sci. USA 98, 11248 (2001).
[17] M. Ostermeier and S. J. Benkovic, Biotech. Lett. 23, 303 (2001).
[18] V. Sieber et al., Nature Biotechnol. 19, 456 (2001).
[19] M. G. Pikkemaat and D. B. Janssen, Nucleic Acids Res. 30, E35 (2002).
[20] T. A. Naumann et al., Nucleic Acids Res. 30, e119 (2002).

[7] Random Multirecombinant Polymerase Chain Reaction

By Toru Tsuji, Michiko Onimaru, Masanobu Kitagawa, Kanehisa Kojoh, Noriko Tabata, and Hiroshi Yanagawa

Introduction

This article describes a method for the construction of combinatorial protein libraries, which have been developed for the directed evolution of proteins. DNA shuffling[1] and/or family shuffling[2] has been used conventionally to create protein libraries. These methods introduce point mutations into the parent sequence(s) by means of homologous recombination among highly related sequences. The major limitation of these methods is that they cannot combine different proteins without homologous sequences, so it is possible to search only a limited sequence space in the vicinity of the parent sequence(s).

Novel functional proteins, which have not appeared in the course of protein evolution, are highly likely to be present in the global sequence space. Recent technologies for the *in vitro* selection of proteins such as ribosome display,[3,4] *in vitro* virus,[5–7] mRNA display,[8,9] and STABLE (DNA display),[10,11] which enable us to test a vast number of independent sequences, make it possible to search sequence spaces far from those of natural proteins. Indeed, novel functional proteins were obtained from random amino acid libraries containing $\sim 10^{12}$ independent sequences.[12,13]

[1] W. P. Stemmer, *Proc. Natl. Acad. Sci. USA* **91**, 10747 (1994).

[2] A. Crameri, S. A. Raillard, E. Bermudez, and W. P. Stemmer, *Nature* **391**, 288 (1998).

[3] J. Hanes and A. Pluckthun, *Proc. Natl. Acad. Sci. USA* **94**, 4937 (1997).

[4] J. Hanes, L. Jermutus, and A. Pluckthun, *Methods Enzymol.* **328**, 404 (2000).

[5] N. Nemoto, E. Miyamoto-Sato, Y. Husimi, and H. Yanagawa, *FEBS Lett.* **414**, 405 (1997).

[6] E. Miyamoto-Sato, N. Nemoto, K. Kobayashi, and H. Yanagawa, *Nucleic Acids Res.* **28**, 1176 (2000).

[7] E. Miyamoto-Sato, H. Takashima, S. Fuse, K. Sue, M. Ishizaka, S. Tateyama, K. Horisawa, T. Sawasaki, Y. Endo, and H. Yanagawa, *Nucleic Acids Res.* **31**, e78 (2003).

[8] R. W. Roberts and J. W. Szostak, *Proc. Natl. Acad. Sci. USA* **94**, 12297 (1997).

[9] R. Liu, J. E. Barrick, J. W. Szostak, and R. W. Roberts, *Methods Enzymol.* **318**, 268 (2000).

[10] N. Doi and H. Yanagawa, *FEBS Lett.* **457**, 227 (1999).

[11] M. Yonezawa, N. Doi, Y. Kawahashi, T. Higashinakagawa, and H. Yanagawa, *Nucleic Acids Res.* **31**, 118 (2003).

[12] D. S. Wilson, A. D. Keefe, and J. W. Szostak, *Proc. Natl. Acad. Sci. USA* **98**, 3750 (2001).

[13] A. D. Keefe and J. W. Szostak, *Nature* **410**, 715 (2001).

An alternative strategy to explore the global sequence space is to use combinatorial protein libraries created by block shuffling, where blocks are amino acid sequences corresponding to particular features of proteins, such as secondary structures,[14,15] modules,[16–20] functional motifs, and so on. This idea is based on the exon shuffling hypothesis, which suggests that proteins acquired their functional diversity by combining different building blocks encoded by ancient exons in the early stages of protein evolution.[21] To realize this strategy, a novel combinatorial method that can combine different DNA fragments without homologous sequences is required.

We have developed random multirecombinant (RM) polymerase chain reaction, (PCR)[22] which permits the shuffling of several DNA fragments without homologous sequences. RM-PCR is based on multirecombinant PCR,[14] which is a modified method of overlap extension PCR.[23] In multirecombinant PCR, several dimer templates having overlapped segments are combined in a single PCR (Fig. 1A). Through the use of dimer templates encoding two peptide sequences, a structural gene consisting of several building blocks is created. However, this is not a combinatorial method because one PCR yields only one structural gene. To create different structural genes simultaneously, different dimer templates are mixed such that at least one segment of a dimer template can overlap with more than two different dimer templates. Figure 1B shows a schematic diagram of RM-PCR in which different structural genes consisting of three building blocks arranged in different orders are synthesized in a single PCR.

In RM-PCR, different dimer templates encoding two building blocks are combined simultaneously and are then flanked by dimer templates containing 5' and 3' consensus sequences (T7 and Ex in Fig. 1). Therefore,

[14] T. Tsuji, K. Yoshida, A. Satoh, T. Kohno, K. Kobayashi, and H. Yanagawa, *J. Mol. Biol.* **286,** 1581 (1999).

[15] T. Tsuji, K. Kobayashi, and H. Yanagawa, *FEBS Lett.* **453,** 145 (1999).

[16] M. Go, *Adv. Biophys.* **19,** 91 (1985).

[17] K. Yoshida, T. Shibata, J. Masai, K. Sato, T. Noguti, M. Go, and H. Yanagawa, *Biochemistry* **32,** 2162 (1993).

[18] H. Yanagawa, K. Yoshida, C. Torigoe, J. S. Park, K. Sato, T. Shirai, and M. Go, *J. Biol. Chem.* **268,** 5861 (1993).

[19] K. Wakasugi, K. Ishimori, K. Imai, Y. Wada, and I. Morishima, *J. Biol. Chem.* **269,** 18750 (1994).

[20] K. Inaba, K. Wakasugi, K. Ishimori, T. Konno, M. Kataoka, and I. Morishima, *J. Biol. Chem.* **272,** 30054 (1997).

[21] W. Gilbert, *Nature* **271,** 501 (1978).

[22] T. Tsuji, M. Onimaru, and H. Yanagawa, *Nucleic Acids Res.* **29,** e97 (2001).

[23] R. M. Horton, H. D. Hunt, S. N. Ho, J. K. Pullen, and L. R. Pease, *Gene* **77,** 61 (1989).

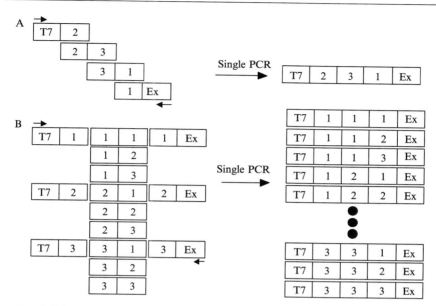

Fig. 1. Schematic diagrams of multirecombinant PCR (A) and random multirecombinant PCR (RM-PCR) (B). In RM-PCR, different sequences consisting of several building blocks arranged in different orders are synthesized in a single PCR. T7 and Ex are 5′ and 3′ consensus sequences where primers anneal to prime the extension.

the frequencies of building blocks in the library can be controlled by changing the amount of each dimer template in a reaction mixture. This allows the design of different types of combinatorial libraries. We have constructed random shuffling libraries and artificial alternative splicing libraries. A random shuffling library contains different block sequences, where every position has an equal probability of encoding any of the original building blocks (Fig. 2A). Six building blocks encoding 25 amino acids from *Escherichia coli* glutaminyl-tRNA synthetase[24] were used to create such a library. Through the use of an appropriate set of dimer templates, an alternative splicing library was created from a structural gene at the DNA level (Fig. 2B). We have constructed alternative splicing libraries using human estrogen receptor α ligand-binding domain (hERα LBD),[25] which was divided into 10 building blocks each with different

[24] F. Yamao, H. Inokuchi, A. Cheung, H. Ozeki, and D. Soll, *J. Biol. Chem.* **257,** 11639 (1982).
[25] G. L. Greene, P. Gilna, M. Waterfield, A. Baker, Y. Hort, and J. Shine, *Science* **231,** 1150 (1986).

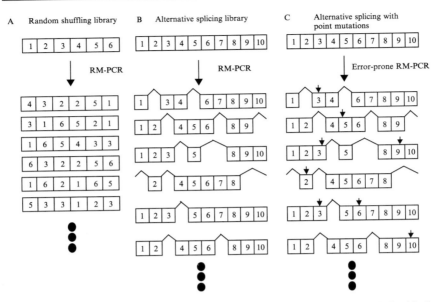

Fig. 2. (A) Construction of a random shuffling library by RM-PCR. The six building blocks are shuffled and combined to yield many different structural genes in RM-PCR. In this library, every position in the block sequences has an equal probability of encoding each building block. (B) Construction of an alternative splicing library by RM-PCR. Ten building blocks from a gene were spliced alternatively at the DNA level to yield different structural genes. (C) Alternative splicing library created by RM-PCR under error-prone conditions where the created block sequences encompass different point mutations (indicated by arrows on the block sequences).

chain lengths. Further, we found that block shuffling and point mutations could be introduced at the same time by performing RM-PCR under error-prone conditions (Fig. 2C). This article describes in detail the methodology used to perform RM-PCR to create combinatorial DNA libraries, which can be used for subsequent protein selection experiments.

Preparation of Dimer Templates for RM-PCR

Dimer templates are prepared from building blocks that encode peptide fragments. Any DNA sequences that can be ligated by overlap extension PCR can be used as building blocks for RM-PCR. The length and GC content of building blocks may affect the efficiency of the overlap extension. Table I lists these values for all building blocks that were combined successfully by RM-PCR in a previous study.[22] DNA sequences with

TABLE I
CHAIN LENGTH AND GC CONTENT OF BUILDING BLOCKS USED FOR RM-PCR[a]

Blocks for random shuffling[b]	Length (bp)	GC (%)	Blocks for alternative splicing[c]	Length (bp)	GC (%)
Block 1	75	54.7	Block 1	102	54.9
Block 2	75	50.7	Block 2	105	53.3
Block 3	75	45.3	Block 3	90	53.3
Block 4	75	52.0	Block 4	48	50.0
Block 5	75	52.0	Block 5	78	52.6
Block 6	75	54.7	Block 6	57	36.8
			Block 7	111	51.4
			Block 8	72	66.7
			Block 9	51	47.1
			Block 10	60	65.0

[a] DNA sequences are described elsewhere.[22]
[b] Six building blocks for random shuffling are from parts of the dinucleotide-binding domain and the acceptor-binding domain of E. coli glutaminyl-tRNA synthetase.
[c] Ten building blocks for alternative splicing are from the human estrogen receptor α ligand-binding domain.

lengths from 48 to 111 bp and GC contents ranging from 36.5 to 66.7% were combined successfully. Another study showed that a 33-bp DNA fragment (GC 58.9%), which is the shortest fragment tested, could also be combined by RM-PCR.

Two building blocks with blunt ends are amplified by PCR with phosphorylated primers, purified using Wizard PCR Preps (Promega), and ligated by DNA ligase. The reaction mixure for the ligation (10 μl) usually consists of 1 μl of each DNA solution, T4 DNA ligase (1 μl = 400 units, New England Biolabs), and the supplemented buffer. Ligation reactions are performed for 1 h at 16°. Ligation products consisted of DNA fragments differing in length and direction. Because the dimer templates are the shortest fragments flanked by two different primers in the ligation products, they are amplified preferentially by PCR. Dimer templates encoding a single peptide sequence in the same direction, such as 1-1, 2-2, and 3-3 shown in Fig. 1B, are obtained together with by-products such as monomeric, trimeric, and/or tetrameric arrangements. For amplification of the dimer templates, PCR is performed with a reaction mixture (50 μl) containing 20 pmol of each primer, 200 μM of each dNTP, 1 μl of the DNA solution, 3 μl of 25 mM MgCl$_2$, 2.5 units of rTaq DNA polymerase (Toyobo), and 5 μl of 10× rTaq DNA polymerase buffer. Pfu DNA polymerase (Stratagene) also amplifies dimer templates efficiently, but Vent DNA polymerase cannot be employed. The program for the PCR consists of one cycle at 95° for 5 min, followed by 30 cycles at 95° for 30 s and 56° for 30 s. The DNA band corresponding to the dimer is purified from the low melting agarose gel (Sigma) after electrophoresis.

Alternatively, dimer templates may be obtained by chemical synthesis as single-stranded DNAs if the building blocks are short sequences. These single-stranded DNAs can be converted to double-stranded DNAs by PCR or hybridization. Mutations causing frame shifts or stop codons in the dimer templates must be removed. We usually select dimer templates without such mutations after cloning and DNA sequencing. Blue–white selection using the TOPO TA cloning kit (Invitrogen) is very useful for this purpose. Most of the blue colonies contain insert DNA; therefore, dimer templates without such mutations are obtained from blue colonies with high probability. At least two out of three colonies contain a dimer template with the correct open reading frame without such mutations under the usual conditions. Dimer templates with the appropriate sequence are amplified by PCR with reaction mixtures (50 μl) consisting of 20 pmol of each primer, 200 μM of each dNTP, 10 ng of DNA template, 3.75 unit of Pfu DNA polymerase, and 5 μl of 10× Pfu DNA polymerase buffer. The program for the PCR is one cycle at 95° for 5 min, followed by 30 cycles at 95° for 30 s and 56° for 30 s. PCR products are recovered from the low melting agarose gels by phenol extraction and ethanol precipitation. Apparent concentrations of all dimer templates are estimated from the intensity of ethidium bromide-stained bands on agarose gels and from the absorbance at 260 nm.

Dimer templates prepared for the construction of the random shuffling library are T7-1, T7-2, T7-3, T7-4, T7-5, T7-6, 1-1, 1-2, 1-3, 1-4, 1-5, 1-6, 2-1, 2-2, 2-3, 2-4, 2-5, 2-6, 3-1, 3-2, 3-3, 3-4, 3-5, 3-6, 4-1, 4-2, 4-3, 4-4, 4-5, 4-6, 5-1, 5-2, 5-3, 5-4, 5-5, 5-6, 6-1, 6-2, 6-3, 6-4, 6-5, 6-6, 1-Ex, 2-Ex, 3-Ex, 4-Ex, 5-Ex, and 6-Ex. T7 is a 5′ consensus sequence and has a T7 promoter sequence, a Kozak sequence,[26] and an initiation codon for *in vitro*-directed evolution systems for proteins, such as the *in vitro* virus[5–7] or STABLE (DNA display)[10,11] developed in our laboratory. Ex is a 3′ consensus sequence and encodes a flexible peptide chain. Dimer templates prepared for the construction of the alternative splicing libraries are T7OM-1, T7OM-2, T7OM-3, T7OM-4, T7OM-5, T7OM-6, 1-2, 1-3, 1-4, 1-5, 1-6, 1-7, 2-3, 2-4, 2-5, 2-6, 2-7, 2-8, 3-4, 3-5, 3-6, 3-7, 3-8, 3-9, 4-5, 4-6, 4-7, 4-8, 4-9, 4-10, 5-6, 5-7, 5-8, 5-9, 5-10, 6-7, 6-8, 6-9, 6-10, 7-8, 7-9, 7-10, 8-9, 8-10, 9-10, 5-CBPHis, 6-CBPHis, 7-CBPHis, 8-CBPHis, 9-CBPHis, and 10-CBPHis. T7OM is a 5′ consensus sequences and encodes a T7 promoter sequence, an

[26] M. Kozak, *Microbiol. Rev.* **47**, 1 (1983).

omega-like sequence,[27] an initiation codon, and the FLAG tag.[28] CBPHis is a 3' consensus sequence and encodes the calmodulin-binding peptide tag[29] and the hexa histidine tag.[30]

RM-PCR to Create Random Shuffling Libraries

Two factors are important for the efficient construction of a random shuffling library by RM-PCR. First, the amounts of dimer templates in the reaction mixture affect the efficiency significantly. The random shuffling library should contain different block sequences whose every position has an equal probability to encode any of the building blocks desired (Fig. 2A). In principle, this is attained by RM-PCR using equal amounts of dimer templates prepared by ligating two building blocks in all possible combinations. In RM-PCR, these dimer templates encoding two building blocks are combined, arranged in different orders, and flanked by other types of dimer templates, which contain 5' or 3' consensus sequences (T7 or Ex) as shown in Fig. 1B. Therefore, it is expected that the molar ratio of these two types of dimer templates will determine the number of building blocks combined in the synthesized structural genes. The appropriate ratio of these two types of dimer templates should be determined experimentally.

Results showed that structural genes consisting of appropriate numbers of building blocks could be constructed from the reaction mixtures, which contained dimer templates encoding two building blocks and dimer templates containing consensus sequences in a molar ratio of 20:1 to 100:1. The total amount of dimer templates encoding two building blocks to be used for efficient construction of the library was also determined experimentally and was in the range of 500 to 1000 fmol for a reaction volume of 50 μl.

The kind of DNA polymerase used for RM-PCR is another important factor. As shown in Fig. 3, *Taq* DNA polymerase could not synthesize long structural genes compared to Vent DNA polymerase (New England Biolabs). Therefore, a reaction mixture (50 μl) containing 20 pmol of each primer, 750 fmol of dimer templates encoding two building blocks, 15 fmol of dimer templates containing the consensus sequence, 5 μl of 10× Vent DNA polymerase buffer, and 1 unit of Vent DNA polymerase is used to create the random shuffling library. The RM-PCR program is one cycle at 95° for 5 min, followed by 20 cycles consisting of 95° for 30 s, 60° for 30 s,

[27] D. R. Gallie and V. Walbot, *Nucleic Acids Res.* **20**, 4631 (1992).
[28] G. Cho, A. D. Keefe, R. Liu, D. S. Wilson, and J. W. Szostak, *J. Mol. Biol.* **297**, 309 (2000).
[29] C. F. Zheng, T. Simcox, L. Xu, and P. Vaillancourt, *Gene* **186**, 55 (1997).
[30] J. Porath, J. Carlsson, I. Olsson, and G. Belfrage, *Nature* **258**, 598 (1975).

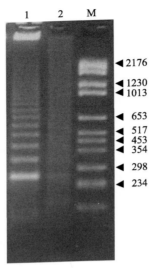

Fig. 3. Random shuffling libraries created by RM-PCR using Vent DNA polymerase (lane 1) and *Taq* DNA polymerase (lane 2). The ladder bands in lane 1 are T7-(block)$_1$-Ex, T7-(block)$_2$-Ex, . . . , from the smallest band. Each lane contained 50 μl of PCR products. M indicates a DNA size marker.

and 72° for 3 min. The yield of long sequences decreases if the extension time is shortened to 1 min.

Finally, a random shuffling library, which was constructed under the conditions described previously, was characterized by DNA sequencing. DNA bands corresponding to T7-(block)$_6$-Ex on an agarose gel were purified, cloned, and sequenced. The DNA sequences analyzed are shown in Table II. Only one sequence contained a deletion, which resided between the identical sequences "gtacgact" in blocks five and six. Thus RM-PCR was employed successfully to construct a random shuffling library, which contained many different block sequences with a correct open reading frame.

RM-PCR to Create Artificial Alternative Splicing Libraries

RM-PCR with reaction mixtures containing dimer templates in appropriate ratios and amounts can also be used to create alternative splicing libraries. The ratios are determined theoretically by the frequencies of dimer templates in the desired library. An example is shown in Fig. 4, where block sequences consisting of three building blocks are created by

TABLE II

BLOCK SEQUENCES OBTAINED FROM THE RANDOM SHUFFLING LIBRARY

112114	245	341514	432335	525155	645465[a]
113134	245256	343	446262	526655	646161
12156	254136	343611	45325	532	646362
12465	256541	345265	455124	5336155[b]	646551
134526	261136	346452	465222	542526	651656
136564	26212	3465125[c]	466325	563656	652636
145412	2641	35354	511533	565145	6552
146521	313542	354531	514231	614231	665115
2116	3212	363262	521314	623652	665145625
216431	324355	412351	521351	625362	
231531	332612[a]	4263152[b]	523515	631234	

[a] Sequences obtained from two colonies.
[b] Sequences without Ex at the 3' end.
[c] Sequence having a deletion between blocks 6 and 5.

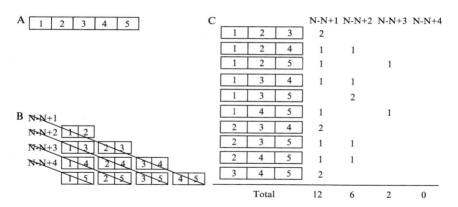

FIG. 4. Dimer templates for the construction of alternative splicing libraries. (A) A parent gene divided into five building blocks. (B) All dimer templates required for construction of the alternative splicing libraries using a parent gene divided into five building blocks. The dimer templates can be grouped into four classes. Each diagonal line covers each class of dimer template. (C) The frequency of each class of dimer template appearing in all possible block sequences consisting of three building blocks is indicated. The sum of the frequencies for each class of dimer template gives the molar ratio of dimer templates in the reaction mixture for RM-PCR.

alternative splicing from a structural gene that is divided into five building blocks (Fig. 4A). All dimer templates required are shown in Fig. 4B and can be grouped into four classes [block (N)-block (N + 1) to block (N)-block (N + 4)]. Figure 4C illustrates all possible block sequences

A

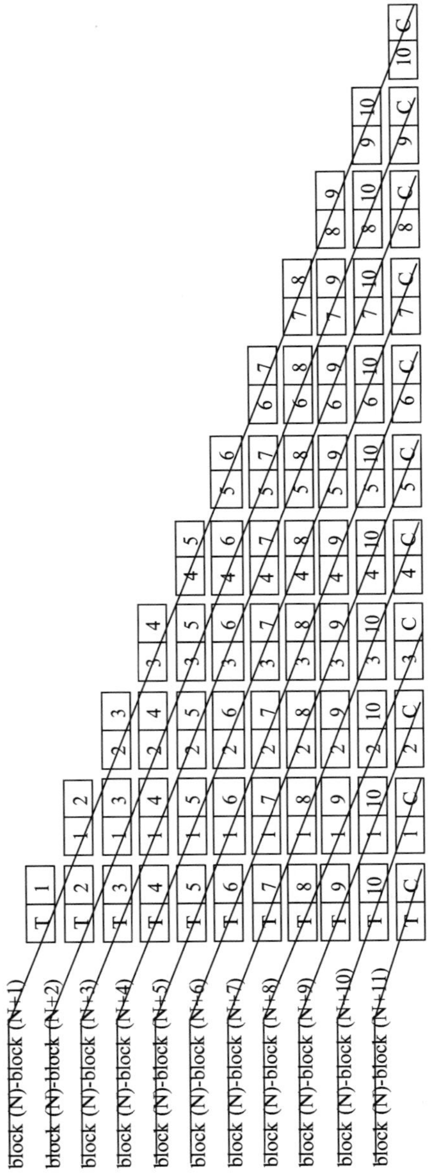

block (N)-block (N+1)
block (N)-block (N+2)
block (N)-block (N+3)
block (N)-block (N+4)
block (N)-block (N+5)
block (N)-block (N+6)
block (N)-block (N+7)
block (N)-block (N+8)
block (N)-block (N+9)
block (N)-block (N+10)
block (N)-block (N+11)

B

	10 mer	9 mer	8 mer	7 mer	6 mer	5 mer	4 mer	3 mer	2 mer
block (N)-block (N+1)	$_9C_0$	$_9C_1$	$_9C_2$	$_9C_3$	$_9C_4$	$_9C_5$	$_9C_6$	$_9C_7$	$_9C_8$
block (N)-block (N+2)	—	$_8C_0$	$_8C_1$	$_8C_2$	$_8C_3$	$_8C_4$	$_8C_5$	$_8C_6$	$_8C_7$
block (N)-block (N+3)	—	—	$_7C_0$	$_7C_1$	$_7C_2$	$_7C_3$	$_7C_4$	$_7C_5$	$_7C_6$
block (N)-block (N+4)	—	—	—	$_6C_0$	$_6C_1$	$_6C_2$	$_6C_3$	$_6C_4$	$_6C_5$
block (N)-block (N+5)	—	—	—	—	$_5C_0$	$_5C_1$	$_5C_2$	$_5C_3$	$_5C_4$
block (N)-block (N+6)	—	—	—	—	—	$_4C_0$	$_4C_1$	$_4C_2$	$_4C_3$
block (N)-block (N+7)	—	—	—	—	—	—	$_3C_0$	$_3C_1$	$_3C_2$
block (N)-block (N+8)	—	—	—	—	—	—	—	$_2C_0$	$_2C_1$
block (N)-block (N+9)	—	—	—	—	—	—	—	—	$_1C_0$

consisting of three building blocks (123 to 345) and shows the frequency of each class of dimer template in the respective block sequences. The sum of the frequencies for each class of dimer template provides the molar ratio of the dimer template in the reaction mixture. In this case, the molar ratio of block (N)-block (N + 1), block (N)-block (N + 2), block (N)-block (N + 3), and block (N)-block (N + 4) is 6:3:1:0. Similarly, the classification and relative frequencies of all dimer templates required for the construction of alternative splicing libraries using a structural gene divided into 10 building blocks are shown in Fig. 5. For example, the 8-mer library can be constructed from a reaction mixture containing dimer templates classified into block (N)-block (N + 1), block (N)-block (N + 2), and block (N)-block (N + 3) mixed in a molar ratio of $36(=_9C_2):8(=_8C_1):1(=_1C_0)$. The total amount of dimer templates required was determined experimentally, and 630 to 1260 fmol of dimer templates was found to be appropriate for a reaction mixture of 50 μl.

The yield of the PCR depended significantly on the kind of DNA polymerase used. The KOD Dash DNA polymerase gave the best efficiency among the DNA polymerases tested. The reaction mixture usually consists of 20 pmol of each primer, 630 to 1260 fmol of total dimer templates, 5 μl of 10× KOD Dash DNA polymerase buffer, and 1.25 units of KOD Dash DNA polymerase (Toyobo). The program of the RM-PCR for the construction of alternative splicing libraries is one cycle at 95° for 5 min, followed by 20 cycles consisting of 95° for 30 s, 54° for 30 s, and 72° for 1 min.

Initially, we constructed 5- and 8-mer libraries and characterized them by DNA sequencing after cloning. The DNA sequences obtained are shown in Tables III and IV. Many different block sequences, were obtained and the 8-mer library contained longer sequences than the 5-mer library, as expected. Most of the block sequences had the correct open reading frame, and only 2 of 71 sequences analyzed had a deletion causing a frame shift. These latter frame shifts resulted from recombination of the identical sequences 'atgatc' in building blocks one and two, and the similar sequences 'ccagtga' and 'ccaggga' in building blocks four and five. Thus, most

---

FIG. 5. (A) Classification of dimer templates required for construction of the alternative splicing library using a structural gene divided into 10 building blocks. T and C represent T7OM and CBPHis, respectively. They are numbered 0 and 11 because they are the 5' and 3' ends of the RM-PCR products. All dimer templates were grouped into 11 classes depending on the building blocks that they encode [block (N)-block (N + 1) to block (N)-block (N + 11)]. Each diagonal line covers one type of dimer template. (B) Relative frequencies of different dimer templates in all possible sequences with a certain number of building blocks are calculated as a binomial coefficient ($_nC_r$ = n! / [r! (n-r)!]).

TABLE III
BLOCK SEQUENCES OBTAINED FROM THE ALTERNATIVE SPLICING LIBRARY (8-mer)

123456789_	123_56789_	12__56__910	1_34_67_9_	_23_567_910
1234567_910	123_567_910	1_3456789_	1_34_67_910	_23_5_7_9_
12345678910	123_5_7__10	1_345678910	1_3_56789_	_2_4_678_10
12345_78910	123_5_78_10	1_345_78__	1_3_5678910	_2_567_910
1234_67_9_	12_45678910	1_345_7_910	1_3_567_910	__3456_8910
1234__78910	12_45678910	1_34_67__10	_234_67__10	__34567__10
123_5678_10	12_4_6_8910	1_34_678__	_234_678__	__3_5__8910

TABLE IV
BLOCK SEQUENCES OBTAINED FROM THE ALTERNATIVE SPLICING LIBRARY (5-mer)

123___7__10	1_3___7__10	_23_5_7_9_	__345_7__10	__3___8__
1_34___910	1_3__7_9_	_23_5__89_	__34_678__	__45_78__
1_3_5_7__10	1__456_8__	_23__67_910	__34_67_910	__4_67_910
1_3_5__89_	1__4_678__	_23___78__	__34___8910	__4_6_89_
1_3__67_10	1__4_67_9_	_2__567__10	__3_567_9_	__5_7_910
1_3__67_9_	1___6__9_	_2__567_910	__3_5_78__	
1_3___7___	_23_56__910	_2__5_7__10	__3_5_7_9_	

of the sequences were found to encode long open reading frames, which were suitable for protein selection experiments.

Error-Prone RM-PCR

For the construction of alternative splicing libraries with greater complexity, we explored conditions where the KOD Dash DNA polymerase introduces random point mutations into block sequences during RM-PCR. We performed RM-PCR with different concentrations of Mn^{2+}, and the frequency of mutations was analyzed by DNA sequencing. As shown in Fig. 6, the fidelity of DNA synthesis by the polymerase decreased as the concentration of Mn^{2+} increased. The results are summarized in Table V, which illustrates the numbers of base pairs analyzed and the frequency of substitutions, transversions, transitions and deletions, or insertions introduced by RM-PCR at varying concentrations of Mn^{2+}. As in the case of *Taq* DNA polymerase, the frequencies of transition and transversion mutations seem to be equal in the presence of Mn^{2+}.[31]

---

[31] D. W. Leung, E. Chen, and D. V. Goeddel, *Technique* **1,** 11 (1989).

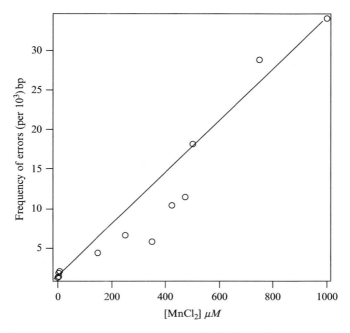

Fig. 6. Frequencies of point mutations during RM-PCR performed under error-prone conditions.

TABLE V

Frequency of Errors Introduced during Error-Prone RM-PCR

[MnCl₂] (μM)	Base pairs analyzed	Substitutions	Substitutions per 1000 bp	Transversions	Transitions	Deletions or insertions
0	47533	60	1.3	33	27	3
3	3669	5	1.4	3	2	0
5	3126	6	1.9	2	4	1
7	2424	5	2.1	3	2	1
150	2751	12	4.4	2	10	1
250	3344	22	6.6	9	13	0
350	3933	23	5.8	11	12	3
425	3663	38	10.4	12	26	0
475	3810	45	11.5	18	27	0
500	3234	59	18.2	29	30	0
750	2916	84	28.8	37	47	3
1000	2202	75	34.0	37	38	4

Figure 7 shows the results of electophoresis of four different alternative splicing libraries on an agarose gel; these libraries were prepared to create block sequences consisting mainly of five to eight building blocks under error-prone conditions. The reaction mixtures used consist of 480 μM $MnCl_2$, 20 pmol of each primer, 960 fmol of total dimer templates, 5 μl of 10× KOD Dash DNA polymerase buffer, and 1.25 units of KOD Dash DNA polymerase. The ratio of each dimer template in each reaction mixture is determined as discussed previously (Fig. 5). The amplification program is identical to that used for RM-PCR to create usual alternative splicing libraries. The regions with strong DNA bands shifted gradually, corresponding to the size of the block sequences that we had selected.

Conclusions

Random multirecombinant PCR which permits the shuffling of several DNA fragments without homologous sequences, can be used to create combinatorial protein libraries containing different block sequences. The frequencies of building blocks in the library can be controlled by changing

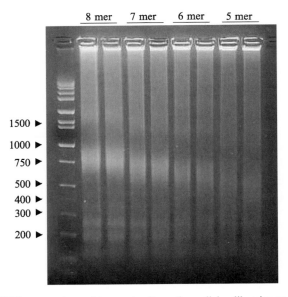

FIG. 7. RM-PCR was performed to create alternative splicing libraries under error-prone conditions. RM-PCR products obtained from reaction mixtures where dimer templates were mixed to yield structural genes consisting of eight, seven, six, and five building blocks were loaded on an agarose gel (2%).

the amounts of dimer templates, making it possible to design different types of combinatorial libraries, such as random shuffling libraries and artificial alternative splicing libraries. The block shuffling and random point mutations can be introduced simultaneously by performing RM-PCR under error-prone conditions. Most of the block sequences in the combinatorial libraries created by RM-PCR encode a long open reading frame and are suitable for protein selection experiments. Thus, RM-PCR promises to be a powerful approach in creating novel proteins, especially simulating early stages of protein evolution on a rapid timescale using an efficient and easily adaptable experimental protocol.

[8] Gene Library Synthesis by Structure-Based Combinatorial Protein Engineering

By Paul E. O'Maille, Ming-Daw Tsai, Bryan T. Greenhagen, Joseph Chappell, and Joseph P. Noel

Introduction

Structure-based combinatorial protein engineering (SCOPE) is a process for the synthesis of gene libraries that lay the genetic foundation for the exploration of the relationship between structure and function in the encoded proteins.[1] The comparative analysis both structurally and functionally of protein primary, secondary, and tertiary structure generates numerous hypotheses with which to probe the relationship between molecular structure and the ensuing functional readout. SCOPE provides a tool for constructing the gene libraries that encode rationally engineered protein variants that provide the raw material for addressing these hypothesis using both structural and functional analyses. Mechanistic hypotheses generated from structures derived from both experiment (crystallographic and nuclear magnetic resonance) and homology modeling is used to design oligonucleotides that code for crossovers between genes encoding structurally related proteins. A series of polymerase chain reactions (PCR), culminating in the selective amplification of crossover products, incorporates spatial information encoded in the oligonucleotide into a full-length gene and the resultant hybrid protein. Iteration of the process enables the synthesis of all possible combinations of desired crossovers, producing a hierarchical collection of chimeras in analogy to a Mendelian population.

[1] P. E. O'Maille, M. Bakhtina, and M. Tsai, *J. Mol. Biol.* **321,** 677 (2002).

The principles of the process are generally applicable and the methodology is easily adapted to a range of experimental objectives. At its inception, SCOPE was developed to provide a means of generating multiple crossover gene libraries from distantly related proteins, constituting a homology-independent *in vitro* recombination approach.[1] This article presents the adaptation of SCOPE to the facile combinatorial synthesis of mutant gene libraries. The newly incorporated refinements to the originally designed SCOPE approach illustrate the underlying principles of the experimental process that make it a robust technique for the parallel exploration of protein sequence space in three dimensions. This tertiary information is embodied within the mechanistic and evolutionary underpinnings of protein structure and function, both of which are fundamental aspects of biochemical adaptive change in organisms. In addition, this information can be exploited by SCOPE for a myriad of applications in biotechnology.

Principles

The construction of gene libraries by SCOPE involves a series of PCRs. Other recombination techniques use multiple primers or random fragments in a single step, thus carrying out multiple reactions in parallel.[2,3] Separation of gene synthesis into discrete steps is an essential feature of SCOPE. This simple but critical property of SCOPE enables one to control recombination through pairing gene fragments and genes that give rise to designed and anticipated combinations of crossovers. As a consequence, libraries are constructed as a series of less complex mixtures, which reduce numerical complexity and the cost and extent of sampling required during screening, including gene sequencing and functional assays. Crossover locations and the frequency of genetically encoded crossovers are established by experimental design and are not dictated or constrained by homology between genes or the linear distance between multiple mutations.

An overview of the process illustrates the basic steps encompassing SCOPE-based recombination (Fig. 1). In step I, standard PCR amplification, using an internal and external primer pair and the appropriate template DNA, produces chimeric gene fragments. Internal primers are designed on the basis of one or more encoded three-dimensional structures viewed with reference to the variable sequence space of protein homologues and code for crossovers in the protein-coding region of genes. External primers correspond to the 5′ and 3′ termini of a given gene, as

[2] W. P. Stemmer, *Nature* **370**, 389 (1994).
[3] F. J. Perlak, *Nucleic Acids Res.* **18**, 7457 (1990).

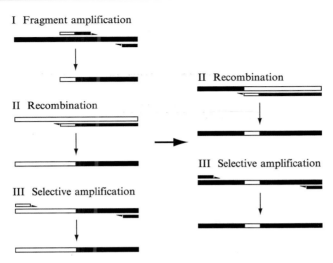

I Fragment amplification

II Recombination

II Recombination

III Selective amplification

III Selective amplification

Fig. 1. Overview of the SCOPE library synthesis process.

with any typical primer pair used in a standard PCR amplification. The template consists of a plasmid or PCR product that contains the gene of interest. In step II, *in vitro* recombination occurs between a gene fragment(s) and a new template; in other words, gene fragments serve as a new set of primers, which anneal and are extended to produce single-stranded full-length chimeras. In step III, a new external primer set directs the selective amplification of recombination products by virtue of the unique genetic identity encoded at their termini. Repetition of steps II and III using various pairs of gene fragments from step I and crossover products from step III allows the production of genetically diverse, multiple crossover libraries in high yield.

Careful oligonucleotide design is central to the SCOPE recombination process. A discussion of the properties of the synthetic oligonucleotides, in relationship to specific applications, illustrates how SCOPE can be adapted to construct multiple crossover libraries from distantly related proteins or combinatorial mutant libraries from mechanistically related proteins.

Internal Primers

Shuffling exons or "equivalent" structural elements between homologues require *chimeric* oligonucleotides. These are composed of approximately equal halves of two distinct genes and code for a crossover region. An example of their use is illustrated in the process overview (Fig. 1).

Without *a priori* knowledge of the optimal point of fusion in regions of low identity or the compatibility of "equivalent" structural elements of low sequence identity, linkage variability can be introduced. This entails designing a set of chimeric oligonucleotides (for a given crossover region), which code for a series of insertions and/or deletions around a fixed point to explore alternative crossovers. The corresponding collection of gene fragments can be used together in recombination reactions (step II). Variable connections between "equivalent" structural elements proved to be an essential aspect of design in producing functional hybrids from distantly related DNA polymerases.[1]

Combinatorial mutagenesis by SCOPE requires *mutagenic* oligonucleotides that code for either a specific or a random set of mutations, insertions, or deletions directed at a given site in the gene of interest. Alternatively, *bridging* oligonucleotides, which code for stretches of native sequence between mutations, can be used to mediate recombination between wild-type and/or mutant genes. Mutagenic and bridging oligonucleotides are employed in PCR in a similar manner as chimeric oligonucleotides, although refinements to the process were required for their efficient incorporation into the desired chimeric sequence as described later.

External Primers

Polymerase chain reaction amplification of mutant or chimeric genes (step III, Fig. 1) is the final step of the SCOPE cycle. Like any conventional amplification, a primer set that flanks the target gene of interest is required during this final amplification step. Additionally, the inclusion of restriction or recombination sites into the final primer set for the efficient cloning of the resultant collection of genes is often desirable. However, a fundamental aspect of SCOPE is that the "proper" primer set be used for the selective amplification of a particular crossover product from a recombination reaction, which may contain a mixture of products. In the chimeragenesis of distantly related proteins, the termini of each gene are unique and can be exploited in this way for selective amplification.

SCOPE, as applied to the combinatorial synthesis of mutant libraries, where the termini of wild-type genes and crossover products are indistinguishable, required the design of alternative external primers and a bookkeeping system for their successful implementation and hierarchical organization and storage. Primary amplification primers (PAPs) code for DNA sequences flanking the gene (like any generic external primer), but contain an additional and unique 5' sequence tag. Their use in gene fragment synthesis (step I) links a unique sequence to a particular mutation. Following recombination, secondary amplification primers (SAPs), which

A   B

I Fragment amplification   IA Mutagenic single-strand synthesis

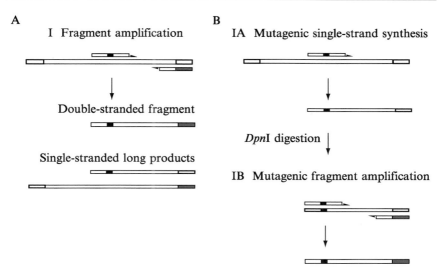

Double-stranded fragment

*Dpn*I digestion

Single-stranded long products

IB Mutagenic fragment amplification

FIG. 2. Refinements to SCOPE. (A) Source of wild-type background. (B) Alternative fragment amplification strategy for the suppression of the wild-type sequence.

correspond to the 5′ unique sequence, are employed in the final amplification (step III) to select for the desired recombination products.

Refinements to the original SCOPE process were necessary for the efficient incorporation of unique sequence tags, their linkage to mutations, and the suppression of wild-type background genes. During step I amplification, single-stranded DNA or "long" product is produced from the extension of each primer on the plasmid template. When derived from PAPs, these single-stranded products code for the wild-type gene (Fig. 2A) and, if carried over into other steps of the process, give rise to a small but significant population of wild-type genes in the background. Separating step I into two reactions alleviates this contamination problem (Fig. 2B). In step IA, the internal primer and template are mixed and single-stranded DNA containing the mutation(s) is synthesized. The product of step IA is treated with *Dpn*I to digest the wild-type plasmid template, leaving only the nascent, single-stranded, mutagenic DNA. This restriction step eliminates the formation of long products that contribute to a wild-type background. A portion of step IA product is then used in step IB, where it serves as a template for conventional PCR amplification with an internal primer and a PAP. In the original development of SCOPE, gel purification of gene fragments (step I) was an essential step of the process. The aforementioned refinements enable the entire sequence of PCR reactions

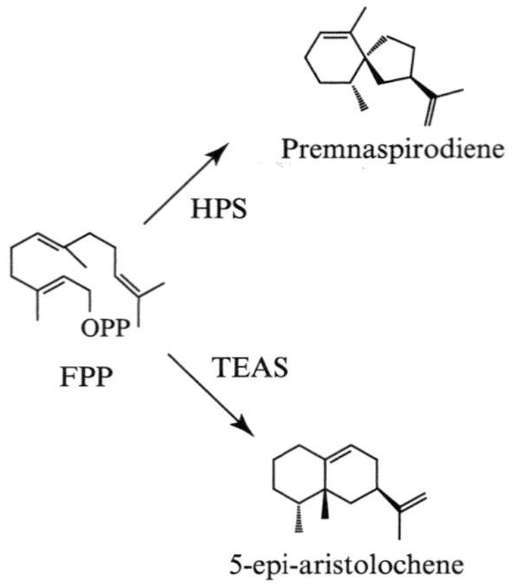

SCHEME 1. Product specificity of two closely related terpene cyclases: henbane premnaspirodiene synthase (HPS) and tobacco 5-epi-aristolochene synthase (TEAS).

(steps I through III) to be conducted without purifying intermediates until the final amplification step.

An illustration of the SCOPE methodology as applied to the combinatorial synthesis of mutant gene libraries of the terpene cyclase enzyme known as tobacco 5-epi-aristolochene synthase (TEAS) is presented as a proof of principle example of the power and ease of this newly revised version of SCOPE. Previously, the product specificity of TEAS was converted from 5-epi-aristolochene to premnaspirodiene, the product of the closely related terpene cyclase premnaspirodiene synthase (HPS) from henbane, by nine sequential mutations.[4] Site-directed mutants were designed using the three-dimensional structure of TEAS[5] and homology modeling of HPS. The products of these cyclases are shown in Scheme 1. Because these mutations were made sequentially, the question remains, are all nine mutations required for product selectivity? What combinations, if any, of these nine mutations are sufficient for this property change? Moreover, in a broader evolutionary sense, what is the mechanistic and energetic landscape that links such specificity switches that are encoded by multiple changes at sites throughout a given protein? Constructing all combinations of nine mutations (2^n combinations, where n is the number of mutations) and analyzing their product specificity and kinetic properties will provide answers to some of these questions.

[4] B. T. Greenhagen, Dissertation thesis. Department of Agriculture at the University of Kentucky, 2003.
[5] C. M. Starks et al., Science 277, 1815 (1997).

Terpene cyclases are an ideal proof of the principal system for exploring the utility of SCOPE given their (1) unusual mechanism employing the conformationally directed production of reactive carbocation intermediates, (2) well-defined three dimensional structures, (3) ease of product identification and quantification using high-throughput GC-MS analysis, (4) evolutionarily diverse distribution of protein sequences and small molecule products across multiple kingdoms, and (5) biotechnological potential for the biosynthesis of unique small molecules representing a currently untapped region of natural product space.

Experimental Procedures

Materials

PCR components: 10× cloned pfu reaction buffer and pfu turbo DNA polymerase (Stratagene, La Jolla, CA), dNTPs (Invitrogen, Carlsbad, CA), and bovine serum albumin (BSA; New England Biolabs, Beverly, MA). PCR reactions are carried out using a PTC 200 Peltier thermal cycler (MJ Research, Waltham, MA). All PCR products are purified by gel extraction (Qiagen, Valencia, CA), cloned into pDONR 207 using Gateway cloning technology (Invitrogen) according to the manufacturer's recommended conditions. Plasmid DNA from gentamicin-resistant transformants is minipreped by the Salk Institute Microarray facility for sequencing at the Salk Institute DNA sequencing/quantitative PCR facility. The cDNA of TEAS is cloned into pH8GW (an in-house gateway destination vector) and this plasmid DNA is used as a template for PCR.

All PCR are carried out using a master mix of a standard set of PCR components for a 50-μl scale reaction:

5 μl of 10× cloned pfu reaction buffer to give 1×

1 μl of pfu turbo DNA polymerase (Stratagene) (2.5 U/μl) to give 0.05 U/μl

0.5 μl of BSA (10 mg/ml) to give 0.1 mg/ml

8 μl of dNTP mix (1.25 mM) to give 200 μM each dNTP

Primers

Oligonucleotides are from Integrated DNA Technologies (IDT) and are listed in Table I. For both mutagenic and chimeric primers, the mutation(s) or crossover point(s) is located in the center of the oligonucleotide, such that the flanking sequence is complementary to a given gene; ideally, this should be 18 to 24 nucleotides (or have a T_m greater than or equal to 50°) for effective PCR. SAPs are designed to consist of 21 nucleotides

TABLE I

OLIGONUCLEOTIDES USED FOR SCOPE COMBINATORIAL MUTAGENESIS

Primer	T_m (°)	Sequence[a]
Internal mutagenic primers[b]		
A275T+	60.2	GAATGCTACTTTTGG**A**CATTAGGAGTTTATTTTGAG
V291A+	62.8	CTCGCGTCATGCTCG**C**TAAGACCATATCAATG
V372I+	60.1	GAAAGAATGAAAGAA**A**TAGTAAGAAATTATAATGTCGAGTCAAC
T402S+	62.8	CCTAAGCAATGCACTAGCAACT**T**CCACATATTAC
Y406L+	59.9	CCACATATTACT**TG**CTCGCGACAAC
T402S/Y406L+	67.9	GCACTAGCAACT**T**CCACATATTACT**TG**CTCGCGACAAC
S436N+	60.3	GAATCCAAAAATTCTTGAAGCTA**A**TGTAATTATATGTCGAG
I438T+	59.9	CTTGAAGCTAGTGTAA**C**TATATGTCGAGTTATCG
I439L+	64.2	CTTGAAGCTAGTGTAATT**CT**G**T**GTCGAGTTATCGATGAC
S436N/I438T+	61.4	AATTCTTGAAGCTA**A**TGTAA**C**TATATGTCGAGTTATCGATG
S436N/I439L+	64.1	AATTCTTGAAGCTA**A**TGTAATT**CTG**TGTCGAGTTATCGATGAC
S436N/I438T/I439L+	64.9	AATTCTTGAAGCTA**A**TGTAA**CTCTG**TGTCGAGTTATCGATGAC
I438T/I439TL+	65.1	CTTGAAGCTAGTGTAA**CTCTG**TGTCGAGTTATCGATGAC
V516I+	60.3	CTCAATCTTGCTCGTATT**A**TTGAGGTTACATATATACAC
Primary amplification primers[c]		
A_b1	72.1	ATGCTTTAAGGCTCTGGGCCGACAAGTTTGTACAAAAAAGCAGGC
B_b1	71.7	GTCACGCATATGATTCGGCGGACAAGTTTGTACAAAAAAGCAGGC
C_b1	74.9	CCTGGCTTGCTCGGATAGAACACAAGTTTGTACAAAAAAGCAGGC
D_b1	73.4	CTACAGAGAATGCCGCGGTCCACAAGTTTGTACAAAAAAGCAGGC
E_b1	68.6	GGTCGTCGACCCCAGCGTGAGACAAGTTTGTACAAAAAAGCAGGC
F_b1	70.8	TCGCAATTCACGGCTTGACCCACAAGTTTGTACAAAAAAGCAGGC
1_b2	70.6	AAATGCAGGTAGCAGAGCTGTACCACTTTGTACAAGAAAGCTGGG
2_b2	71.6	TTAGGACGGGACTGCTGTAGCACCACTTTGTACAAGAAAGCTGGG
3_b2	73.4	ACATTCTGACGTGAAACGCGCACCACTTTGTACAAGAAAGCTGGG
4_b2	72.0	GGACACACTTAGCCTTCCAGGACCACTTTGTACAAGAAAGCTGGG
5_b2	74.1	GGCCTGGAGTAGGATCTTTGCACCACTTTGTACAAGAAAGCTGGG
6_b2	70.2	CCCGTCCCACTTCGTGACCGCACCACTTTGTACAAGAAAGCTGGG
Secondary amplification primers[d]		
A	61.3	ATGCTTTAAGGCTCTGGGCCG
B	59.7	GTCACGCATATGATTCGGCGG
C	58.7	CCTGGCTTGCTCGGATAGAAC
D	60.9	CTACAGAGAATGCCGCGGTCC
E	65.1	GGTCGTCGACCCCAGCGTGAG
F	61.7	TCGCAATTCACGGCTTGACCC
1	58.2	AAATGCAGGTAGCAGAGCTGT
2	59.7	TTAGGACGGGACTGCTGTAGC
3	59.9	ACATTCTGACGTGAAACGCGC
4	58.4	GGACACACTTAGCCTTCCAGG
5	58.6	GGCCTGGAGTAGGATCTTTGC
6	65.7	CCCGTCCCACTTCGTGACCGC

[a] Bold and underlined characters indicate the sites of designed mutations. Shading is used to indicate part of the attB1 and attB2 recombination sequences; complete attB sites are generated by amplification of the target gene(s) from the destination vector pH8GW.

[b] Mutageneic primers are named according to the amino acid substitutions they code for.

and have a T_m greater than or equal to 55°. PAPs contain 24 bases (in addition to their unique sequence), which correspond to partial Gateway attB sites; the remaining attB sequence becomes incorporated into PCR products by amplification from pH8GW. T_m values are calculated based on nearest-neighbor thermodynamic parameters.[6]

Gel Electrophoresis

Analysis of PCR fragments and separation of products for gel purification are performed using 2% (w/v) agarose gels in 1× TAE buffer containing 0.1 μg/ml ethidium bromide. Concentrations of PCR products (steps IB and III) are estimated by comparison to a standard of known concentration, such as the low DNA mass ladder (Invitrogen) using densitometry software such as ImageJ (http://rsb.info.nih.gov/ij/).

Method

Prior to library construction, all primers are tested to ensure that they result in unique amplification products of the expected size. Like any standard PCR amplification, optimization of cycling parameters for specific template and primer sets may be necessary.

Step IA: Mutagenic/Chimeric Single-Stranded DNA Synthesis

Procedure. Reactions are mixed on ice using the following:
14.5 μl of PCR master mix (as defined earlier)
1 μl internal primer (5 μM stock) to give 0.1 μM
1 μl plasmid DNA template (10 nM stock) to give ~200 pM
33.5 μl filter-sterilized H_2O added to give a 50-μl reaction volume
The master mix is added last and the resultant reaction is mixed by pipetting. Cycling program: 96° for 5 min, followed by 50 cycles of 96° for 30 s, 55° for 30 s, and 72° for 1 min/kb of product followed by incubation at 4° at the completion of cycling.

Comments. The amount of single-stranded product formed is limited by the amount of template DNA and the number of cycles performed.

---

[6] H. T. Allawi and J. Santa Lucia, Jr., *Biochemistry* **36,** 10581 (1997).

---

[c] Primary amplifications primers are named according to their unique sequence tag (A through F for forward and 1 through 6 for reverse as listed in Fig. 4) and gateway recombination sequence (b1 for attB1 and b2 for attB2).
[d] Unique sequence tags are labeled according to their corresponding primary amplification primer (A through F for forward and 1 through 6 for reverse as listed in Fig. 4).

Estimated yields for the aforementioned reaction (using 50 cycles and ~200 pM plasmid) are about 10 fmol of the final single-stranded product; this amount is well in excess of what is required for subsequent amplification reactions. A 0.1 μM concentration of internal primer ($>10^3$ molar excess of plasmid template) is sufficient; higher primer concentrations may promote the formation of alternative products in subsequent amplification steps (step IB).

Dpn*I Digestion of Plasmid DNA*

Procedure. The addition of 1 μl of *Dpn*I (20 U/μl, New England Biolabs) with mixing is followed by incubation at 37° for 1 h for digestion of the original DNA template and 20 min at 80° for heat inactivation.

Step IB: Mutagenic/Chimeric Double-Stranded DNA Fragment Amplification

Procedure. Reactions are mixed on ice using the following:
   14.5 μl of PCR master mix (as defined earlier)
   2 μl internal primer (5 μM stock) to give 0.2 μM
   1 μl primary amplification primer (5 μM stock) to give 0.1 μM
   1 μl of step IA reaction as template to give ~1–10 pM single-stranded DNA
   31.5 μl filter-sterilized H$_2$O added to give a 50-μl reaction volume
The master mix is added last with pipetting to mix reactions. Cycling program: 96° for 5 min, followed by 40 cycles of 96° for 30 s, 55° for 30 s, and 72° for 1 min/kb of product followed by incubation at 4° at the completion of cycling. Amplification products are verified by agarose gel electrophoresis.

Comments. Internal primers should be in excess of external primers; keeping the concentration of external primers below saturation and increasing the number of cycles ensure their depletion. This is sufficient to suppress wild-type background arising from "long" products generated during subsequent amplification steps from carryover of the external primer. The step IA product can be diluted up to 10,000-fold and still provide enough template for robust amplification.

Step II: Recombination

SINGLE MUTANTS/CROSSOVERS

Procedure. Reactions are mixed on ice using the following:
   5.8 μl of PCR master mix (as defined earlier)
   1 μl of step IB reaction to give ~10 nM (or 1–5 ng/μl) gene fragment

1 μl plasmid DNA template (10 nM stock) to give $\sim$200 pM final (1 ng/μl for a 7-kb plasmid)

12.2 μl filter-sterilized H_2O added to give a 20-μl reaction volume

The master mix is added last with pipetting to mix reactions. Cycling program: 96° for 5 min, followed by 15 cycles of 96° for 30 s ($+2''$/cycle), 55° for 30 s, and 72° for 1 min/kb of product followed by incubation at 4° at the completion of cycling.

Multiple Mutants/Crossovers

Procedure. Same as just described, except the gel-purified full-length mutant/chimeric gene (step III product) at $\sim$1.0 ng/μl ($\sim$1 nM final concentration) is substituted for plasmid DNA.

Multiplex Recombination

Procedure. A mixture of gene fragments (step IB products) corresponding to a collection of mutations or alternative crossovers is pooled, and 1 μl (to give $\sim$10 nM) is used with either a plasmid or a full-length mutant/chimeric gene (step III product) as the template in a recombination reaction.

Comments. The amount of full-length, single-stranded recombination product produced in step II is limited by the amount of gene fragment from step IB added to the reaction mixture. Gene fragments should ideally be 1- to 10-fold molar excess of the plasmid or mutant gene that it is recombining with. This is particularly important in the case of *single mutants/crossovers*, where only one terminus can be exploited in the following step for selective amplification. The plasmid concentration should be kept to a minimum; about 10 pM is the lowest concentration that can be used to give the amplifiable recombination product in step III.

Step III: Selective Amplification of Recombination Products

Amplification of Single Mutants/Crossovers

Procedure. Reactions are mixed on ice using the following:

14.5 μl of PCR master mix (as defined earlier)

2 μl secondary amplification primer (5 μM stock) to give 0.2 μM

2 μl primary amplification primer (5 μM stock) to give 0.2 μM

1 μl of step II reaction as template to give $\sim$100–200 pM single-stranded DNA

30.5 μl filter-sterilized H_2O added to give a 50-μl reaction volume

The master mix is added last with pipetting to mix reactions. Cycling program: 96° for 5 min, followed by 30 cycles of 96° for 30 s, 55° for 30 s, and 72° for 1 min/kb of product followed by an additional 10 min at 72° and incubation at 4° at the completion of cycling. Amplification products are verified by agarose gel electrophoresis.

Amplification of Multiple Mutants/Crossovers

Procedure. Same as *amplification of single mutants/crossovers*, but only secondary amplification primers are used.

Comments. The final step in a cycle of SCOPE is a standard PCR amplification of full-length mutant/chimeric genes with unique sequence tags at both 5' and 3' ends. In synthesis of the first generation of mutants, only one SAP can be used for selective amplification. This corresponds to the unique sequence of the PAP used in step IB. A PAP is directed at the opposite terminus, where it incorporates unique sequence at this terminus. Because this primer is directed to the flanking sequence of the gene, it can efficiently prime any carryover long product (single-stranded wild-type DNA) from step IB or any plasmid from step II. Therefore, it is important to eliminate the long product from step IB and minimize the amount of plasmid in step II, as the single-stranded product generated at this step has the potential to carry over into subsequent rounds of synthesis. In the step III amplification of multiple mutants/crossovers, SAP combinations are chosen to allow selective amplification of desired recombination products.

Product Isolation and Cloning

Full-length mutant genes from step III are gel purified using the Qiagen gel extraction kit according to the manufacturer's recommended procedures. Gel-purified attB PCR products are cloned into pDONR207 via the gateway BP reaction according to the manufacturer's recommendations.

Controls

DpnI Treatment. Digestion is omitted and the step IB reaction is performed using the undigested step IA product as the template. Because plasmid DNA is carried over into the step IB reaction, PAPs can be extended to produce wild-type single-stranded DNA as described previously (Fig. 2A). As a result, wild-type genes can be amplified efficiently using a 1-μl portion of step IB as the template and a PAP and SAP primer pair. If the step IB reaction is performed using a 10-fold molar excess of mutagenic primer, then the amount of amplifiable wild-type gene decreases markedly. In fact, the combination of increasing the number of cycles in step IA to 100 (resulting in 2-fold more template) and using a 10-fold excess of internal mutagenic primer in step IB enables the suppression of wild-type background and a mutagenesis efficiency of 80%, as apparent from terpene cyclase libraries produced in this manner.

Step IB Products. The selectivity of amplification (or the suppression of wild-type sequences) is evaluated using the step IB reaction as the template. If *Dpn*I digestion is complete, no amplifiable wild-type product is observed. In the case where restriction digestion is omitted, wild-type product is observed.

Results and Discussion

SCOPE was applied to create a library representing all possible combinations of nine point mutations in the terpene cyclase, TEAS. The location of mutations in the amino acid and nucleotide sequences of TEAS are indicated in Fig. 3. The nine positions were recombined as six units (shown in boxes). Some mutations were clustered, requiring a plurality of internal primers. For example, amino acid positions 436, 438, and 439 required a collection of seven internal primers to code for all permutations: three single, three double, and a triple mutant. A system was developed to introduce unique sequence tags; PAPs link mutation (or a collection thereof) to a unique sequence during gene fragment amplification and SAPs enable selective amplification of the desired combinations of mutations. An illustration of their use and the nomenclature system is described in Fig. 4.

An attribute of SCOPE synthesis is the fractionation of complex mixtures into many simpler ones. This has the benefit of reducing the numerical complexity and hence the screening requirements necessary to verify and identify the collection of desired changes. This effect arises from sampling probability as described by the following mathematical expression:

$$p(n) = 1 - \sum_{i=1}^{n-1} (-1)^{i+1} \frac{n!}{i!(n-i)!} \left[\frac{(n-i)}{n} \right]^k \qquad (1)$$

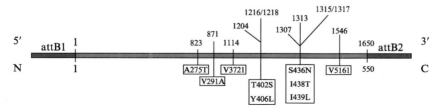

FIG. 3. Location of mutations in TEAS recombined by SCOPE. The nucleotide positions and corresponding amino acid changes are shown above and below, respectively

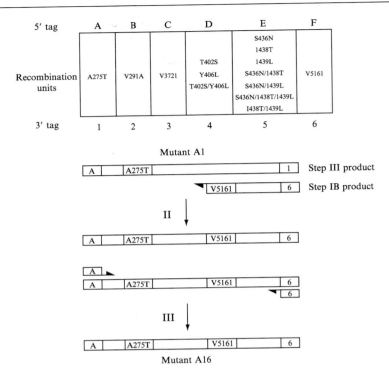

Fig. 4. Recombination units and tagging system. The recombined positions and their associated unique sequence tags give rise to a naming system to describe the recombination product created by SCOPE.

where k is the sample size, n is the number of unique members, and p is the probability that a sample of size k contains at least one representative of each unique member. As complexity increases, the amount of oversampling required to achieve the same probability of screening the library increases, where oversampling refers to sample size (k) in multiples of library complexity (n). This can be shown graphically in Fig. 5.

Each iteration of the process ends with a conventional PCR amplification step, and after multiple iterations, additional mutations accumulate. The overall frequency of nondesired additional mutations in the population analyzed is 5.5%. No strong bias for the type of error or its location within the gene was observed. The nondesired mutation rate after the first round was 2.67%, which matches previous measures of pfu error frequency.[7] However, the random mutation rate increases as a function of

[7] J. Cline, J. C. Braman, and H. H. Hogrefe, *Nucleic Acids Res.* **24,** 3546 (1996).

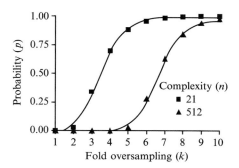

Fɪɢ. 5. Sampling probability as a function of oversampling. The probability that a sample contains one copy of each unique clone for a given complexity *(n)* is calculated using Eq. (1). Probability is calculated for a range of sample sizes *(k)* that are in multiples of a fixed library complexity *(n)* and the results are fit to a sigmoidal curve.

iterations of SCOPE and, after four iterations, reached 8.9%. Using a higher fidelity polymerase can minimize random mutation rates. Alternatively, products from step III amplification reactions can be cloned and the SCOPE cycle started anew (from step IA). Bridging oligonucleotides may be useful in this case to recombine various mutations. Also, gene fragments (from step IB) can be made to include multiple mutations from previous cycles.

In the example used here, 512 mutants were made from a series of simpler mixtures, the most complex of which contained 21 unique members. To achieve a 50% probability ($p = 0.5$) of identifying by screening every unique member in a mixture of that complexity requires 3.38-fold oversampling. To achieve the same probability of identifying all unique members of a library by screening a mixture of 512 unique possibilities requires 6.6-fold oversampling. Given the exponential relationship between sample size and library complexity, this difference equates to a reduction in numerical complexity of a factor of 25 for the entire library.

Library Analysis

Over 600 colonies from discrete mixtures, representing about half of the complexity of the TEAS library (241 unique members), were picked and their sequences determined. A summary of the results is listed in Table II. Of the clones sequenced, only 24 wild-type genes (3.5%) were found. This library was synthesized prior to addition of the *Dpn*I restriction step (as described earlier), and while the efficiency of the first round of mutagenesis was ~80%, the overall efficiency of the entire process reached 96.5%

TABLE II
SEQUENCE ANALYSIS RESULTS

Library statistics	
Clones sequenced	692
Wild-type genes	24
Percentage of mutants	96.5%
Complexity screened	241
Unique clones identified	193
Fold oversampled	2.8
Complexity covered	80.1%
Total library complexity	512
Percentage of verified mutants	37.70%
Additional mutations	
Silent	9
Frameshift	16
Point mutants	13
Total	38
Mutation rate	5.49%

(Table II). Mutations become incorporated into the wild-type sequence during recombination reactions in subsequent iterations of the process. As a result, wild-type sequences vanish in multiple crossover populations.

Aside from the low-level appearance of wild-type sequence and random mutations likely arising from PCR errors, the actual distribution of mutations in a given mixture was as designed experimentally. Some recombination reactions produce a single product having several designed mutations, such as A1236. In reactions containing multiple mutations, the reaction distribution appears random.

Concluding Remarks

Adaptation of SCOPE to combinatorial mutant library design and construction demonstrates the broader utility of these library construction principles. While various techniques have been developed for either homology-independent recombination or combinatorial mutagenesis, none can efficiently do both. SCOPE provides an effective means for the creation of both global or local sequence space as demonstrated by the synthesis of DNA libraries representing the genetically encoded information spanned by distant homologues[1] or closely related members of a gene family.

Acknowledgments

We are grateful to the National Institutes of Health for the grants that supported this work (GM43268 to M.D.T. and GM54029 to J.C. and J.P.N.). P.E.O. is an NIH Postdoctoral Research Fellow (GM069056-01). Additionally, we thank Marina Bakhtina and Brandon Lamarch for valuble consultations during the early phases of this work.

[9] New Enzymes from Combinatorial Library Modules

By Werner Besenmatter, Peter Kast, and Donald Hilvert

Introduction

Current strategies for the construction of combinatorial gene libraries for directed evolution experiments generally make use of cassette mutagenesis[1,2] to insert library modules[3–6] into plasmids. We have applied this technique in a variety of formats to investigate chorismate mutase, a key enzyme in the biosynthesis of aromatic amino acids.[7] Active variants are directly selected from gene libraries transformed into a chorismate mutase-deficient *Escherichia coli* strain (Fig. 1).[8] Because catalytic activity is an extremely sensitive probe for protein integrity, a wealth of information on structural and functional aspects of this enzyme can be derived from sequence patterns in selected variants.[9]

The extent of randomization of the gene library cassettes depends on the questions asked. For instance, to investigate the roles of individual active site residues, one or two codons were randomized at a time.[8,10] When loops connecting secondary structural elements were (re-)designed, we opted for formats mutagenizing three to seven codons

[1] J. A. Wells, M. Vasser, and D. B. Powers, *Gene* **34,** 315 (1985).

[2] J. F. Reidhaar-Olson and R. T. Sauer, *Science* **241,** 53 (1988).

[3] S. Kamtekar, J. M. Schiffer, H. Xiong, J. M. Babik, and M. H. Hecht, *Science* **262,** 1680 (1993).

[4] G. Cho, A. D. Keefe, R. Liu, D. S. Wilson, and J. W. Szostak, *J. Mol. Biol.* **297,** 309 (2000).

[5] S. V. Taylor, K. U. Walter, P. Kast, and D. Hilvert, *Proc. Natl. Acad. Sci. USA* **98,** 10596 (2001).

[6] T. Matsuura, A. Ernst, and A. Plückthun, *Protein Sci.* **11,** 2631 (2002).

[7] E. Haslam, "Shikimic Acid: Metabolism and Metabolites." Wiley, Chichester, UK, 1993.

[8] P. Kast, M. Asif-Ullah, N. Jiang, and D. Hilvert, *Proc. Natl. Acad. Sci. USA* **93,** 5043 (1996).

[9] S. V. Taylor, P. Kast, and D. Hilvert, *Angew. Chem. Int. Ed.* **40,** 3310 (2001).

[10] P. Kast, J. D. Hartgerink, M. Asif-Ullah, and D. Hilvert, *J. Am. Chem. Soc.* **118,** 3069 (1996).

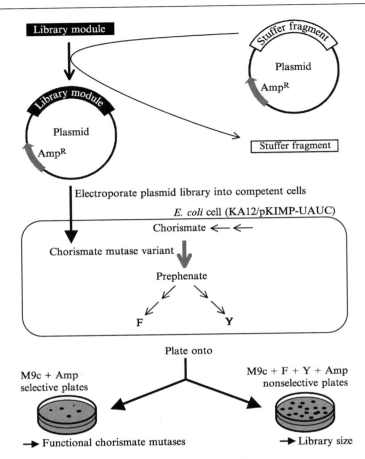

Fig. 1. From a library module to functional enzymes. The library module is inserted into the acceptor plasmid by replacing a stuffer fragment. The resulting plasmid library is transformed into the selection strain KA12/pKIMP-UAUC, which has deletions of both endogenous chorismate mutase genes.[8] After transformation, the pool of cells is divided. One portion is plated onto M9c minimal medium agar plates[11] (additionally containing 150 μg/ml sodium ampicillin and 30 μg/ml chloramphenicol)[5] to select for active chorismate mutase gene variants from the library.[8] Chorismate mutase, which catalyzes the conversion of chorismate to prephenate in the biosynthesis of the aromatic amino acids phenylalanine (F) and tyrosine (Y), is essential for survival of the strain on minimal medium in the absence of F and Y. Another portion of the cells is plated onto nonselective plates (which—in addition to containing the antibiotics—are supplemented with F and Y) to determine the number of transformants. Amp[R], ampicillin resistance.

[11] P. Kast, C. Grisostomi, I. A. Chen, S. Li, U. Krengel, Y. Xue, and D. Hilvert, *J. Biol. Chem.* **275,** 36832 (2000).

simultaneously.[12–14] To assess the importance of the structurally ill-defined region at the C terminus of the AroH class chorismate mutase from *Bacillus subtilis*, the randomized stretch was increased to 12 and even 17 codons; additionally, it was biased to enrich for stop codons to obtain a set of truncated but still active variants.[15] By taking the cassette mutagenesis approach to extremes, we have created binary patterned proteins that still possess chorismate mutase activity but are composed of a reduced alphabet of just four hydrophobic and four hydrophilic amino acids at the randomized positions.[5] This article discusses in greater detail the strategies currently used in our laboratory to generate large patterned libraries from long synthetic oligonucleotides.

Methods

Although many proteins can be mutagenized extensively without loss of fold or function, it is generally believed that native-like proteins occur extremely infrequently in random sequence space.[5,16,17] To explore the extent to which sequence can be varied without loss of catalytic activity, we have replaced all the secondary structural elements in the homodimeric, helical bundle AroQ class chorismate mutase from *Methanococcus jannaschii* with modules of random sequence and selected functional variants by complementation of our chorismate mutase-deficient *E. coli* strain (Fig. 1).[5] The modules themselves can be fully randomized or biased in various ways, for example, to follow the inherent binary pattern of hydrophobic and hydrophilic residues in the parent enzyme or by using restricted sets of building blocks. The helices can be replaced individually or in combination, and separately constructed library modules can be crossed to obtain more comprehensively randomized enzymes (Fig. 2).

Our basic strategy is illustrated by the construction of two different H1 library modules for an *E. coli* chorismate mutase (Fig. 2A and C) for studies on the influence of the templating scaffold and different amino acid sets. In analogy to earlier work on the homologous *M. jannaschii* protein,[5] the catalytic residues at positions 11, 28, and 39 are held constant but the rest of the sequence is randomized according to a binary pattern. The modules differ in the number of polar and apolar amino acids permitted.

[12] G. MacBeath, P. Kast, and D. Hilvert, *Protein Sci.* **7**, 325 (1998).

[13] G. MacBeath, P. Kast, and D. Hilvert, *Science* **279**, 1958 (1998).

[14] G. MacBeath, P. Kast, and D. Hilvert, *Protein Sci.* **7**, 1757 (1998).

[15] M. Gamper, D. Hilvert, and P. Kast, *Biochemistry* **39**, 14087 (2000).

[16] A. D. Keefe and J. W. Szostak, *Nature* **410**, 715 (2001).

[17] D. A. Moffet and M. H. Hecht, *Chem. Rev.* **101**, 3191 (2001).

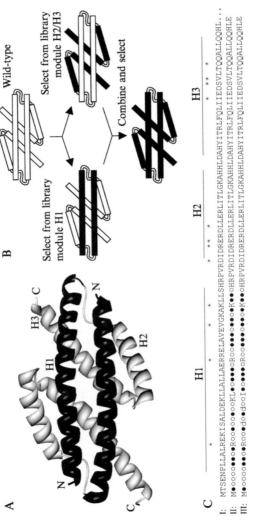

H1

```
        *            *        *       *           *        *       *       *
I:   MTSENPLLAIREKISALDEKLLALLAERRELAVEVGKAKLLSHRPVRDIDRERDLLERLITLGKAHHLDAHYITRLFQLIIEDSVLTQQALLQOHL...
II:  M●ooooo●●●Roo●oo●oKL●●●oRoo●●●o●K●●oHRPVRDIDRERDLLERLITLGKAHHLDAHYITRLFQLIIEDSVLTQQALLQOHLE
III: M●ooooo●●●Roo●d●dooI●●●●oRoo●●●oK●●oHRPVRDIDRERDLLERLITLGKAHHLDAHYITRLFQLIIEDSVLTQQALLQOHLE
        *
```

H2

```
              *   **     *
```

H3

```
              *   **      *
```

FIG. 2. (A) Structure of the all α-helical, homodimeric *E. coli* enzyme chorismate mutase.[18] The H1 helices that span the dimer are shown in black. (B) Combinatorial strategy to select active enzymes containing new library modules (in black) using the wild-type sequence (in white) as the template. (C) Sequence of wild-type *E. coli* chorismate mutase (I) and binary-patterned libraries based on eight (FILMDENK, II) or four (IKEA, III) amino acid building blocks. Filled and open circles correspond to apolar and polar positions, respectively. Capital letters indicate amino acids held constant by design. The lower case d indicates positions where I, K, E, A, and D were offered. Lines above the sequence show the lengths of the H1, H2 and H3 helices; active site residues are denoted with asterisks.

The first library utilizes Asp, Glu, Asn, and Lys as the hydrophilic residues and Phe, Ile, Leu, and Met as the hydrophobic residues. The former are encoded by the RAV codon (where R represents a mixture of A and G; and V = A + C + G) and the latter by the HTS codon (where H = A + C + T; and S = C + G). Equimolar mixtures of nucleotides at the variable positions lead to a probability distribution of 1:2:1:2 for Asp:Glu:Asn:Lys and 1:1:3:1 for Phe:Ile:Leu:Met.

In the second library, only two polar amino acids, Glu and Lys, and two apolar amino acids, Ile and Ala, are allowed. The polar amino acids are specified by the degenerate RAA codon in a 1:1 ratio. As there is no consensus codon sequence that specifies only Ile and Ala, a different strategy is required to create mixtures of these two residues. We employed the "split and mix" technique for oligonucleotide synthesis.[19] In this approach, the resin is split in a 3:1 ratio at positions corresponding to apolar amino acids, and the ATT or GCG codons, which respectively encode Ile and Ala, are synthesized separately. The resin is then remixed and the synthesis continued. The advantage of this approach is that all synthetic steps are conducted on a standard DNA synthesizer without special reagents, such as trinucleotides. Moreover, the number, proportion, and choice of codons can be varied at will.[19]

The dimer-spanning H1 helix of the *E. coli* chorismate mutase is 37 amino acids long (Fig. 2). Because an oligonucleotide corresponding to this segment plus the flanking regions required for cloning is too long for current synthetic methodology, we construct the H1 modules from two oligonucleotides, each about 100 nucleotides long (Fig. 3). The degenerate oligonucleotides are hybridized via complementary nonvariable sequences at their 3′ ends that also serve as primers for DNA polymerase-mediated, second-strand synthesis. By designing nonvariable regions that cannot form hairpin structures, problems associated with self-priming can be avoided (Fig. 4). The double-stranded DNA fragment is then inserted into a plasmid to obtain a fusion with the 3′ wild-type portion of the gene for *E. coli* chorismate mutase. The DNA overlap region in the middle of the H1 module is subsequently excised by restriction digestion and religation (Figs. 3 and 4). This procedure yields a degenerate gene library in which all but the codons for the catalytic residues Arg11, Arg28, and Lys39 (Fig. 2C) and residue 21 at the splice site (Fig. 4B) have been randomized.

[18] A. Y. Lee, P. A. Karplus, B. Ganem, and J. Clardy, *J. Am. Chem. Soc.* **117**, 3627 (1995).
[19] W. D. Huse, *United States Patent* US005808022A (1998).

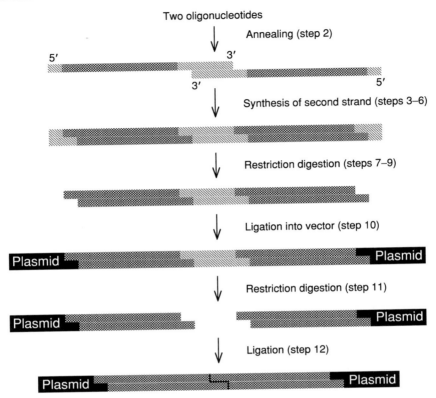

Fig. 3. Synthesis of a large library module from long degenerate oligonucleotides. DNA from the (circular) acceptor plasmid is shown in black, DNA that is removed during library construction is striped, and all other DNA is checkered. Steps are numbered according to the experimental protocol in the text.

Protocol

The following protocol details the construction of a library module assembled from two large degenerate oligonucleotides. The individual steps are illustrated in Fig. 3.

1. Synthesize oligonucleotides by standard methods or obtain them from a commercial supplier, preferably already purified by polyacrylamide gel electrophoresis (PAGE).

2. Combine 0.6 nmol (ca. 18 μg) of each oligonucleotide, 80 μl dNTP mix (each dNTP at 2 mM), 100 units Klenow fragment of *E. coli* DNA

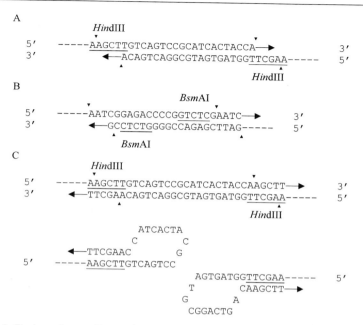

FIG. 4. Design of annealing region sequences. The direction of DNA synthesis after annealing is shown by arrows. Recognition sites for restriction enzymes are underlined and their cleavage sites are indicated by small arrowheads. Designs A and B were reduced successfully to practice. Design C is very similar to A, but it results in unwanted self-priming due to hairpin formation during second-strand synthesis. In design A, two fixed codons are required at the splice site because of the constraints imposed by the HindIII restriction site. In design B, a restriction enzyme (BsmAI) was chosen that cuts outside the recognition site. As a consequence, only a single amino acid codon needs to be held constant at the splice site. The identity of this encoded amino acid can be changed easily because BsmAI has no sequence preference at the site of cleavage.

polymerase I (New England Biolabs, Beverly, MA), and 40 μl of the supplied 10× buffer stock solution in a total volume of 400 μl.

3. Incubate at 37° for 2 h.

4. Add 100 μl 5× DNA heat stabilizer solution (5 M NaCl, 65 mM EDTA, pH 7.0).

5. Incubate at 75° for 20 min to inactivate the polymerase.

6. Purify by ultrafiltration, e.g., with a Nanosep 3K Omega centrifugal device (Pall, Ann Arbor, MI). Replenish with the buffer supplied with the restriction endonuclease to be used in step 7. Repeat ultrafiltration and redilution three times.

7. Digest with appropriate restriction endonucleases. We used 330 units of each *Nde*I and *Eag*I for 3h.

8. Add 5× DNA heat stabilizer solution and heat inactivate the restriction endonucleases.

9. Purify by nondenaturing PAGE (in TBE buffer). We used a precast gel (Criterion 345-0052) from Bio-Rad (Hercules, CA) and desalted by ultrafiltration prior to electrophoresis. After staining the gel with ethidium bromide, cut out the slice containing DNA of the correct size under UV light (use long wavelength lamp, e.g., $\lambda = 365$ nm). Mash the gel slice in a petri dish with the pistil from a single-use syringe. Extract the DNA with about 2.5 ml elution buffer (10 mM Tris–HCl, pH 7.5, 50 mM NaCl, 1 mM EDTA) per gram of gel slice. Shake at 30° overnight. Centrifuge the suspension and remove the supernatant. Wash the crushed gel with elution buffer and centrifuge again. Concentrate the combined supernatants by ultrafiltration. Replenish with 2 mM Tris–HCl buffer (pH 8.5). Repeat ultrafiltration and replenishing two times.

10. Use T4 DNA ligase to insert the gene fragment from step 9 into a gel-purified acceptor vector cleaved previously with the same restriction endonucleases as in step 7. Roughly 4–8 μg of a 3-kb acceptor and 1–3 molar equivalents of insert in a total volume of 400 μl give the best results.[20] We purified the ligation product with the QIAquick PCR purification kit (Qiagen, Hilden, Germany).

11. Digest with appropriate restriction endonuclease to remove the oligonucleotide annealing region (see Figs. 3 and 4). We used *Bsm*AI and purified the linearized plasmid with the kit described in step 10.

12. Ligate all of the purified product of step 11 in a total volume of 1.5 ml (the low DNA concentration favors intramolecular ligation over intermolecular concatemer formation). Afterwards, concentrate the solution by ultrafiltration. For purification we used the kit described in step 10.

13. Electroporate[21,22] the plasmid library into your selection strain and grow transformed cells under selective conditions.

Results and Discussion

We have used the aforementioned protocol successfully with several different sets of oligonucleotides to prepare randomized modules corresponding to the H1 helix in AroQ chorismate mutases (unpublished

[20] A. Dugaiczyk, H. W. Boyer, and H. M. Goodman, *J. Mol. Biol.* **96,** 171 (1975).
[21] D. Hanahan, J. Jessee, and F. R. Bloom, *Methods Enzymol.* **204,** 63 (1991).
[22] W. J. Dower, J. F. Miller, and C. W. Ragsdale, *Nucleic Acids Res.* **16,** 6127 (1988).

and published[5] results). Analogous procedures have been employed to randomize the other two helices in the enzyme and to combine these modules combinatorially (Fig. 2B).[5] Modules derived from oligonucleotides containing up to 108 nucleotides were assembled successfully to yield such libraries.[5]

To obtain the frequency of functional variants in our randomized chorismate mutase libraries, the ligation products (e.g., of step 12 in the protocol) were electroporated into the chorismate mutase-deficient *E. coli* strain KA12/pKIMP-UAUC.[8] Control experiments showed that electro-competent cells of this strain typically yield 1×10^8 transformants per μg of supercoiled plasmid DNA [pGEM-3Zf(+); Promega, Madison, WI] with large amounts of DNA (1 μg plasmid transformed). Applying the same electroporation procedure to the library plasmids prepared as described in the protocol, we typically obtain between 10^6 and 10^7 independent transformants per library (see library size determination in Fig. 1).

The probability of finding functional clones able to grow under selective conditions obviously depends on library design. For example, for an eight amino acid H1 library based on the *M. jannaschii* chorismate mutase, active enzymes occurred at a frequency of approximately 1 in 4500 clones.[5] In this case, the typical library sizes were more than adequate to obtain large numbers of sequences to analyze for sequence patterns that contribute to enzyme stability and function. Similar frequencies were obtained for the *M. jannaschii* H2/H3 library and when the active members of the two libraries were crossed to obtain chorismate mutases in which all the secondary structural elements were replaced by randomized modules (Fig. 2B).[5]

Instead of combining only the functional members of the H1 and H2/H3 library modules in the combinatorial library of Fig. 2B, another strategy could be employed that inserts the second library module directly into positives from the first library. As an alternative, both library modules could also be combined directly in one step; however, depending on the number of functional sequences in the individual modules, the frequency of active enzymes in the combinatorially assembled libraries may be too low to be experimentally accessible.[5,16]

Tricks and Pitfalls

Before engaging in the construction of large combinatorial gene libraries, which generally represents a considerable investment of time and resources, it is useful to reflect about potential problems and how to circumvent them.

Contamination with Wild-Type Gene

One possible reason for false positives is contamination with a wild-type sequence, either from a natural source or from a precursor to the random library. To reduce the risk of the latter, an acceptor plasmid with a stuffer fragment is used in step 10 of the protocol. The stuffer fragment replaces the part of the wild-type sequence that is ultimately replaced with the library module (Fig. 1). The length of the stuffer DNA (we often use a ca. 1-kb fragment from phage λ) should be chosen such that the desired doubly cut acceptor fragment can be purified easily from the singly cut plasmid and that the correct recombinants can later be distinguished from the acceptor vector by a simple analytical restriction digest.

Plasmid Mixtures

Another possible source of false positives is the coexpression of an active and an inactive library gene in one cell. The addition of salt, which stabilizes double-stranded DNA,[23] in steps 4 and 8 of the protocol can alleviate this problem. In the absence of salt, short double-stranded library DNA might denature easily upon heating. After cooling, individual DNA strands that are not perfectly complementary could reanneal and form heteroduplex DNA, which might be cloned into the acceptor vector. This DNA would be transformed as one plasmid, but would subsequently be propagated as two plasmid variants in one cell. Our protocol also limits exposure of the library module to chaotropic salts, which are generally employed in gel extraction and DNA purification kits but denature short double-stranded DNA, until the module is safely ligated to the acceptor plasmid (step 10).

Furthermore, to avoid wasting time on false positives, it is crucial to confirm that phenotype and genotype match. Plasmids isolated from selected library clones need to be retransformed into the selection strain to verify *in vivo* activity. However, it is of utmost importance to dilute the plasmid preparation appropriately prior to retransformation, since at high DNA concentrations it is again possible that two different library plasmids will be electroporated into a single cell.

Quality and Diversity of the Library

The outlined method for the construction of library modules does not require DNA amplification by PCR, which would increase the number of DNA copies but not the diversity of the library. In addition, PCR could

---

[23] J. G. Wetmur, *Crit. Rev. Biochem. Mol. Biol.* **26,** 227 (1991).

skew the library composition due to copying artifacts. It is nevertheless recommended to check the DNA sequences of a representative number of clones picked randomly from nonselective media (we usually look at 20 variants) to confirm the absence of a bias at the randomized positions. It is not uncommon to find that only a minority of the library members (for the randomized H1 module between 30 and 40%) has the correct length.[3–5,24] In fact, we observe all types of point mutations—insertions, deletions, and substitutions—in library members that grow in the absence of selection pressure. These may have arisen during chemical synthesis of the oligonucleotides, *in vitro* synthesis by a DNA polymerase, or *in vivo* DNA synthesis in *E. coli*. From the observation that such unplanned mutations are generally much more frequent in stretches derived from synthetic oligonucleotides, chemical synthesis can be singled out as the main source of such errors.

Oligonucleotide Artifacts

Unplanned mutations are commonly observed in any approach that relies on chemically synthesized oligonucleotides.[1,3–5,24] Single base deletions can originate during chemical synthesis from incomplete coupling and capping as well as from inefficient removal of the 5′ protection group prior to coupling.[25] Insertions can result from cleavage of the 5′ protection group on the phosphoramidite building blocks before and during coupling. Contamination of the phosphoramidites, potentially due to the use of common supply lines in the DNA synthesizer, can lead to base substitutions.[25] In our experience, longer oligonucleotides are damaged more extensively than shorter ones, presumably because of longer exposure to the reagents (e.g., resulting in nonspecific acylations).

Nevertheless, mutations introduced into DNA libraries through oligonucleotide synthesis generally do not represent a serious problem if library members are subsequently subjected to a selection procedure such as the one described here. For instance, the requirement for function automatically eliminates frameshifts leading to dysfunctional gene products.

Small Library Sizes

Many factors could be responsible for a low yield of library clones. Some of these include the use of inappropriate or inefficient restriction endonucleases or competent cells. Contamination by nucleases or by

[24] J. E. Ness, S. Kim, A. Gottman, R. Pak, A. Krebber, T. V. Borchert, S. Govindarajan, E. C. Mundorff, and J. Minshull, *Nature Biotechnol.* **20,** 1251 (2002).
[25] K. H. Hecker and R. L. Rill, *Biotechniques* **24,** 256 (1998).

inhibitors of the enzymes employed constitutes another problem, which may require more thorough DNA purification. However, numerous purification steps may also lead to small library sizes due to loss of DNA, and it is therefore recommended that the progress of library construction be monitored by analytical agarose gels. In particular, smaller fragments are more difficult to handle than longer DNAs (that is one reason why the annealing site in the H1 module is removed *after* ligation with the large vector fragment in step 11; see Fig. 3). A reduced library size may also result if the DNA is damaged. Damage can occur at any stage, but particularly critical is exposure to UV light after gel electrophoresis. UV radiation-induced lesions in synthetic oligonucleotides or double-stranded DNA may block subsequent replication in the commonly used RecA-deficient *E. coli* strains.[26]

Summary

Directed evolution of biological macromolecules involves the generation of gene libraries followed by selection or screening for molecules with the desired function. Using chorismate mutase as a target, we have explored different cassette mutagenesis formats to create libraries from which structurally and functionally interesting variants can be directly selected by *in vivo* complementation of a metabolic defect. In the most extensively randomized libraries, all the secondary structural elements in the enzyme have been replaced by modules constructed from long degenerate oligonucleotides. This article discussed methods for designing and assembling these large library modules based on different patterned sets of hydrophilic and hydrophobic building blocks. Strategies for optimizing the efficiency of library construction and for avoiding common artifacts, such as false positives, plasmid mixtures, and unplanned mutations, were addressed.

Acknowledgments

We are grateful to Ying Tang and Applied Molecular Evolution (San Diego, CA) for oligonucleotides prepared by the "split and mix" technique and to Sean Taylor and Ken Woycechowsky for helpful discussions.

---

[26] K. C. Smith and T. C. Wang, *Bioessays* **10**, 12 (1989).

[10] Construction of Protein Fragment Complementation Libraries Using Incremental Truncation

By David E. Paschon and Marc Ostermeier

Introduction

Many proteins can have their peptide backbone cut by proteolytic or genetic means, yet the two fragments can associate to make an active heterodimer. This "monomer-to-heterodimer conversion" is referred to as protein fragment complementation (PFC). Such complementation is the reverse of evolutionary processes in which domains are recruited and fused at the genetic level.[1] Classic examples of protein fragment complementation include ribonuclease S[2] and β-galactosidase.[3] Protein fragment complementation can be used to examine theories of protein evolution,[4-6] protein folding,[7] macromolecular assembly,[8] structure–function relationships,[9,10] and mapping contacts in membrane proteins.[11] Sites for successful protein bisection for protein fragment complementation are quite varied. Bisection sites need not fall between well-defined domains or structural units and fall within conserved and nonconserved regions, as well as within secondary structure elements such as α helixes.[9,12] Overlapping sequences at the bisections point are often tolerated and, in some cases, even required.

For most locations, bisection does not lead to protein fragment complementation, presumably due to inefficient assembly or improper folding of the fragments. For some sites, this can be overcome by fusion of the fragments to dimerization domains to facilitate correct assembly. This is known as assisted protein reassembly (APR). A bisection point for APR

[1] M. Ostermeier and S. J. Benkovic, *Adv. Protein Chem.* **55,** 29 (2000).

[2] I. Kato and C. B. Anfinsen, *J. Biol. Chem.* **244,** 1004 (1969).

[3] A. Ullmann, F. Jacob, and J. Monod, *J. Mol. Biol.* **24,** 339 (1967).

[4] B. L. Bertolaet and J. R. Knowles, *Biochemistry* **34,** 5736 (1995).

[5] G. de Prat-Gay, *Protein Eng.* **9,** 843 (1996).

[6] B. Hocker, S. Beismann-Driemeyer, S. Hettwer, A. Lustig, and R. Sterner, *Nature Struct. Biol.* **8,** 32 (2001).

[7] A. G. Ladurner, L. S. Itzhaki, G. P. Gray, and A. R. Fersht, *J. Mol. Biol.* **273,** 317 (1997).

[8] M. L. Tasayco and J. Carey, *Science* **255,** 594 (1992).

[9] K. Shiba and P. Schimmel, *Proc. Natl. Acad. Sci. USA* **89,** 1880 (1992).

[10] J.-M. Betton and M. Hofnung, *EMBO J.* **13,** 1226 (1994).

[11] H. Yu, M. Kono, T. D. McKee, and D. D. Oprian, *Biochemistry* **34,** 14963 (1995).

[12] M. Ostermeier, A. E. Nixon, J. H. Shim, and S. J. Benkovic, *Proc. Natl. Acad. Sci. USA* **96,** 3562 (1999).

must meet two requirements: (a) the respective protein fragments do not assemble efficiently into an active protein in the absence of a dimerization domain and (b) when these fragments are each fused to one half of a dimer (a dimerization domain), the activity of the protein is restored. In the case where the bisected protein is an enzyme, assembly of the dimerization domains can be detected as the reconstitution of enzyme activity. Such enzymatic two-hybrid systems [also referred to as "protein fragment complementation assays" (PCAs)] have a number of biotechnological applications.[13] Enzymes that have been used as PCAs include dihydrofolate reductase,[14-16] glycinamide ribonucleotide transformylase,[17] green fluorescent protein,[18] ubiquitin,[19,20] β-galactosidase,[21,22] β-lactamase,[23,24] and aminoglycoside and hygromycin B phosphotransferases.[17] The primary use of these PCAs has been as an enzymatic two-hybrid system to evaluate protein–protein interactions[16] and to select for interacting proteins,[25,26] including antibody/antigen pairs.[27]

Incremental Truncation

Incremental truncation,[12] a method for rapidly generating a DNA library of every 1 base pair deletion of a gene or gene fragment (Fig. 1), can be used to evaluate "protein fragment complementation space." Separate,

[13] S. W. Michnick, *Chem. Biol.* **7,** R217 (2000).
[14] J. N. Pelletier, F. X. Campbell-Valois, and S. W. Michnick, *Proc. Natl. Acad. Sci. USA* **95,** 12141 (1998).
[15] I. Remy and S. W. Michnick, *Proc. Natl. Acad. Sci. USA* **96,** 5394 (1999).
[16] I. Remy, I. A. Wilson, and S. W. Michnick, *Science* **283,** 990 (1999).
[17] S. W. Michnick, I. Remy, F. X. Campbell-Valois, A. Vallee-Belisle, and J. N. Pelletier, *Methods Enzymol.* **328,** 208 (2000).
[18] I. Ghosh, A. D. Hamilton, and L. Regan, *J. Am. Chem. Soc.* **122,** 5658 (2000).
[19] N. Johnson and A. Varshavsky, *Proc. Natl. Acad. Sci. USA* **91,** 10340 (1994).
[20] G. Karimova, J. Pidoux, A. Ullmann, and D. Ladant, *Proc. Natl. Acad. Sci. USA* **95,** 5752 (1998).
[21] F. Rossi, C. A. Charlton, and H. M. Blau, *Proc. Natl. Acad. Sci. USA* **94,** 8405 (1997).
[22] F. M. Rossi, B. T. Blakely, C. A. Charlton, and H. M. Blau, *Methods Enzymol.* **328,** 231 (2000).
[23] A. Galarneau, M. Primeau, L. E. Trudeau, and S. W. Michnick, *Nature Biotechnol.* **20,** 619 (2002).
[24] T. Wehrman, B. Kleaveland, J. H. Her, R. F. Balint, and H. M. Blau, *Proc. Natl. Acad. Sci. USA* **99,** 3469 (2002).
[25] K. M. Arndt, J. N. Pelletier, K. M. Muller, T. Alber, S. W. Michnick, and A. Pluckthun, *J. Mol. Biol.* **295,** 627 (2000).
[26] J. N. Pelletier, K. M. Arndt, A. Pluckthun, and S. W. Michnick, *Nature Biotechnol.* **17,** 683 (1999).
[27] E. Mossner, H. Koch, and A. Pluckthun, *J. Mol. Biol.* **308,** 115 (2001).

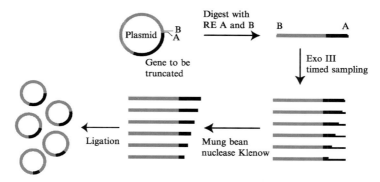

Fig. 1. Incremental truncation. The gene or gene fragment is cloned into a plasmid designed for incremental truncation. The vector is then digested with restriction enzymes A and B. Restriction enzyme A leaves a 3' recessed end, which is susceptible to Exo III digestion, whereas restriction enzyme B leaves a 3' overhang, which is resistant to Exo III digestion. Following the addition of Exo III, small samples are taken after discrete time intervals. Single-stranded tails left from Exo III digestion on the plasmids are removed by incubating with mung bean nuclease and the ends are blunted by Klenow fragment. The plasmids are then recircularized using ligase under dilute conditions and are thus ready for transformation into the desired host.

comprehensive libraries of N-terminal and C-terminal truncations of any gene can be constructed in compatible vectors and cotransformed (Fig. 2). Provided that a sufficient selection or screen is available for the activity of the bisected protein, all possible protein fragment pairs can be evaluated in a single experiment. Sites for PFC or APR can be thus identified.

There are two basic methods for constructing incremental truncation libraries: time-dependent truncation (described here) and a method involving the polymerase-catalyzed incorporation of α-phosphothioate nucleotides (THIO truncation).[28,29] Although the latter has the advantage of experimental convenience and has its advantages for the construction of ITCHY libraries,[30] for purposes of creating libraries for PFC or APR, it suffers from a bias toward short truncations and limited control over the truncation range. Time-dependent truncation is preferred, as it produces a much more even distribution of truncation products.[30]

A schematic depiction of time-dependent truncation is shown in Fig. 1. Plasmid DNA containing the gene to be truncated is prepared by digesting closed circular plasmid DNA with restriction enzymes such that (1) a 3'

[28] S. Lutz, M. Ostermeier, and S. J. Benkovic, *Nucleic Acids Res.* **29,** e16 (2001).
[29] M. Ostermeier and S. Lutz, *Methods Mol. Biol.* **239,** 129 (2003).
[30] M. Ostermeier, *Biotechnol. Bioeng.* **82,** 564 (2003).

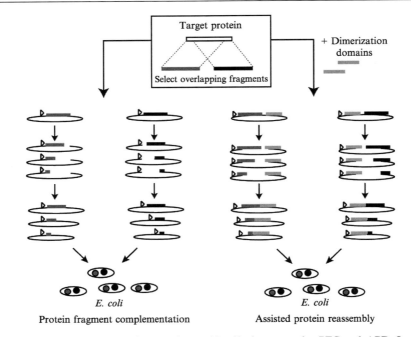

FIG. 2. Use of incremental truncation to identify fragments for PFC and APR. Large, overlapping (and ideally, inactive) N-terminal and C-terminal fragments are selected from the gene of the target protein. For PFC, these fragments are cloned into vectors designed for performing incremental truncation. The N-terminal fragment is truncated from its C terminus, and the C-terminal fragment is truncated from its N terminus. N-terminal fragments are fused to a series of stop codons in all three frames, and C-terminal fragments are fused to an ATG start codon. Individual N-terminal and C-terminal truncation libraries are first constructed and then cotransformed into E. coli cells. For APR, the same procedure is followed except that the vectors contain dimerization domains, which are fused to the truncated genes after truncation has been performed.

recessed end is left at the end of the DNA to be truncated and (2) a 3′ overhang is left on the opposite end of the vector to protect it from degradation (the end not to be truncated can also be protected by the incorporation of α-phosphothioate nucleotides[31]). Exonuclease III (Exo III) is used to digest linear double-stranded DNA. Exo III is a 3′–5′ exonuclease that can digest 3′ recessed or blunt ends efficiently, but cannot digest from a 3′ overhang of four or more bases.[32] The incremental

[31] S. D. Putney, S. J. Benkovic, and P. R. Schimmel, *Proc. Natl. Acad. Aci. USA* **78,** 7350 (1981).

[32] J. D. Hoheisel, *Anal. Biochem.* **209,** 238 (1993).

truncation library is created by time-dependent sampling during Exo III digestion. The rate of Exo III digestion is controlled carefully by a combination of low temperature and salt. During digestion, small aliquots are removed and the reaction is quenched in a low-pH, high-salt buffer. The resulting single-stranded DNA tails are removed by digestion with mung bean nuclease and the ends are then blunted using the Klenow fragment. The blunt ends of the linear plasmid are then ligated, and the resulting library is transformed into the desired host organism.

Aside from constructing libraries to identify fragments for PFC and APR, the following protocols can also be used to optimize peptide linkers in gene fusions and to construct libraries of promoter variants that vary the distance between important sites within the promoter. In addition, the truncation of a protein can result in altered properties. Removal of the first 70 N-terminal amino acids of a phospholipase resulted in the change of substrate specificity from a phospholipase to a lipase.[33] Incremental truncation can also be used to create fusion libraries between genes (called ITCHY libraries). Detailed protocols for the construction of ITCHY libraries have been published elsewhere[29] and are not described here.

Materials and Methods

Vectors for Creating PFC or APR Libraries

The vectors used for incremental truncation are shown in Fig. 3; other vectors are suitable as well, provided that they meet certain criteria outlined later. Plasmid pDIM-N2 is used for making libraries of N-terminal fragments, whereas pDIM-C8 is used for making libraries of C-terminal fragments. Each plasmid contains a different antibiotic resistance marker: ampicillin for pDIM-N2 and chloramphenicol for pDIM-C8. Each also has a different origin of replication so that both plasmids may be stably maintained in the same bacterial cell. The f1 origin of replication allows for the possibility of packing DNA into phage as another method for transformation into *Escherichia coli.* IPTG (Isopropyl-BD-Thiogalactoside) inducible *lac*-based promoters allow for leaky expression of the resulting protein fragments under noninducing conditions, with higher expression under inducing conditions.

The gene to be truncated is cloned between the *Nde*I and the *Bam*HI (or *Spe*I) sites in pDIM-N2 and between the *Bgl*II and the *Spe*I sites in pDIM-C8. For use in creating libraries for PFC, pDIM-N2 contains stop codons in all three frames between the *Nsi*I and the *Kpn*I restriction sites. This ensures that all N-terminal protein fragments will have a stop codon

[33] J. K. Song, B. Chung, Y. H. Oh, and J. S. Rhee, *Appl. Environ. Microbiol.* **68,** 6146 (2002).

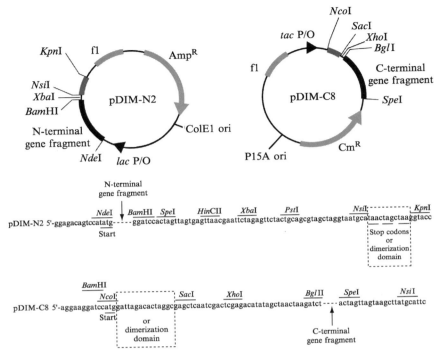

F<small>IG</small>. 3. Circular maps and sequences of cloning and truncation regions of incremental truncation vectors. All vectors have the filamentous phage origin of replication (f1), as well as plasmid origins of replication (ColE1 ori and p15A ori). In pDIM-N2, the N-terminal gene to be truncated is cloned between the *Nde*I and the *Bam*HI sites downstream from an IPTG-inducible lac promoter (*lac* P/O). The vector also has an antibiotic resistance gene (ampicillin, Amp^R). In pDIM-C8, the C-terminal gene to be truncated is cloned between the *Bgl*II and the *Spe*I sites downstream from an IPTG-inducible trp/lac hybrid promoter (*tac* P/O). The vector also has an antibiotic resistance gene (chloramphenicol, Cm^R). For PFC, pDIM-N2 contains a series of stop codons between the *Nsi*I and the *Kpn*I sites. For APR, dimerization domains are located between the *Nsi*I and the *Kpn*I sites of pDIM-N2 and the *Nco*I and the *Sac*I sites of pDIM-C8.

proximal to the truncated gene and contain C-terminal extensions of no more than three residues not in the original protein (one-third of all N-terminal protein fragments will contain no extensions). Plasmid pDIM-C8 contains an *Nco*I site, which, upon digestion and blunt end formation, leaves an ATG start codon that is fused to the truncated C-terminal gene fragment. One-third of this library will have this start codon in-frame with the truncated gene, the other two-thirds will not produce truncation fragments except in the case of spurious internal translation initiation.

For constructing libraries for APR, the dimerization domains must be positioned properly in the vectors. This is accomplished by cloning the gene for one subunit of the dimer between *Nsi*I and *Kpn*I in pDIM-N2. If the first codon of the dimerization domain begins with a T, then this can be the last T in the *Nsi*I site. Alternatively, the last T in the *Nsi*I site can serve as the wobble base for genes in which truncation stopped after removal of the third base of a codon; however, this will necessarily result in some truncations having a nonsynonymous mutation at the fusion point. The gene for the other half of the dimer is cloned between the *Nco*I and the *Sac*I sites in pDIM-C8. In most cases the last codon of the dimerization domain either already ends with a G or can be mutated silently to a G, which can be designed to be the first G of the *Sac*I site.

Preparation of Plasmid DNA for Truncation

Plasmid DNA to be truncated must be prepared to ensure that the majority of plasmid molecules do not contain a nick. Exo III can digest from single-stranded nicks in double-stranded DNA, leaving single-stranded gaps. These single-stranded gaps will be digested by mung bean nuclease and result in the undesired effect of random deletions throughout the entire plasmid. It is also important to limit the amount of restriction enzyme used to avoid nicking of the DNA for the same reasons. We routinely isolate plasmid DNA of pDIM-N2 and pDIM-C8 vectors from *E. coli* strain DH5α using commercial plasmid prep kits, although for pDIM-C8 vectors, we have found that the fraction of nicked molecules and the total yield of plasmid DNA are highly dependent on growth conditions.[34] Note that *Sac*I and *Xho*I have difficulty digesting supercoiled DNA, thus necessitating higher amounts of enzyme during digestions to ensure the complete digestion of the pDIM-C8 vector.

Digestion mixtures are as follows. pDIM-N2 consists of 10 μg of pDIM-N2, 10 μl of 10× NEB buffer 3, 1 μl of 100× bovine serum albumin (BSA), 1.5 μl of *Pst*I (30 units), 2 μl of *Nsi*I (20 units), and water to 100 μl. pDIM-C8 consists of 10 μg of pDIM-C8, 15 μl of 10× NEB buffer 1, 1 μl of 100× BSA, 7.5 μl of *Sac*I (150 units), 10 μl of *Xho*I (200 units), and water to 150 μl.

1. Incubate both digestions at 37° for 1.5–2 h.
2. Incubate at 65° for 20 min to inactivate enzymes.
3. Add 5× volume of digestion of QIAquick buffer PB.

---

[34] M. Ostermeier, S. Lutz, and S. J. Benkovic, *in* "Protein–Protein Interactions: A Molecular Cloning Manual" (E. Golemis, ed.). Cold Spring Harbor Laboratory Press, Cold Spring Harbor, NY, 2002.

4. Follow QIAquick protocol.
5. Elute DNA from the column with 50 μl of QIAquick buffer EB.

Construction of Individual Truncation Libraries

The following protocol makes repeated use of the QIAquick DNA purification kit from QIAGEN. We have found that QIAquick buffer PB is a suitable buffer for inactivating Exo III.

1. Equilibrate 250 μl QIAquick buffer PB at room temperature in a 1.5-ml tube (tube A).
2. Into a 0.5-ml tube (tube B) add 2 μg DNA, 6 μl of 10× Exo III buffer, the desired amount of 1 M NaCl, and water to 60 μl. At 22°, the rate of Exo III digestion can be estimated as a function of NaCl concentration using the equation:

$$\text{Rate(bp/min)} = 48 \times 10^{-0.00644[\text{NaCl}]}$$

where the concentration of NaCl is in mM.

3. Equilibrate tube B in a minifridge or thermocycler at 22°.
4. At time = 0, add 200 units of Exo III to tube B and mix immediately.
5. Begin removing 1-μl samples from tube B every 20–30 s and add to tube A. Mix tube A well. Because the rate of Exo III digestion is highly dependent on temperature, leave tube B open during sampling to avoid warming the tube by repeated handling. QIAquick buffer PB is slightly viscous; therefore, some liquid will stick to the side of the pipette tip when adding the truncation samples to tube A. Thus the final volume will be lower than the starting 250 μl of buffer PB plus the 60 μl of truncation samples.
6. After all samples are taken, estimate the volume remaining in tube A and add 5× volume of QIAquick buffer PB.
7. Follow the QIAquick protocol.
8. In the final step of the QIAquick protocol, elute DNA from the column with 44 μl of QIAquick buffer EB.
9. Add 5 μl of 10× mung bean nuclease buffer and 3 μl (3 units) of mung bean nuclease and incubate tube at 30° for 30 min. Occasionally, the amount of mung bean nuclease added has to be adjusted to ensure the formation of a high-quality library. The amount of mung bean nuclease to add can be determined experimentally using a single truncation time point. For example, digest 10 μg of DNA with Exo III for a sufficient amount of time to digest 300 bp. This single time point is then divided into five tubes with 2 μg of DNA per tube. Add 5 μl of 10× mung bean nuclease buffer to each tube and then add 5 μl of various amounts of mung bean nuclease (say 0, 1, 3, 6, and 12 units) to each sample. A portion of the DNA is then

run on a 0.8% agarose gel. DNA not digested by mung bean nuclease will run as a slightly diffuse band larger than the expected size due to the large single-stranded overhangs. As the level of mung bean is increased, the DNA will start to smear between the size obtained with no mung bean nuclease and the expected size. When the level of mung bean nuclease is optimum, the DNA will run as a focused band at the expected size. When there is too much mung bean nuclease, the DNA will smear to smaller sizes as the mung bean nuclease begins to make double-stranded breaks.

10. Add 250 μl of QIAquick buffer PB.

11. Follow the QIAquick protocol.

12. Elute truncated DNA from the QIAquick column with 82 μl of QIAquick buffer EB except for constructing pDIM-C8 PFC libraries, in which case elution is with 90 μl of QIAquick buffer EB.

For pDIM-N2 (both PFC and APR libraries) or for pDIM-C8 in the case of constructing APR libraries skip to step 17. Only perform steps 13–16 for pDIM-C8 in the case of constructing PFC libraries. These steps are necessary in preparing the ATG start codon for fusion to the truncated gene.

13. For pDIM-C8 only, add 10 μl of 10$\times$ NEB buffer 1, 1 μl of 100$\times$ BSA, 18 units of NcoI, and water to 100 μl. Incubate at 37° for 2 h.

14. Add 0.5 ml of QIAquick buffer PB.

15. Follow the QIAquick protocol to purify digested pDIM-C8.

16. Elute digested pDIM-C8 from the column with 82 μl of QIAquick buffer EB.

17. Equilibrate pDIM-N2 and pDIM-C8 at 37°.

18. For pDIM-N2, add 10 μl of Klenow mix [20 mM Tris–HCl, pH 8.0, 100 mM MgCl$_2$, 0.25 units/μl Klenow (exo$^+$)], incubate at 37° for 3 min, add 10 μl of dNTPs (0.125 mM each), and incubate at 37° for 5 min. For pDIM-C8, add 10 μl of dNTP, 10 μl of Klenow mix, and incubate at 37° for 5 min.

19. Inactivate Klenow by incubating at 72° for 20 min.

20. Cool to room temperature and add 0.4 ml of ligase mix (320 μl of water, 40 μl of 10$\times$ T4 DNA ligase buffer, 40 μl of 50% PEG 8000, and 18 Weiss units of T4 DNA ligase).

21. Incubate at room temperature $\geq$12 h.

22. Concentrate by ethanol precipitation. To 250 μl of ligation mixture add 125 μl of 7.5 M ammonium acetate, 750 μl of 100% ethanol (at $-20°$), and 2 μl Pellet Paint (Novagen) and incubate on ice for 30 min. Centrifuge at 10,000g at 4° for 10 min. Wash the DNA pellet with 750 μl of 70% ethanol (at $-20°$). After an additional 2-min spin and thorough removal of

all liquid by pipetting, air dry the DNA pellet for 10 min. Resuspend the DNA pellet in 15 μl water.

23. Electroporate $\leq$5 μl of DNA into 50 μl of electrocompetent DH5α-E *E. coli* cells (e.g., Bio-Rad Gene Pulser II set to 25 μF capacitance, 200 Ω resistance, and 2.5 kV). Allow cells to recover for 1 h in 1 ml of SOC media while shaking at 200 rpm at 37°.

24. Plate 1 μl of the cells on a small plate with growth media containing either 100 μg/ml ampicillin for pDIM-N2 or 50 μg/ml chloramphenicol for pDIM-C8. Plate the remaining electroporation mixture ($\sim$1 ml) on a large LB plate (e.g., 245 $\times$ 245-mm bioassay dish; Nalgene-Nunc) with the appropriate antibiotics. Colonies that grow on the small plate are used to determine the number of transformants (total transformants = number of colonies on the small plate $\times$ 1000) and will also be used for evaluating the truncation libraries.

25. Incubate at 37° overnight.

Evaluating the Individual Truncation Libraries

After creating a truncation library for pDIM-N2 or pDIM-C8, verify that the library has a random distribution in the expected size range. This can be performed by mini-prepping the DNA from individual transformants and performing restriction enzyme digests followed by analysis using agarose gel electrophoresis. It is more convenient to perform PCR on individual transformants using primers flanking the insertion site as described next.

1. Combine 70 μl of 10$\times$ *Taq* buffer (containing 15 mM MgCl$_2$, Promega), 56 μl of dNTPs (2.5 mM each), 42 μl of 10 μM forward primer, 42 μl of 10 μM reverse primer, 4 μl *Taq* DNA polymerase (20 units, Promega), and 486 μl water to create the PCR mixture.

2. Add 25 μl of PCR mixture per tube.

3. Transfer one colony per tube from the 1-μl plate using a sterile pipette tip.

4. Perform PCR for 30 cycles of 1 min at 94°, 1 min at 56°, and 1 min at 72°. Also do a control PCR on the original pDIM-N2 and pDIM-C8 (plasmid before truncation) using the same primers as described earlier. This will give an upper limit on the size of fragments obtained from the PCR.

5. Analyze the PCR reactions by agarose gel electrophoresis (Fig. 4 shows a sample result).

If results show that the library is biased toward too little truncation, reconstruct the library with a lower concentration of NaCl. If the library is

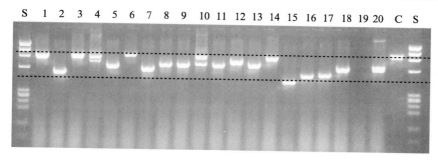

Fig. 4. PCR analysis of truncation-length diversity. Individual C-terminal library members were amplified by colony PCR using primers flanking the gene fragment to be truncated. The two standard lanes (S) are the ϕX174-*Hae*III-digested DNA marker (NEB). The control lane (C) is a PCR amplification of the vector before truncation has been performed. This untruncated fragment is the largest size that should appear in the library. Dotted horizontal lines give the desired size range for this particular truncation library. Note that lanes 4 and 10 each have two bands. Two bands can result from a single cell receiving two plasmids with different truncation lengths.

biased toward too much truncation, reconstruct the library with a higher concentration of NaCl. Note that too much truncation can also be caused by overdigestion with mung bean nuclease.

Recovering and Storing the Individual Truncation Libraries

Libraries are recovered from the large plate. One portion of the library is saved as frozen cell stocks and from another portion the library is isolated in plasmid form.

1. Recover cells from a 245 × 245-mm bioassay dish by adding 2 × 15 ml storage media [18 ml LB, 9 ml 50% glycerol, 3 ml 20% (w/v) glucose] to the top of the plate, scrape cells from media using a cell spreader, and then pipette cells into a 50-ml polypropylene centrifuge tube.

2. Spin cells in a centrifuge at 5000 rpm (maximum allowable speed for polypropylene centrifuge tubes) at 4° for 10 min.

3. Decant supernatant and add 2 ml of storage media.

4. Resuspend the pelleted cells by gentle shaking.

5. Store 4 × 200-μl aliquots in 1.5-ml tubes at −80°.

6. Pellet the remaining cells in a centrifuge at 5000 rpm at 4° and follow the protocol for the QIAGEN HiSpeed plasmid midiprep kit (or other midiprep kit) to recover the library in plasmid form.

Cotransformation of the Individual Truncation Libraries

The individual N-terminal and C-terminal truncation libraries are now ready to be cotransformed into *E. coli* strain DH5α–E. It should be noted that using too much DNA for the transformation could lead to problems when screening the final library. If too much DNA is used, multiple library members of either pDIM-N2 or pDIM-C8 can transform a single bacterial cell, thus complicating the analysis and identification of positives. Using 20 ng of DNA from each library for the cotransformation appears to minimize the occurrence of multiple copies of library members in the same cell while still resulting in cotransformed libraries of significant size.

1. Combine 20 ng of DNA from each library in a 0.5-ml tube and chill on ice. Add 50 μl of electrocompetent DH5α-E *E. coli* cells and electroporate (e.g., Bio-Rad Gene Pulser II set to 25 μF capacitance, 200 Ω resistance, and 2.5 kV). Allow cells to recover for 1 h in 1 ml of SOC media shaking at 200 rpm at 37°.

2. Plate 1 μl of the cells on each of three small LB plates (one with Cm, one with Amp, and one with Cm and Amp). Plate the remaining electroporation mixture ($\sim$1 ml) on a large LB plate (e.g., 245 × 245-mm bioassay dish; Nalgene-Nunc) with Cm and Amp. Colonies that grow on the small plate with both antibiotics are used to determine the number of transformants (total transformants = number of colonies on the small plate × 1000). Small plates with only one antibiotic are used to determine which plasmid is limiting the number of double transformants, should the number of Amp^{R}Cm^R transformants be small.

3. Recover the library from the large plate as before and store as a frozen cell stock.

Confirmation of True Positives

The library of heterodimers can now be screened or selected for the desired function. However, once positives are identified, these must be analyzed further to ensure that recombination has not occurred. It is important to separate the plasmids and retransform them together for each possible positive to ensure that (1) recombination producing a full-length gene has not occurred, (2) one of the protein fragments alone is not functional, and (3) the function is exhibited only when both halves of the heterodimer are present.

Recombination is the most frequent cause of false positives. Because the individual truncation libraries originated from large overlapping fragments, there is a high probability of finding large overlaps in the N-terminal and C-terminal genes in individual bacterial cells. Presumably these overlapping DNA sequences sometimes lead to homologous recombination between the pDIM-N2 and the pDIM-C8 plasmids to reconstitute the original, full-length gene. This can occur at a significant frequency (about 3 in 10^4) even in $recA^-$ strains of E. coli. Therefore, positives must be screened further to ensure that recombination has not occurred. Although PCR (with forward and reverse primers for the starting gene), minipreps (most recombinants result in larger or abnormal size plasmids), and restriction digests can be used as a quick screen to eliminate recombinants, the most definitive test to ensure that a positive is a true heterodimeric positive is to isolate the individual plasmids, retransform each plasmid individually into fresh cells, and confirm the absence of the function. After it is confirmed that the individual plasmids do not produce a protein with the desired function, these plasmids are cotransformed together again and the presence of the function is confirmed. The truncated genes residing in the plasmids are then sequenced as a final confirmation.

1. Miniprep the plasmid DNA from 5 ml of an overnight inoculum of a positive clone.

2. Prepare a 1:100–1:1000 dilution of the miniprep. The miniprep needs to be diluted so that there is a high probability of transforming only one of the plasmids into a given cell.

3. Electroporate 5 μl of this diluted DNA into DH5α-E (e.g., Bio-Rad Gene Pulser II set to 25 μF capacitance, 200 Ω resistance, and 2.5 kV). Allow cells to recover for 1 h in 1 ml of SOC media shaking at 200 rpm at 37°.

4. Spread 5 μl of the electroporation mixture on a plate containing 100 μg/ml Amp and 50 μl of the electroporation mixture on a plate containing 50 μg/ml Cm. pDIM-C8 has a lower copy number than pDIM-N2, resulting in a lower frequency of cells transformed with pDIM-C8. Therefore, a greater volume needs to be plated to obtain a similar number of colonies.

5. Incubate plates at 37° overnight.

6. Pick individual colonies from the Amp^R transformant plate and streak first onto a Cm plate and then immediately onto an Amp plate. If the colony contains cells with only the pDIM-N2 plasmid, nothing should grow when streaked onto the Cm plate. If there is growth on the Cm plate, either the cells contain both plasmids or a recombination has occurred (recombinants often contain a large plasmid with both antibiotic markers). Generally, when transforming this level of dilution, virtually 100% of Amp^R transformants will contain only pDIM-N2.

7. Perform the analogous streak test with colonies selected from the Cm^R transformant plate, except now streak first onto an Amp plate. If

there is growth on the Amp plate, either the cells contain both plasmids or a recombination has occurred. In general, greater than 90% of the Cm^R transformants will contain only pDIM-C8. This frequency is lower than the frequency of transformants containing only pDIM-N2, as pDIM-C8 is the lower copy number plasmid.

8. Pick a transformant that only has one antibiotic resistance and test for the desired function of the heterodimer. If the desired function is present, either the protein fragment maintains the function by itself or recombination has occurred such that a full-length gene is present on an individual plasmid.

9. If the function is not present for both separated plasmids, the final step is to retransform them together to confirm that the function requires the presence of both plasmids. Grow an inoculum and miniprep the plasmid DNA from about 5 ml of culture. Mix together 2 μl each of DNA from pDIM-N2 and pDIM-C8 minipreps, electroporate into the desired strain, and plate onto plates containing 100 μg/ml Amp and 50 μg/ml Cm.

10. Test these cotransformants for the desired function.

Section II

Applications: Optimization and Screening

[11]   GigaMatrix: A Novel Ultrahigh Throughput Protein Optimization and Discovery Platform

By MIKE LAFFERTY and MARK J. DYCAICO

Introduction

The myriad of microbes inhabiting this planet represent a tremendous repository of biomolecules for pharmaceutical, agricultural, industrial, and chemical applications. The great majority of these microbes, estimated at greater than 99.5%, have remained uncultured by modern microbiological methods due in large part to the complex chemistries and environmental variables encountered in extreme or unusual biotopes. The discovery of novel enzymes from the biodiversity present in nature using culture-independent, recombinant approaches has been demonstrated successfully.[1] Large, complex ($>10^7$ member) gene libraries are constructed by the direct isolation of DNA from selected microenvironments around the world. These libraries are then expressed in various host systems and are subjected to high throughput screens specific for an activity of interest. Enzymes discovered by this or other methods can then be optimized if necessary to function under environmental extremes or specific industrial application conditions. Laboratory evolution methods can be used to create large $>10^{10}$ member libraries of phenotypic variants. These variants are created and screened in an attempt to identify enzymes or other proteins with improved properties. The GigaMatrix* screening platform (patents pending worldwide) was developed to exploit the opportunities provided by biodiversity and protein optimization techniques by addressing many of the shortcomings of current screening technologies.

Current Screening Methods

Liquid-phase enzyme assays, often in robotically manipulated 96-, 384-, or 1536-well microplate formats, have been used successfully to detect and/or quantify the presence and expression of heterologous or mutated genes in environmental and evolved libraries. The development of much of the automated microplate screening equipment utilized to date has been driven by the needs of pharmaceutical companies to perform

---

[1] D. E. Robertson, E. J. Mathur, R. V. Swanson, B. L. Marrs, and J. M. Short, *SIM News* **46**, 3 (1996).

* Gigamatrix is a trademark of Diversa Corporation.

traditional chemical compound screening. Typically, in these screens, each well contains a single specific compound that is screened against a single specific target. The identity of each compound and its physical location are known and the activity against a known target is measured. Although these microplate-based screening technologies are being used successfully, limitations do exist. The primary limitation is throughput, as these techniques generally allow the screening of only about 10^4 to 10^5 wells per day for each instrument. While these throughputs may be more than sufficient for screening relatively low-complexity compound libraries or gene libraries derived from isolated organisms, it could take more than a year to thoroughly screen one complex 10^8 member gene library. Clearly, higher throughput screening technologies are necessary to take complete advantage of the opportunities provided by biodiversity and protein optimization techniques.

In contrast to typical chemical compound library screening, the identity of the variants from these large discovery or optimization libraries is not known unless each clone is sequenced prior to screening, which in most cases is not feasible. For this reason, there is no drawback to randomly placing clones into any well within the collection of screening plates. Researchers using solid-phase colorimetric screening of bacterial colonies or phage plaques (e.g., Kuritz[2]) have been distributing their samples randomly for years. Typically, libraries of clones are distributed randomly onto agar plates where the assay is performed and putative hits are recovered for further characterization. However, these solid-phase methods often suffer from limitations in assay flexibility. In most solid-phase assays, the products of enzyme turnover must either be sequestered within the cell or form a precipitate to prevent excessive diffusion across the plate over the course of incubation. Thus, this type of screening is usually amenable only to certain chromogenic substrates. Moreover, because there is a point at which increasing the clone density on an agar plate results in lowered assay sensitivity, the throughput is often limited to about 20,000 clones per 150-mm petri dish. Liquid-phase methods that utilize random clone distribution have also been used for the discovery and optimization of novel bioactivities from gene libraries. These methods include growth selection,[3] *in vitro* expression cloning,[4] cell surface or phage display,[5] and FACS-based strategies.[6] However, each of these systems has limitations that have been described previously.

[2] T. Kuritz, *Lett. Appl. Microbiol.* **28,** 445 (1999).

[3] K. S. Lundberg, P. L. Kretz, G. S. Provost, and J. M. Short, *Mutat. Res.* **301,** 99 (1993).

[4] R. W. King, K. D. Lustig, P. T. Stukenberg, T. J. McGarry, and M. W. Kirschner, *Science* **277,** 973 (1997).

[5] I. Benhar, *Biotechnol. Adv.* **19,** 1 (2001).

[6] C. Schmidt-Dannert, *Biochemistry* **40,** 13125 (2001).

The GigaMatrix Screening Platform

GigaMatrix screening combines many of the best features of the microplate and solid-phase plate screening paradigms while improving throughput by two to three orders of magnitude. Further, it takes advantage of an abundance of commercially available fluorescent substrates used commonly in liquid-phase assays. The processes and assay protocols used in GigaMatrix screening closely parallel those used in microplate screening, yet with significant simplification in required instrumentation, reduction in plate storage capacity requirements and reagent costs, and the inclusion of integrated "cherry picking." These advantages are realized through miniaturization, automation, and massively parallel processing.

At the heart of the platform are special high-density plates, into which samples are distributed randomly, and a custom detection and recovery system used to identify and isolate putative (primary) hits. GigaMatrix plates (Fig. 1), which are fabricated using manufacturing techniques originally developed for the fiber optics industry, consist of approximately 100,000 cylindrical compartments, or wells, contained within a reusable plate the size of a standard microplate (85.48×127.76 mm). Each 6-mm-long well has an inner diameter of 200 μm and is open at both ends. These wells act as discrete 190-nl microenvironments in which isolated clones can be grown and screened.

GigaMatrix plates are loaded with a suspension of library clones and other assay components (e.g., growth medium and fluorogenic enzyme substrates) by simply placing the surface of the plate in contact with the assay solution. Surface tension at the liquid/plate interface causes the assay components to be drawn or wicked into all of the through-hole wells simultaneously without the need for complicated dispensing equipment.

Fig. 1. A GigaMatrix plate.

Clones are deposited randomly among the available wells. Once loaded, the plates are then incubated typically 24 to 48 h to allow for both clonal amplification and enzymatic turnover.

After incubation, GigaMatrix plates are transferred to a custom-built automated detection and recovery station (Fig. 2) where fluorescence imaging is used to identify and isolate clones expressing the desired bioactive molecules. The detection and recovery system combines fluorescence imaging and precision motion control technologies through the use of machine vision and image-processing techniques. Images are generated by focusing light from a metal halide arc lamp (EXFO, Vanier, Québec,

Fig. 2. (A) The automated detection and recovery system. (B) Exploded three-dimensional model view of the major detection and recovery station components. (1) GigaMatrix plate, (2) recovery microplate, (3) recovery needles, (4) illumination optics, (5) filter wheel, (6) cooled CCD camera with telecentric lens, and (7) needle removal device. (See color insert.)

Canada) onto the GigaMatrix plate through a set of fluorescence excitation filters (Omega Optical, Brattleboro, VT). The resulting fluorescence emission is filtered and then imaged by a telecentric lens (Edmund Scientific, Barrington, NJ) onto a high-resolution cooled CCD camera (Roper Scientific, Tucson, AZ) in an epifluorescent configuration. The GigaMatrix plates are scanned to generate a total of 56 slightly overlapping images in approximately 45 s. The images are digitized and processed on the fly to detect and locate positive wells or primary hits. Clones that have converted the substrate to a fluorescent product appear as bright spots on a dark background.

Automatic discrimination between positive wells and inevitable dirt and dust is performed in order to simplify the use of the screening platform and avoid the necessity for extraordinary cleanliness (i.e., operation in a clean room). Discrimination is accomplished using standard image processing algorithms (Matrox, Dorval, Québec, Canada) to extract and measure a number of features of connected regions of pixels (commonly known as blobs) within an image. Blobs, or fluorescent wells in this case, are segmented from the background by adaptive intensity thresholding. The resulting feature measurements for each blob, which include shape, size, and intensity profile, are then used to distinguish positive wells from objects such as dirt and dust as well as locate the mechanical center of the well.

Once detected and located, primary hits are recovered from the Giga-Matrix plate and transferred to a microplate (the recovery plate) for secondary confirmation and isolation. The recovery process is a five-step automated process that takes place on the same machine used for imaging and detection. These five steps are (1) mounting and locating the tip of a sterile recovery needle, (2) moving the GigaMatrix plate to precisely align the tip of the needle to the well containing the primary hit, (3) applying vacuum to aspirate the contents of the well into the needle, (4) flushing the well contents from the needle into the recovery plate with an appropriate medium, and (5) removing the used needle in preparation for the next recovery. Closed loop positioning with image-based feedback provides the positional accuracy required to allow the aspiration of individual wells without contamination from neighboring wells. Each 30-gauge recovery needle (EFD, East Providence, RI) is fitted with a 0.22-μm filter (Millipore, Billerica, MA) that serves as an upstream barrier to avoid contamination from the vacuum and flushing tubes. Approximately 0.2 μl of medium is then used to back-flush the sample from the needle into the well of the recovery plate.

A number of key details must be addressed in order to fully exploit the potential of the GigaMatrix platform hardware to deliver ultrahigh throughputs. Issues related to miniaturization include evaporation, cell growth, and assay sensitivity. Random clone distribution introduces the possibility of having multiple clones per well. This raises issues with hit identification and recovery due to growth competition between clones. These issues, along with methods utilized to perform assays with the GigaMatrix platform, are described.

GigaMatrix Screening Methods

Library Construction and Clone Preparation

Most of the GigaMatrix screens performed to date have been carried out using gene libraries constructed in λZAP-based cloning vectors (Stratagene, La Jolla, CA). Library insert sizes in this type of vector typically range from 2 to 10 kb in length. These libraries are propagated in the form of bacteriophage λ and are amplified to produce high-titer stocks for use in screening. A unique advantage of λZAP-based vectors is that the λ library can be converted easily into a phagemid library through a process known as *in vivo* excision.[7] Infection of a second bacterial host strain by the phagemid library transforms the host efficiently, allowing the library clones to be propagated internally as plasmids. The benefit of this cloning strategy is that λZAP-based libraries can be screened in either of two forms: a (plasmid-like) phagemid library or its parental λ phage library. The decision of whether to use the phagemid or the λ form of a library in activity-based screening hinges on several factors. Because a phagemid replicates as a plasmid once in the host, the gene products tend to accumulate within the cell at high concentrations. This can be advantageous for amplifying enzymatic turnover from weakly active clones. However, it also can be disadvantageous in screens where the desired gene product is likely to be toxic to the host cell. In such cases, it is preferable to screen the λ form of a library, as during lytic replication the growth of a λ clone is not dependent on the continued survival of one particular host cell. Another advantage of screening λ libraries is that during λ replication, each infected host cell is fated to burst during the bacteriophage life cycle. Thus, unlike phagemid library screening, the observation of enzymatic activity is not dependent on membrane permeability to the enzyme and/ or substrate. The procedures used for preparing library clones for growth in GigaMatrix differ depending on which form of the library is used.

---

[7] J. M. Short, J. M. Fernandez, J. A. Sorge, and W. D. Huse, *Nucleic Acids Res.* **16,** 7583 (1988).

Preparation of Phagemid Libraries. For enzyme screening, the host strain may need to be altered genetically to reduce or eliminate endogenous background. To prepare clones for screening, the phagemid library is combined with the host strain and allowed to adsorb for 15 min, after which the sample is supplemented with growth medium and incubated at 37° for 35–45 min to allow expression of the antibiotic resistance gene present on the cloning vector. Following centrifugation and suspension in storage medium, a sample of this slurry is plated on agar plates containing antibiotic and is incubated overnight in order to determine the titer of infected cells. The results of this titer are used to determine the extent to which the suspension is to be diluted with a mixture of growth medium, fluorescent substrate, and antibiotic, producing the final assay suspension used to load the GigaMatrix plates. A sample of the assay suspension is plated on solid agar plates containing antibiotic to monitor the *seed density* (see later). The remainder of the phagemid library suspension is used to fill the GigaMatrix plates.

Preparation of λ Libraries. As described earlier for phagemid library hosts, endogenous genes may also need to be interrupted to reduce background in certain enzyme screens. To prepare clones for screening, the titer of the λ library is measured using standard techniques.[8] Using this titer information, the λ library is allowed to adsorb to the screening host using a library-to-host ratio such that the final assay suspension will satisfy conditions for lytic growth and result in the desired concentration of library clones. After the 15-min adsorption period, a mixture of growth medium and fluorescent substrate is added to produce the final assay suspension. A sample of this suspension is plated with soft agar to monitor *seed density* (see next section). The remainder of the λ library suspension is used to fill the GigaMatrix plates.

Plate Loading and Seed Density

When loading a GigaMatrix plate, its surface is brought into contact with the suspension containing library clones diluted in growth medium and substrate. This loading technique relies on surface tension between the plate material and the suspension to pull the fluid into all wells simultaneously. This parallel loading model is advantageously rapid and avoids the need for specialized dispensing apparatus.

Seed density is defined here as the average number of library clones deposited per well at the time of plate loading. Clone deposition within

[8] J. Sambrook and D. W. Russell, "Molecular Cloning: A Laboratory Manual." *in* Cold Spring Harbor Laboratory Press, Cold Spring Harbor, NY, 2001.

wells is random and appears to generally follow the statistical characteristics of the Poisson distribution (Fig. 3). At a seed density of one, the
Poisson formula predicts that roughly 37% of available wells will be empty,
37% will be loaded with only one clone, and 26% will be loaded with more
than one clone. Raising the seed density to five ensures that greater than
99% of available wells will receive at least one clone. However, high seed
densities introduce complications inherent to growing several clones together in one well. A bacterium carrying a "positive" clone that inhibits
cell growth is at a disadvantage among faster-growing cells carrying "negative" clones. Similarly, a poorly expressed positive clone is more likely to
escape detection if it shares the same well with several other negative
clones. Most library screens in GigaMatrix are performed at seed densities
ranging from 0.5 to 1 as a compromise between two opposing motivations:
maximizing throughput through the use of all available wells and lessening
the chance that some valuable positives will be overlooked due to intrawell

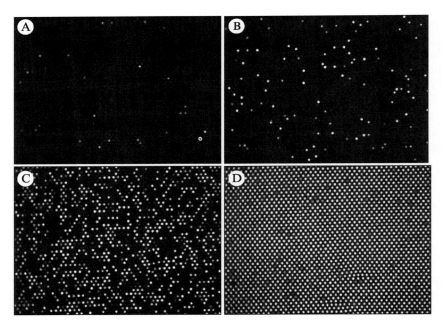

Fig. 3. Sample images from GigaMatrix plates loaded at seed densities of (A) 0.005,
(B) 0.05, (C) 0.5, and (D) 5. *E. coli* host bacteria harboring a phagemid clone expressing a
β-galactosidase gene were loaded at the aforementioned seed densities using LB medium and
25 μM resorufin β-D-galactopyranoside (Sigma, St. Louis, MO). GigaMatrix plates were
incubated at 37° for 24 h using techniques described in the text. Fluorescence imaging was
performed using the Model XF38 filter set (Omega Optical, Brattleboro, VT).

competition. In addition, seed density can also impact the downstream process of isolating each positive clone from the negative clones recovered from the same well (see later).

Incubation

Once loaded, the plates are typically incubated for 24 to 48 h at temperatures between 30 and 37°. Incubation conditions must be developed to achieve effective levels of clonal amplification, protein maturation, and enzymatic turnover of the fluorescently labeled substrate. The average number of live bacteria or λ phage recovered and transferred to the recovery plate in a typical phagemid screen is on the order of 10^4 cells per well.

Sealed Plate Incubation Method. A distinguishing requirement of Giga-Matrix incubation is careful control of moisture. Whereas an acceptable water balance for a microplate assay is accomplished simply with the use of a standard humidified incubator and plate lids, the microscopic well volumes in GigaMatrix mandate a much higher level of moisture control. Seemingly minor evaporation/condensation events can have disastrous effects both on the growth and containment of cells during incubation and on the subsequent fluorescence imaging of individual wells. In our laboratory, consistent and reliable moisture control is attained by an enclosed system, where incubation is carried out on individually encapsulated GigaMatrix plates. The rate of evaporative loss is less than 1% (by fluid weight) per day when using the technique as follows. After loading, plates are prewarmed individually in a dry oven just long enough for the temperature to reach approximately 40°. The purpose of the prewarming step is to prevent condensation on the plate surface during the first hour of incubation when plate temperature and air humidity approach equilibrium. Each plate is then sandwiched quickly between two clean microplate lids, sealed tightly with plastic wrap, and placed in the incubator.

An important requisite of this technique is precise temperature stability within the incubator. Air layers trapped by the plate lids remain saturated with water vapor throughout the incubation process. The vapor pressure at which air is saturated, however, is a function of temperature. Because this value drops sharply with decreasing temperature, a drop in surface temperature at one region of a GigaMatrix plate will force excess water vapor to condense onto the plate at that region. An even and constant temperature must therefore be maintained throughout incubation in order to prevent pools of fluid forming over multiple adjacent wells. The best incubators for incubating GigaMatrix plates maintain temperature stability with rapid forced air flow, providing temperature stability to within ±0.1°.

Growth Medium. Although standard microbiological media such as LB and NZY are used routinely in GigaMatrix, consideration should be given to the way in which libraries are propagated and screened when choosing a growth medium. As aeration issues are similar to those for microplates, the growth of *Escherichia coli* in GigaMatrix is presumed to be semiaerobic. Unlike an actively shaking liquid culture, gas exchange at the air/medium interface of each well is limited. The ratio of surface area to volume in a GigaMatrix well is about 3.3 cm^{-1} (both ends of each well exposed), which is roughly equivalent to that of a standard 384-well microplate loaded at 30 μl per well. Also limited is the supply of molecular oxygen within the plate enclosure. As the shift from aerobic to anaerobic growth progresses, the medium can become acidic, resulting in not only slowed cell division, but a weakened fluorescent signal from several pH-affected fluorophores. The use of a buffer in the medium can protect against these adverse pH-related effects. Highly supplemented forms of buffered defined media such as that described by Neidhardt *et al.*[9] have been used successfully for screening in GigaMatrix. In addition, the use of glucose in growth medium is generally avoided due to the levels of acid produced by its metabolism and because of its effects on cAMP-dependent gene regulation.[10] A further consideration when designing a GigaMatrix screen is the level of background fluorescence contributed by the growth medium during the subsequent detection step, as discussed next.

Detection and Recovery

Following incubation, each plate is taken individually from the incubator and is unwrapped and placed quickly in the detection and recovery station. The plate is scanned using the appropriate filter set, and selected hits are recovered robotically as described earlier. Background fluorescence and optical effects are important obstacles to overcome for the successful detection of positive wells.

Background Fluorescence. Background fluorescence must be taken into consideration when designing any screen based on fluorescence imaging. The imaging system described earlier was designed to be suitable for typical enzyme discovery assays in which fluorogenic substrate is added at concentrations between 10 and 100 μM. For example, the system currently in use can detect fluorescein in water at approximately 10 nM with a signal-to-background ratio greater than two. However, when imaged under similar conditions, a cell suspension in LB growth medium has a natural

[9] F. C. Neidhardt, P. L. Bloch, and D. F. Smith, *J. Bacteriol.* **119,** 736 (1974).
[10] A. Kolb, S. Busby, H. Buc, S. Garges, and S. Adhya, *Annu. Rev. Biochem.* **62,** 749 (1993).

fluorescence equivalent to approximately 100 nM fluorescein in water. Thus the level of background fluorescence limits overall sensitivity of the GigaMatrix screen. Background fluorescence in these screens usually derives from the growth medium. Fluorescence from standard growth media such as LB and NZY increases as fluorescence imaging wavelengths approach the UV region of the electromagnetic spectrum. Figure 4 demonstrates the background contributed by LB at excitation/emission wavelengths used to detect three common fluorophores. Background fluorescence can be reduced significantly using rich defined media or hybrid formulas combining defined and undefined media. Another potential

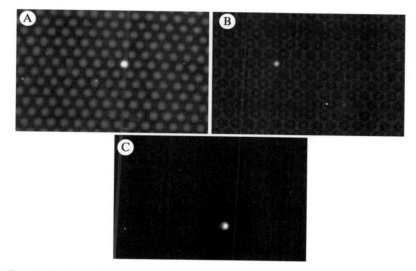

Fig. 4. Background fluorescence from LB growth medium as a function of imaging wavelength. Visualization of three fluorophores is shown using the relevant filter sets. Fluorophores, optimal absorbance/emission wavelengths (in nanometers), and part numbers (Omega Optical, Brattleboro, VT) for the filter sets used are as follows: (A) 7 amino-4-methylcoumarin (AMC), 351/430, and XF02-2; (B) fluorescein, 490/514, and XF136-2; and (C) resorufin, 571/585, and XF38. In each case, *E. coli* cells harboring a phagemid clone expressing the enzyme of interest were loaded sparsely in GigaMatrix plates using LB medium and the fluorescently labeled substrate. All plates were incubated at 37° for 24 h as described previously. The positive well shown in each frame contains cells carrying an active clone; the surrounding wells contain no cells and demonstrate the extent of background fluorescence contributed by the LB medium. Enzymes encoded by active clones, as well as substrates used to detect them, are (A) secondary amidase with a proprietary D-2-aminoadipic acid 6-AMC substrate; (B) amylase with DQ starch labeled with BODIPY FL dye (Molecular Probes, Inc., Eugene, OR); and (C) β-galactosidase with resorufin β-D-galactopyranoside (Sigma, St. Louis, MO).

source of significant background fluorescence is the fluorescently labeled substrate. The extent of substrate autofluorescence depends on various aspects, such as the fluorophore, the molecule to which it is linked, and the specific batch of manufacture. An awareness of excessive fluorescence from a particular substrate is useful both in the initial design of an enzyme screen and in discriminating primary hits during the detection step.

Optical Effects. Certain optical effects can lead to variability in both signal and background fluorescence and result in false positives or false negatives. The surface tension of the assay suspension tends to form a meniscus at the air/fluid interface with extreme curvature within the microscopic diameter of a GigaMatrix well. The refractive effects of the meniscus affect the optical efficiency of both excitation and emission. Additional variability in signal and background levels can result from well-to-well variation in the position of the assay solution within the well. In general, higher signals are obtained when evaporation is minimized and the meniscus is closer to the optical system. These effects further demonstrate the need for a uniform (from well-to-well) control of evaporation and condensation during the incubation process.

Breakout

During the robotic recovery process, primary hits are identified, aspirated, and deposited into a standard microplate (recovery plate) for further analysis. Each well of this recovery plate holds the contents of a positive well, or primary hit, taken from the GigaMatrix screen. The primary hit is often a mixed population of library clones, especially if high seed densities were used in the screen. Each primary hit is taken through a process (called *breakout*) of secondary screening and clonal isolation. In a phagemid library screen, for example, each primary hit is plated onto agar plates and incubated to produce colonies. Several isolated colonies from each primary hit are picked and arrayed individually into microplates, where they are evaluated for enzymatic activity in a liquid assay that mimics the conditions of the primary (GigaMatrix) screen.

The seed density used in the primary screen impacts the breakout process directly. In comparisons of seed densities over the range of 0.5 to 5, we have observed that modest increases in seed density can result in a dramatic drop in the average fraction of positive isolates identified from each primary hit. In screens where the seed density is approximately two, it is common to observe only 1 positive isolate among 384 isolates picked from a single primary hit. This effect can be explained, at least in part, to differences in growth rate among heterogeneous populations within a single GigaMatrix well, where cells harboring clones that actively express enzyme are sometimes outcompeted. Using seed densities below one in the primary screen simplifies the subsequent breakout phase because fewer isolates are needed in order to find the positive clone within each recovered sample. For example, most of the recovered samples from a screen initially seeded at an average of 0.1 clones per well will be clonal populations of active hits. However, routine screening at a seed density of 0.1 severely decreases the throughput of the primary screen. As mentioned previously, the preferred seed density of approximately 0.5 to 1 appears to maximize the efficiency and throughput of the entire screening process, taking into account the effort needed for both primary screening and breakout. Alternatively, sparsely seeded GigaMatrix plates can be substituted for microplates in the breakout method. This alternative can be especially effective in situations where the seed density of the primary screen is high and/or the fraction of positive clones among primary hits is low.

GigaMatrix Plate Cleaning

As GigaMatrix plates are designed to be reusable, an effective and reliable cleaning method is required. A method has been developed that produces plates with a degree of cleanliness sufficient for all aspects of the screening process. This method, which has become automated, involves a combination of decontamination, ultrasonic cleaning, and two high-velocity rinsing steps. Rinse water is removed efficiently from all wells of the cleaned plates using a specially designed vacuum device. Clean plates are sterilized by either autoclaving or baking in a 115° oven. Plates are spot checked randomly using a three-part cleanliness evaluation method: (1) testing the ease of fluid wicking, (2) measuring *E. coli* growth within the wells during an overnight incubation under typical conditions, and (3) scanning the plate at multiple wavelengths for fluorescent signals that could be construed as a positive well.

GigaMatrix Screening Examples

Enzyme Discovery

The implementation of GigaMatrix in an enzyme discovery project is illustrated by a recent effort to identify novel proteases having minimal substrate specificity. The discovery of such proteases is generally problematic, the traditional means of which is limited to protease-active organisms that can be cultured in the laboratory. The recombinant strategy used in this project provided access to the vast majority of microbes that remain uncultured. Although new enzymes can be identified from recombinant libraries through techniques based on DNA sequence similarity, a screening approach based on enzyme activity has the potential for the discovery of novel enzymes with little homology to known protease genes. Using GigaMatrix for activity-based screening offers the combined advantages of employing sensitive fluorescent substrates with the high throughput necessary to overcome limitations imposed by heterologous gene expression of recombinant DNA libraries. A highly quenched, fluorescein-labeled gelatin substrate (Molecular Probes, Inc., Eugene, OR) was used in these screens. A combined total of nearly 5×10^7 library members from a combination of λ and phagemid libraries were interrogated with this substrate. Of the resulting confirmed protease clones identified in GigaMatrix, 54 were determined to be both unique and novel.[11]

Protein Optimization

In addition to enzyme discovery projects, the GigaMatrix platform has also been used for protein optimization screening. In one project, the goal was to modify the excitation/emission characteristics of a proprietary fluorescent protein. An optimized phagemid library was produced from three parental genes using the GeneReassembly* technique (U.S. Patent No. 6,537,776 and patents pending). For this project, fluorescent substrates were not employed. Plates were simply filled with the diluted clone suspension, allowed to grow and express the fluorescent protein, and then screened for the desired variants. The imaging software was modified in order to distinguish subtle changes in the fluorescent properties of the protein. To accomplish this, three filter sets with unique spectral properties were used to image each frame of the plate. The filters were selected to highlight shifts in wild-type green fluorescent emission. The grayscale images obtained from each of the three filter sets were mapped into discrete bands of an RGB color display buffer to create a composite false-colored image (Fig. 5) used for hit discrimination and display to the operator. Under this scheme, protein variants with emission spectra shifted toward shorter wavelengths would appear as blue or cyan, the wild type as green, and variants with longer wavelength emission spectra would appear red. Utilizing these methods, 2×10^6 clones were screened and numerous protein variants with modified spectral properties were isolated and characterized.[12]

---

[11] C. Hansen, A. McClure, M. Dycaico, M. Lafferty, M. Sun, A. Anderson, D. Nunn, and M. Cayouette, submitted for publication.

[12] L. Parra-Gessert, C. Abulencia, M. Dycaico, M. Lafferty, C. Tweedy, E. Tozer, F. Zhang, E. Mathur, J. Short, and G. Frey, submitted for publication.

* Gene Reassembly is a trademark of Diversa Corporation.

FIG. 5. Composite false-color image from a GigaMatrix plate in which four control clones encoding different fluorescent proteins were grown. The fluorescent protein encoded by clone A has absorbance/emission peaks at 388/440 nm and is identified here with the color blue. Clones B and C encode proteins with peaks at 463/488 and 488/507 nm, respectively. Clone C is identified by bright green, whereas clone B is a darker combination of blue and green. Clone D encodes a protein with peaks at 530/540 nm and is visualized by the color red in this image. (See color insert.)

Concluding Remarks

The GigaMatrix platform has been used routinely for screening since 2002. Over the course of several projects, hundreds of unique novel enzymes have been identified from both λ and phagemid DNA libraries using this platform. These comprise a range of protein types and enzyme functions, including β-xylosidases, β-glucosidases, β-galactosidases, α-amylases, cellulases, α-L-arabinofuranosidases, proteases, other secondary amidases, and fluorescent proteins. As with all screening platforms, the success of GigaMatrix is strongly influenced by the performance of the particular screening assay. Those with high signal-to-noise ratios tend to be very robust, delivering success under varied screening conditions and identifying clones covering a wide range of enzymatic strengths. Less sensitive assays require more care in both developing and performing the screens and are limited to identifying clones that exceed a higher activity threshold. Nevertheless, GigaMatrix far surpasses microplate-based methods in both throughput and economy, making it a valuable screening platform for protein discovery and optimization.

Acknowledgments

The development of the GigaMatrix platform was a multidisciplinary effort requiring the skills of many engineers and scientists. The authors gratefully acknowledge Betsy George who first demonstrated the concept and the members of the research and development team: Patti Kretz, Dan Kline, Chuck Tweedy, Scott Beaver, Ian Storer, and Tom Todaro. We thank our colleagues whose work provided the examples presented; Connie Hansen, Amy McClure, and Michelle Cayouette for their work on protease discovery, and Lillian Parra-Gessert and Gerhard Frey for their work on fluorescent protein optimization and GeneReassembly. We also thank Dan Robertson and Jay Short for their vision and support.

[12] High Throughput Microplate Screens for Directed Protein Evolution

By Melissa L. Geddie, Lori A. Rowe,
Omar B. Alexander, and Ichiro Matsumura

Introduction

Protein engineers seek to enhance the utility of wild-type proteins through rational design or directed evolution.[1-3] In the latter approach, molecular diversity is generated by random mutagenesis or chimeragenesis of protein-coding genes. The resulting libraries of sequence variants are subcloned and expressed in microorganisms. Clones exhibiting improvement in the desired property are isolated in high throughput screens (HTS) or selections. This approach enables the direct alteration of protein function without a complete understanding of protein structure.

In recent years, molecular biologists have developed a bewildering number of methods to generate sequence variation.[4-6] High throughput screening generally requires a substrate that is spectroscopically distinct from the corresponding product[7] and requires some way to express and

[1] E. T. Farinas, T. Bulter et al., Curr. Opin. Biotechnol. 12, 545 (2001).
[2] K. A. Powell, S. W. Ramer et al., Angew Chem. Int. Ed. Engl. 40, 3948 (2001).
[3] H. Zhao, K. Chockalingam et al., Curr. Opin. Biotechnol. 13, 104 (2002).
[4] S. Harayama, Trends Biotechnol. 16, 76 (1998).
[5] S. Lutz and S. J. Benkovic, Curr. Opin. Biotechnol. 11, 319 (2000).
[6] A. A. Volkov and F. H. Arnold, Methods Enzymol. 328, 447 (2000).
[7] D. Wahler and J. L. Reymond, Curr. Opin. Biotechnol. 12, 535 (2001).

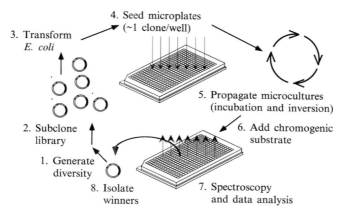

FIG. 1. A semiautomated high throughput microplate screen. (1) The protein-coding gene is mutated randomly by mutagenic polymerase chain reaction or recombined randomly with homologues. (2) The resulting library is ligated into a constitutive expression vector and (3) transformed into *Escherichia coli*. (4) The transformants are distributed into 384-well microtiter plates with a microplate dispenser. (5) The microplates are sealed manually and inverted end over end in an environmental rotator. (6) The chromogenic substrate is added to each of the saturated microcultures with the dispenser. (7) The reaction is monitored with a microplate spectrophotometer. (8) Clones exhibiting the most activity are isolated and propagated manually.

display the protein so that the most active variants can be identified and isolated. Ultrahigh throughput screens have been reported,[8,9] but astronomic throughputs (10^7–10^9 clones/day) are often associated with modest sensitivity, dynamic range, and/or versatility. Microtiter plates (or microplates) enable high throughput (~10,000 clones per day) assays that are precise, sensitive, and broad in dynamic range. Industrial laboratories have generally opted for microplate-based screens,[8,10] but the automation associated with these assays is generally too expensive for most academic laboratories.

This article provides protocols for a versatile but relatively inexpensive high throughput microplate enzyme assay. Our general strategy is illustrated in Fig. 1. *Escherichia coli* cells constitutively expressing a library of proteins are seeded into liquid media contained in 384-well microplates. These plates are sealed and the microcultures are propagated to saturation (by agitation at 37° for 16 h). A chromogenic substrate is dispensed into

[8] M. Olsen, B. Iverson *et al.*, *Curr. Opin. Biotechnol.* **11**, 331 (2000).
[9] W. J. Dower and L. C. Mattheakis, *Curr. Opin. Chem. Biol.* **6**, 390 (2002).
[10] J. E. Ness, M. Welch *et al.*, *Nature Biotechnol.* **17**, 893 (1999).

each microculture, and the plates are incubated. The formation of product within each well is measured by microplate spectrophotometry. The absorbance values associated with each microculture are sorted by a Microsoft Excel macro. Clones exhibiting the most improvement over ancestral control cultures are isolated and used as templates for the next round of directed evolution.

We have employed this screen to direct the evolution of *E. coli* β-glucuronidase.[11] The screening methods, however, are versatile enough to enable the evolution of "difficult" enzymes, including those that are toxic, those that react with membrane-impermeable substrates, and those expressed heterologously in microorganisms with competing endogenous activities. We have constitutively expressed several toxic proteins, including HIV protease[12] and T4 endonuclease VII,[13] at modest but detectable levels. The HTS can accommodate substrates that do not permeate cell membranes by replica plating and lysis of the cultures (as described later). Replica plating and lysis could also enable the purification of six histidine-tagged proteins in nickel chelate-coated microplates (Qiagen, Chatworth, CA; Xenopore, Hawthorne, NJ); this procedure would eliminate inhibitors or competing endogenous activities.

Heterologous Expression

The heterologous expression system is a major determinant of the sensitivity and precision of any HTS. We generally employ *E. coli* and expression vectors based on the pBAD (Invitrogen, Carlsbad, CA), *lac*, *trc* (Invitrogen), or pPROTet (BD Biosciences/Clontech, Palo Alto, CA). We produce liquid *E. coli* microcultures from single transformed cells (as described later). Under these conditions, constitutive expression at moderate transcription levels produces more activity (with less variation between isogenic cultures) than the induction of a strong promoter. The pET expression system (Novagen, Madison, WI) is not appropriate for constitutive expression because the expression of the T7 RNA polymerase alone is apparently toxic.[14,15]

Because proteins differ with respect to codon bias, toxicity, and conformational stability, the development of an efficient expression system often requires some trial and error of different vectors and induction conditions.

[11] L. A. Rowe, M. L. Geddie *et al.*, *J. Mol. Biol.* **332**, 851 (2003).

[12] E. Z. Baum, G. A. Bebernitz *et al.*, *Proc. Natl. Acad. Sci. USA* **87**, 5573 (1990).

[13] H. G. Kosak and B. W. Kemper, *Eur. J. Biochem.* **194**, 779 (1990).

[14] F. W. Studier and B. A. Moffatt, *J. Mol. Biol.* **189**, 113 (1986).

[15] J. W. Dubendorff and F. W. Studier, *J. Mol. Biol.* **219**, 61 (1991).

pBAD and pPROTet promoters can be fine-tuned within appropriate *E. coli* strains by controlling the inducer concentration. In our hands, pBAD is apparently weaker and more tightly regulated than pPROTet. As noted earlier, decreased transcription can sometimes lead to greater enzyme activity. We have also reported an alternative method to optimize transcription from promoters that cannot be modulated such as *lac* and *trc* (Invitrogen).[16] In short, the conserved "−10" regions of these promoters are "randomized" by cassette mutagenesis. This produces a library of promoter variants, most of which are presumably weaker (and therefore more appropriate for constitutive expression from multicopy plasmids) than the parental promoter. High throughput screens (described later) are employed to identify inducer concentrations (for pBAD or pPROTet) or promoter variants (of *lac* or *trc*) that confer genetically stable, high level constitutive expression upon transformed cells.

Seeding and Propagation of Clonal Microcultures

The seeding of transformed cells into microplate wells can be automated in stages as described in Table I. Automation is generally governed by the law of diminishing returns. Robots are very good at simple repetitive processes, but the cost of automation rises sharply as the procedures become more sophisticated. We employ a Thermo LabSystems Multidrop384 dispenser (Thermo LabSystems, Waltham, MA) retrofitted with a Titan microplate stacker (Titertek, Huntsville, AL). The transformed cells are aliquoted such that each well receives an average of one transformed cell. The Poisson distribution predicts that ~37% of the wells will receive no cells, ~37% will receive exactly one cell, and the remainder will receive two or more. Colony pickers enable the seeding of exactly one colony in each well, but that level of automation is too expensive for most individual academic investigators.

TABLE I
AUTOMATION OF MICROPLATE SEEDING

Instrument	Approximate cost (2003)	Function
Multichannel pipetter	$500–1200	Fills one row at a time
Dispenser	$6200– >7500	Fills one plate at a time
Handler or stacker	+ $11,500– >17,000	Feeds 60–80 plates at a time
Colony picker	+ $ 90,000– >120,000	Seeds exactly clone well

16 I. Matsumura, M. J. Olsen *et al.*, *Biotechniques* **30**, 474 (2001).

Seeding and Propagation Protocol

1. *Transform cells with expression library ("mutants").* Separately transform cells with the best clone from the previous round of evolution ("ancestral"). Add SOC for a final volume of 1 ml after heat shock[17] or electroporation[18] and allow cells to recover for 1 h at 37°.

2. *Freeze transformants.* Add sterile glycerol to 15%, and vortex lightly. Aliquot into sterile microfuge tubes on dry ice to snap freeze. Store at −80°.

3. *Titer library and ancestral control population.* Unfreeze one aliquot each of the ancestral and mutant populations. Plate serial dilutions onto Luria broth (LB) agar plates supplemented with the appropriate antibiotic(s). Incubate agar plates overnight at 37°. Count colonies and calculate the titer of transformants in each of the undiluted frozen populations.

4. *Sterilize the dispensing cassette.* We run 95% ethanol and then autoclave water through the tubing of the microplate dispenser.

5. *Distribute transformants in 384-well microplates.* Thaw aliquots of frozen ancestral and mutant cells and dilute each in LB (plus appropriate antibiotic) to 1 cell/5 μl. Use the sterilized dispenser to aliquot 5 μl of ancestral cells into 3 × 384 square well microplates (1152 wells) and then 5 μl of mutant cells into 77 microplates (29,568 wells).

6. *Seal plates.* We use a rolling pin to press autoclaved silicone 384-well microplate seals (Specialty Silicone Products, www.ssp.com) into each of the wells (Fig. 2A).

7. *Propagate the microcultures.* Stack the microplates into an environmental rotator (Environmental Express, www.envexp.com) in a 37° room or incubator. Insert pieces of Styrofoam or cardboard into the rotator to keep the microplates from rattling around (Fig. 2B). Close and lock the top of the rotator and invert the microplates end over end (30 rpm) overnight for at least 16 h. The cell densities of the resulting microcultures should be comparable (∼75%) to those of cultures aerated in regular culture tubes shaken at 250 rpm.

Notes

3. We use Nunc 384 square well microplates (Nunc, Rochester, NY) for the following reasons. Square wells provide a corner into which cells

---

[17] H. Inoue, H. Nojima *et al.*, *Gene* **96,** 23 (1990).
[18] W. J. Dower, J. F. Miller *et al.*, *Nucleic Acids Res.* **16,** 6127 (1988).

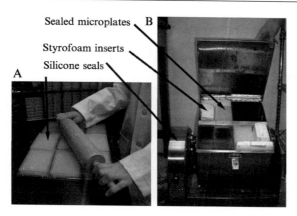

FIG. 2. Propagation of *E. coli* in 384-well microplates. (A) Tightly fitting silicone seals are pressed into 384-well microplates containing 5-μl microcultures containing an average of one transformed cell. (B) The sealed microplates are stacked up and placed into an environmental rotator. Styrofoam spacers are placed into the rotators to prevent the plates from crashing around. The lid is closed and locked, and the rotator turns the microplates end over end at 30 rotations/minute and enables *E. coli* growth to saturation.

can settle. The wells are compatible with the SSP silicone seals. The plastic lids are compatible with the Assist microplate handler; lids from other manufacturers sometimes slip the grasp of the handler.

4. The tubing of some dispensers, such as the Thermo LabSystems Multidrop384, can be autoclaved.

5. We designate each of the microplates (numbers 1–80) with a permanent Sharpie marker. The operation of the dispenser will depend on the make and model. We fill microplates #1, #40, and #80 with the ancestral control cells. After the microplates have been filled, we place the ancestral controls within the top, middle, and bottom of the stack.

6. Box packing tape (3M Scotch 3750) can also be used to seal microplates, but it leaves behind residues that cause microplates to stick to each other; this diminishes the performance of plate handlers. Microplate sealing tape (Nunc) can be used, but it is ultimately more expensive than the reusable silicone seals. Automated microplate sealers are also available commercially, but are expensive.

7. Standard orbital shakers (250 rpm) do not agitate microplates enough for propagation of *E. coli*. The ATR Multitron shaker (ATR, Laurel, MD) can agitate microplates sufficiently, but this instrument costs roughly 10 times as much as an environmental rotator.

High Throughput Assays

1. *Remove the silicone seals manually.* Rinse the seals, wrap them in aluminum foil (10 seals per package), and autoclave them for the next round of screening.

2. *Sterilize the dispensing cassette.* We run 95% ethanol and then autoclave water through the tubing of the microplate dispenser.

3. *Add substrate.* Use the sterilized dispenser to fill the wells ($+75 \mu l$ for 384 square well Nunc microplates) with buffer containing an appropriate concentration of a chromogenic or fluorogenic substrate.

4. *Incubate the reactions.* Use a Kimwipe to remove any condensation that has formed on the bottoms of the microplates. Place a plastic cover on each microplate. Put the microplates into a rack (Custom BioGenic Systems, Shelby Township, MI) and incubate the rack at an appropriate temperature at a $>45°$ angle for 1–24 h (depending on how quickly the product is formed). The cells should settle into an edge of each well after 4 h and will not interfere with subsequent spectrophotometry.

5. *Spectrophotometry.* Remove each lid just prior to placement in the microplate spectrophotometer (or fluorimeter); this eliminates condensation.

6. *Analyze data.* Identify wells containing microcultures exhibiting the greatest improvement in the desired phenotype.

7. *Secondary screen.* Transfer 2 μl of the best microcultures into tubes containing 1 ml of fresh LB (plus appropriate antibiotics). Aliquot 20 μl of each dilution into 4–8 duplicate wells of a sterile 96-well microplate. Shake overnight on an orbital shaker (250 rpm, $37°$). Add 300 μl of substrate in buffer and monitor the formation of product in a kinetic run of the microplate spectrophotometer.

8. *Archive the best clones.* Streak cultures onto LB agar plates and incubate them overnight at $37°$. Prepare a rectangular LB agar Omniplate (Nunc). Use a 96-pin microplate replicator to prick the surface of the agar. Pick individual colonies with sterile toothpicks and stab them into marked locations of the LB agar Omniplate. Incubate overnight at $37°$, wrap the plate with Parafilm, and store the stab cultures at $4°$.

Notes

1. An absorbance reading could be taken after overnight growth so that enzyme activity data can be corrected for cell density.

3a. We usually detect enzyme activity in the absence of any cosolvent or detergent,[19] even with substrates that do not permeate the cell membrane.

---

[19] U. Schwaneberg, C. Otey *et al.*, *J. Biomol. Screen* **6,** 111 (2001).

Because most of the enzyme activity remains in the supernatant after the cells have been removed, some cell lysis therefore apparently occurs during cell growth.

3b. Timing is important because each reaction will be measured spectrophotometrically only once. The substrate should be added at the same rate at which the microplates will be loaded into the spectrophotometer. It takes 4 h for the cells to settle (see step 3). If the substrate is generally depleted by that time, it will be necessary first to add most of the buffer, let the cells settle, and than add concentrated substrate in a small volume.

4–5. The Thermo LabSystems Assist microplate stacker can remove each lid just before the plate is put into the microplate spectrophotometer. Make sure that all of the microplates are in the correct order and orientation. The software associated with the Thermo LabSystems Multiskan Ascent spectrophotometer can write data from a series of microplates into a single Excel spreadsheet (Table IIA).

6. We find it convenient to rearrange spectrophotometer data to compare all of the data points. For example, data collected on the Thermo LabSystems Multiskan Ascent microplate spectrophotometer can be exported as Microsoft Excel files. The absorbance values are formatted in consecutive 24 × 16 matrices, one for each 384-well plate (Table IIA). We have written a Visual Basic macro (Table III) that rearranges data into a single column, annotated with address information (plate/column/row for each point) (Table IIB). Different software solutions will be required for data arrayed in other formats. The sort function of Excel enables rapid identification of the microcultures exhibiting the strongest activities (Table IIC).

7. We generally reevaluate the performance of the best 12–24 microcultures. The elimination of clones that fail to exhibit reproducible improvement accelerates the rate of adaptation greatly.

8. Stab cultures survive longer than colonies grown on agar surfaces. The 96-pin format of stab cultures enables replication onto fresh agar plates. Alternatively, the clones can be inoculated into 100-μl liquid cultures and propagated. The saturated cultures can be diluted twofold in 70% LB/30% glycerol (15% glycerol final) and stored at −80°.

Washing and Sterilizing Microplates

We save money and reduce plastic waste by washing, sterilizing, and reusing microplates. Unfortunately 384-well microplates are difficult to wash because the wells are narrow. We developed the following procedure by washing colored dyes out of 384-well microplates.

TABLE II
DATA PROCESSING

A. Raw data

Measurement count: 1 Filter: 405

	1	2	3	4
A	0.107	1.582	0.107	2.103
B	0.109	0.104	0.125	0.112
C	0.102	1.479	0.163	0.273
D	0.105	0.104	0.107	0.166
E	0.105	0.104	0.106	1.886
F	0.104	1.56	0.104	0.108
G	1.395	0.112	0.109	0.111
H	0.106	0.106	0.108	0.108
I	0.108	0.108	0.106	0.107
J	1.86	0.107	0.107	0.106
K	0.105	0.108	0.107	0.106
L	0.112	0.108	1.722	0.105
M	0.117	0.109	0.108	0.111
N	1.761	0.108	0.107	0.108
O	0.11	0.11	0.11	0.108
P	0.119	0.114	0.11	0.109

B. Rearranged

Abs405	Plate	Column	Row
0.107	1	1	A
0.109	1	1	B
0.102	1	1	C
0.105	1	1	D
0.105	1	1	E
0.104	1	1	F
1.395	1	1	G
0.106	1	1	H
0.108	1	1	I
1.86	1	1	J
0.105	1	1	K
0.112	1	1	L
0.117	1	1	M
1.761	1	1	N
0.11	1	1	O
0.119	1	1	P
0.107	2	1	A
1.894	2	1	B
1.421	2	1	C

C. Sorted

Abs405	Plate	Column	Row
4.385	14	18	N
3.58	20	20	G
3.532	22	24	M
3.516	14	21	O
3.476	7	12	B
3.328	8	4	N
3.324	24	18	G
3.287	16	24	D
3.241	25	1	B
3.205	28	15	F
3.125	48	7	M
3.116	52	18	L
3.116	16	18	D
3.106	17	13	F
3.095	32	21	M
3.088	8	20	P
3.055	8	12	P
3.047	52	20	J
3.044	26	13	N

Measurement count: 1 Filter: 405

	1	2	3	4										
A	0.107	0.28	0.275	0.76	1.119	2	1	D	3.034	10	24	N
B	1.894	1.259	1.301	0.714	1.764	2	1	E	3.032	33	24	H
C	1.421	0.276	1.28	0.641	1.805	2	1	F	3.012	41	1	G
D	1.119	1.116	1.333	0.231	2.133	2	1	G	3.011	32	15	I
E	1.764	1.682	0.757	0.459	0.956	2	1	H				
F	1.805	1.135	1.565	1.656								
G	2.133	1.827	1.458	2.113								
H	0.956	0.2	0.235	0.331								

Spectroscopic data were collected by a Multiskan Ascent and exported into an Excel file. Raw data are listed in 24 × 16 arrays; a small subset of the 30,700 data points is shown (column A). Data were rearranged (column B) by a Visual Basic macro (listed in Table III), and the fittest clones were identified using the Excel sort command (column C).

TABLE III
VISUAL BASIC MACRO TO REARRANGE MICROPLATE SPECTROPHOTOMETER DATA

```
Sub Macro5 ()
platenum = InputBox ("How many 384 well plates were read?")
'data trimming: deletes rows with labels in rows 1,2,3, 21,22,23 etc.
For platecount = 1 To 100
    platenumb = 101 - platecount
    rocount = (platenumb - 1) * 20
    ro1 = rocount + 3
    Rows (ro1) .Delete
    ro2 = rocount + 2
    Rows (ro2) .Delete
    ro3 = rocount + 1
    Rows (ro3) .Delete
    Cells (1, 26) = platecount
Next platecount

'moves columns 3-25 into column 2
'finalro is the number of rows before cutting and pasting the 23 columns

finalro = 17 * platenum
For col = 3 To 25
    For fromro = 1 To finalro
        toro = ( ( (col - 2) * finalro) + fromro)
        Cells (fromro, col) .Cut Cells (toro, 2)
    Next fromro
    Cells (1, 26) = col
Next col

'copies the plate row labels

For origcol = 1 To 24
For origplate = 1 To platenum
ro = ( ( (origcol - 1) * finalro) + ( (origplate - 1) * 17) + 1)

For rowletter = 65 To 80
Cells (ro, 5) = Chr (rowletter)
ro = ro + 1
Next rowletter

Next origplate
Cells (1, 26) = origcol
Next origcol

'attaches original column and plate numbers to each datapoint

finalro = 17 * platenum
For origcol = 1 To 24
    For platecount = 1 To platenum
        For ro = 1 To 16
            toro = ( ( ( (origcol - 1) * finalro) + ( (platecount - 1) * 17) + ro)
            Cells (toro, 3) = platecount
            Cells (toro, 4) = origcol
Next ro
    Next platecount
    Cells (1, 26) = origcol
Next origcol

End-Sub
```

1. *Shake out* the bacterial, buffer, and substrate from each microplate down the sink drain.
2. *Dunk the plate in 1% Alconox detergent*. The wells should fill up without additional agitation.
3. *Dunk the microplate into a bucket of water*. Shake the inverted microplates up and down several times. Shake the water out into the sink drain.
4. *Dunk the microplate in 95% ethanol*. Again, the wells should fill up without agitation. Shake the ethanol back into the reservoir.
5. *Stack the microplates into a rack*. Heat in a drying oven at 50° until the liquid has evaporated. Overheating (or autoclaving) will cause the polystyrene microplates to warp.

Acknowledgments

We thank the National Science Foundation (MCB0109668) and the Emory University Research Committee (URC 00–01 Matsumura) for support. We are grateful to Monal Patel for her helpful comments on the manuscript.

[13] Periplasmic Expression as a Basis for Whole Cell Kinetic Screening of Unnatural Enzyme Reactivities

By Grazyna E. Sroga and Jonathan S. Dordick

Introduction

The application of evolutionary and combinatorial techniques to generate improved enzymes, achieve novel enzyme selectivities, and probe enzyme mechanisms has become one of the most dynamic tools of chemistry and biology. Two principally different strategies can be applied to alter enzyme properties by targeting the enzyme gene(s)—rational design, which has been applied extensively, and directed evolution. While the former has seen some specific successes,[1–6] the lack of detailed structural

[1] T. Yano, S. Oue, and H. Kagamiyama, *Proc. Natl. Acad. Sci. USA* **95,** 5511 (1998).
[2] G. MacBeath, P. Kast, and D. Hilvert, *Science* **279,** 1958 (1998).
[3] B. Van den Burg, G. Vriend, O. R. Veltman, G. Venema, and V. G. H. Eijsink, *Proc. Natl. Acad. Sci. USA* **95,** 2056 (1998).
[4] J. R. Cherry, M. H. Lamsa, P. Schneider, J. Vind, A. Svendsen, A. Jones, and A. N. Pedersen, *Nature Biotechnol.* **17,** 379 (1999).

and mechanistic information about many enzymes and our insufficient understanding of the structure–function relationship of proteins in general hinder the rational design of biomolecules with new functions. Hence, evolutionary approaches employing mutation and selection strategies have grown dramatically. In addition, information gained from evolutionary experiments can help improve our understanding of how proteins function, and thus may ultimately support rational design efforts.

Although the basic concept of directed evolution has been known for many years,[7] recent advances in automation and miniaturization facilitate the screening of millions of mutants in a reasonable time frame[8–10] and, therefore, have increased the utility of this technique. Directed evolution relies on creating genetic diversity, which is largely independent of the enzyme to be optimized. However, filtering the variants for desired function is unique for each enzyme and follows the principle of either *in vivo* selection or screening. While a screen that targets physical properties (e.g., pH or temperature optima) of an enzyme is relatively straightforward, screening for catalytic activity and specificity, particularly when the desired catalytic property overlaps with the native reactivities of the host cell, is more difficult. A case in point is proteolytic vs esterolytic activities of proteases, the former being the native function of the enzyme, whereas the later is unnatural yet synthetically relevant. The key to developing a successful screen for proteolytic and esterolytic reactivities using whole cell biocatalysts is to segregate and/or differentiate between native and unnatural activities. To address esterolytic specificity of enzyme variants by screening whole cell biocatalysts, we developed a strategy combining periplasmic expression with cell-based kinetic assays. This resulted in a powerful approach for the selection of esterolytic enzymes.

The great majority of protein engineering studies use cytoplasmic expression in *Escherichia coli*.[11,12] However, quite often this approach does

---

[5] P. C. Cirino and F. H. Arnold, *in* "Directed Molecular Evolution of Proteins or How to Improve Enzymes for Biocatalysis" (S. Brakmann and K. Johnsson, eds.), p. 215. Wiley-VCH Verlag GmbH, Weinheim, 2002.

[6] G. Xia, L. Chen, T. Sera, M. Fa, P. G. Schultz, and F. E. Rosemberg, *Proc. Natl. Acad. Sci. USA* **99**, 6597 (2002).

[7] B. G. Hall, *Biochemistry* **20**, 4042 (1981).

[8] G. Georgiou, *in* "Advances in Protein Chemistry—Evolutionary Protein Design" (F. H. Arnold, ed.), Vol. 55, p. 293. Academic Press, San Diego, 2000.

[9] M. Olsen, B. Iverson, and G. Georgiou, *Curr. Opin. Biotechnol.* **11**, 331 (2000).

[10] A. Schweinhorst, *in* "Directed Molecular Evolution of Proteins or How to Improve Enzymes for Biocatalysis" (S. Brakmann and K. Johnsson, eds.), p. 159. Wiley-VCH Verlag GmbH, Weinheim, 2002.

[11] G. Hannig and S. C. Makrides, *Trends Biotechnol.* **16**, 54 (1998).

[12] S. Witt, M. Singh, and H. M. Kalisz, *Appl. Environ. Microbiol.* **64**, 1405 (1998).

not allow the expressed proteins to be used in a whole cell screen because either the protein (inclusion bodies) or the substrate (metabolized inside the cytoplasm) is inaccessible to each other. An alternative is to display heterologous proteins on the surface of microorganisms, including *E. coli*,[13,14] or phage display.[15–18] Unfortunately, displaying proteins onto the cell or phage surface may result in an undesirable environment for the enzyme, thereby resulting in a possible poor assembly of multiprotein complexes or low stability of the expressed enzyme.

A growing number of proteins have been expressed successfully in the *E. coli* periplasm.[19–25] During periplasmic expression, recombinant proteins are continuously transported and displayed *near* the bacterial surface (Fig. 1). The unanchored, yet cell-bound recombinant enzyme is protected from denaturation that could occur upon enzyme display onto the negatively charged cell surface. Moreover, the relatively unhindered accessibility of substrate to the recombinant enzymes in subsequent screening of the whole cells is facilitated by periplasmic expression (Fig. 1). Finally, the periplasmic expression of mutant enzymes results in a dramatically reduced proteolysis[26] and a reduced likelihood of inclusion body formation.

Statement of the Problem: Periplasmic Expression Strategy

This article describes work on the modification of esterolytic activities of the serine protease subtilisin E by directed molecular evolution using a periplasmic expression strategy. In addition to its natural amidase activity,

[13] G. Georgiou, C. Stathopoulos, P. S. Daugherty, A. R. Nayak, B. L. Iverson, and R. Curtiss III, *Nature Biotechnol.* **15**, 29 (1997).

[14] J. Jose and S. Handel, *ChemBioChem.* **4**, 396 (2003).

[15] G. P. Smith, *Science* **228**, 1315 (1985).

[16] D. J. Rodi, L. Makowski, and B. K. Kay, *Curr. Opin. Chem. Biol.* **6**, 92 (2002).

[17] S. S. Sidhu, W. J. Fairbrother, and K. Deshayes, *ChemBioChem.* **4**, 14 (2003).

[18] K. Weisehan and D. Willbold, *ChemBioChem.* **4**, 811 (2003).

[19] J. R. Vasquez, L. B. Evnin, J. N. Higaki, and C. S. Craik, *J. Cell. Biochem.* **39**, 265 (1989).

[20] C. Wulfing and A. Pluckthun, *J. Mol. Biol.* **242**, 655 (1994).

[21] L. A. Collins-Racie, J. M. McColgan, K. L. Grant, E. A. DiBlasio-Smith, J. M. McCoy, and E. R. LaVaille, *Biotechnology* **13**, 982 (1995).

[22] C. Wulfing and R. Rappuoli, *Arch. Microbiol.* **167**, 280 (1997).

[23] Y. Zhang, D. R. Olsen, K. B. Nguyen, P. S. Olson, E. T. Rhodes, and D. Mascarenhas, *Protein Expr. Purif.* **12**, 159 (1998).

[24] X. Zhan, M. Schwaller, H. F. Gilbert, and G. Georgiou, *Biotechnol. Progr.* **15**, 1033 (1999).

[25] Z. Ignatova, S. O. Enfors, M. Hobbie, S. Taruttis, C. Vogt, and V. Kasche, *Enzyme Microb. Technol.* **26**, 165 (2000).

[26] K. H. Swamy and A. L. Goldberg, *J. Bacteriol.* **149**, 1027 (1982).

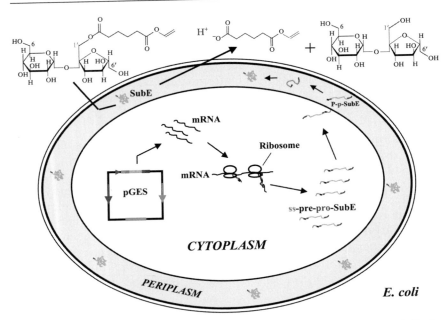

Fig. 1. The principle of periplasmic expression of recombinant proteins in *E. coli*. During expression, subtilisin E is synthesized in its pre-pro form along with the signal sequence of gene III that encodes the minor capsid protein pIII of a filamentous phage fd. The signal sequence directs the recombinant protein to the periplasm, where it matures. Substrates have easy access to the enzyme expressed near the bacterial surface. Hydrolysis of S1'A is shown as an example. Screening is based on the release of vinyl adipate, which causes a pH decrease detected by bromothymol blue. (See color insert.)

subtilisin E possesses a relatively weak esterolytic activity. The key questions addressed were as follows.

1. To what extent could the amidase and esterase activities of subtilisin be modified?
2. Would it be possible to evolve a broad esterolytic subtilisin with high activity on several structurally different esters?

To answer these questions, we employed a directed molecular evolution strategy coupled with periplasmic expression. Typical screens of subtilisin mutants rely on the bacterial host *Bacillus subtilis*[27–29] and thus employ

[27] L. You and F. H. Arnold, *Protein Eng.* **9**, 77 (1994).
[28] H. Zhao and F. H. Arnold, *Protein Eng.* **12**, 47 (1999).
[29] J. E. Ness, M. Welch, L. Giver, M. Bueno, J. R. Cherry, T. V. Borchert, W. P. C. Stemmer, and J. Minshull, *Nature Biotechnol.* **17**, 893 (1999).

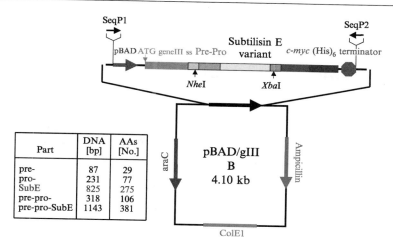

Part	DNA [bp]	AAs [No.]
pre-	87	29
pro-	231	77
SubE	825	275
pre-pro-	318	106
pre-pro-SubE	1143	381

FIG. 2. General plasmid map of DNA libraries carrying subtilisin E variants. The pBAD/ gIII B (Invitrogen) vector served as the basis for the generation of libraries of subtilisin E variants. (See color insert.)

commonly available protease-deficient *B. subtilis* mutants, such as the DB428 strain or strain 168. Unfortunately, these hosts display diverse esterolytic activities, which prevented us from using them in whole cell biocatalysts screens. Conversely, we determined that several *E. coli* strains do not have observable esterolytic activities on different ester substrates added to the culture medium. We focused on the *E. coli* TOP10 and TOP10 F′ strains, which were engineered to facilitate the periplasmic expression of recombinant proteins from the pBAD/gIII vector (Fig. 2) using arabinose as the inducer. Among other benefits,[30] this vector permits control of the level of recombinant protein expression through the modulation of arabinose concentration. We determined that the periplasmic fraction of *E. coli* TOP10 and TOP10 F′ could not catalyze ester hydrolysis and, therefore, was free of interfering enzyme activities that would prevent the effective screening of mutant enzymes expressed in the periplasm.

Cell-based kinetic assays require the simultaneous expression of similar amounts of protein variants. To achieve uniform expression of subtilisin libraries in *E. coli* cells growing in wells of microtiter plates, we introduced growth "pseudo-synchronization." Growth pseudo-synchronization was achieved by culturing the cells for a prolonged time in minimal medium

[30] G. E. Sroga and J. D. Dordick, *Biotechnol. Bioeng.* **78,** 761 (2002).

at low cell density to ensure that the cells were in steady state, balanced growth.[31–33]

We examined the reactivity of subtilisin E on N-succinyl-Ala-Ala-Pro-Phe-p-nitroanilide (AAPF-NA, a common amidase substrate) and N-acetyl-DL-Phe-p-nitrophenyl ester (Phe-NPE, a common esterase substrate). Three parent genes were used, including wild-type subtilisin E (GES201), a DMF-tolerant mutant (GES202), and a TeR-tolerant mutant (GES203). In addition to pseudo synchronization, we introduced *the ratio* of the initial reaction rates (v_0) for different enzyme reactivities as a key criterion for selecting desired enzyme variants (Table I). Using ratios overcomes the problem of natural fluctuations in differential expression levels of mutant enzymes, which are promoter and gene dependent. Wild-type subtilisin hydrolyzes Phe-NPE approximately 15-fold slower than AAPF-NA (Table I). The dimethylformamide (DMF)-tolerant and temperature (TeR)-resistant mutants are also less reactive on the esterase substrate. Although all evolved mutants showed improved reactivity on Phe-NPE, GES208 displayed over 7-fold higher activity on the ester substrate than on the amidase substrate as compared to the wild-type subtilisin E, which corresponded to an approximately 65-fold swing in the ratio of esterase/amidase activity. Thus, our first question was answered.

Although the primary screening was performed with Phe-NPE, we found that enhanced esterolytic activity of the evolved mutants could be extended to structurally different esters. Specifically, sucrose adipates, which have been used in sugar-based hydrogels,[34–37] were screened using an assay based on the generation of protons during ester hydrolysis (Fig. 1), which were detected using the pH indicator bromothymol blue (BTB). Sucrose 1'-adipate and sucrose 6-adipate were effective substrates of the three mutant variants, and hence, our second question was answered in the affirmative. Interestingly, subtilisin can differentiate between sucrose 1'-adipate (S1'A) and sucrose 6-adipate (S6A) with the former approximately 60% more reactive than the latter (Table I). This specificity was not altered significantly in the mutant enzymes, which was expected as the screen did

[31] T. E. Shehata and A. G. Marr, *J. Bacteriol.* **103,** 789 (1970).

[32] A. L. Koch and M. L. Higgins, *J. Gen. Microbiol.* **128,** 2877 (1982).

[33] H. Bremer and P. P. Dennis, *in* "*Escherichia coli* and *Salmonella typhimurium* Cellular and Molecular Biology" (F. C. Nieidhardt, ed.), p. 1527. ASM, Washington, DC, 1987.

[34] B. D. Martin, S. A. Ampofo, R. J. Lindhardt, and J. S. Dordick, *Macromolecules* **25,** 7081 (1992).

[35] N. S. Patil, J. S. Dordick, and D. G. Retchwisch, *Biomaterials* **17,** 2343 (1996).

[36] N. S. Patil, Y. Li, D. G. Retchwisch, and J. S. Dordick, *J. Polym. Sci. A: Polym. Chem.* **35,** 2221 (1997).

[37] S. J. Novick and J. S. Dordick, *Chem. Mater.* **10,** 955 (1998).

TABLE I

AMIDASE ($v_{0[\text{AAPF-NA}]}$) AND ESTERASE ($v_{0[\text{S1'A}]}$, $v_{0[\text{S6A}]}$, $v_{0[\text{Phe-NPE}]}$) INITIAL REACTION RATES AND RESPECTIVE RATES RATIOS FOR WILD-TYPE SUBTILISIN E AND THE IDENTIFIED MUTANTS

Mutant	AAPF-NA	Phe-NPE	S1'A	S6A	$\dfrac{v_{0[\text{AAPF-NA}]}}{v_{0[\text{Phe-NPE}]}}$	$\dfrac{v_{0[\text{AAPF-NA}]}}{v_{0[\text{S1'A}]}}$	$\dfrac{v_{0[\text{AAPF-NA}]}}{v_{0[\text{S6A}]}}$
	$V_0{}^a$ [μM/100 μl cell$_{\text{ind}}$ min^{-1}]						
Parents[b]							
GES201	1.273 ± 0.013	0.086 ± 0.008	0.243 ± 0.016	0.151 ± 0.012	14.73	5.24	8.43
GES202	0.997 ± 0.011	0.130 ± 0.003	0.182 ± 0.015	0.303 ± 0.006	12.80	5.48	10.89
GES203	1.099 ± 0.017	0.091 ± 0.007	0.240 ± 0.009	0.113 ± 0.012	12.08	4.58	9.73
Mutants							
GES206	0.237 ± 0.013	0.201 ± 0.006	0.430 ± 0.016	0.214 ± 0.017	1.18	0.55	2.01
GES207	0.186 ± 0.011	0.166 ± 0.002	0.433 ± 0.017	0.165 ± 0.007	1.13	0.49	1.12
GES208	0.157 ± 0.014	0.678 ± 0.021	0.349 ± 0.019	0.131 ± 0.011	0.23	0.45	1.20

[a] $\varepsilon_{405} = 7.51 \times 10^{-3}$ μM^{-1} cm^{-1} for BTB; $\varepsilon_{405} = 9.33 \times 10^{-3}$ μM^{-1} cm^{-1} for p-nitroaniline; $\varepsilon_{405} = 1.57 \times 10^{-2}$ μM^{-1} cm^{-1} for p-nitrophenol.

[b] The parental strains carry the following plasmids: pGES201 (the WT-$sprE$ gene in pBAD/gIII), pGES202 (the DMF-$sprE$ gene in pBAD/gIII), and pGES203 (the TeR-$sprE$ gene in pBAD/gIII). Parental genes were a gift from Dr. Frances H. Arnold (California Institute of Technology, Pasadena, CA).

not focus on sucrose ester hydrolysis. Based on the work described earlier, the combination of periplasmic expression and cell-based kinetic screening led to successful identification of subtilisin variants with broad reactivity on diverse esters.[38]

Protocols

This section describes specific methods and protocols for the periplasmic expression of subtilisin E from the pBAD/gIII series of vectors (Invitrogen).

Complete Minimal Medium

The expression of subtilisin E mutant libraries in the *E. coli* periplasm is conducted in complete minimal (CM) medium (g/liter: K_2HPO_4, 10.50; KH_2PO_4, 1.13; $(NH_4)_2SO_4$, 1.00; sodium citrate $\times$ $2H_2O$, 0.50; $MgSO_4$ $\times$ $7H_2O$, 0.25; $CaCl_2$ $\times$ $2H_2O$, 0.015) supplemented with 20 amino acids (40 μg/ml), vitamin B_1 (5 mg/l), glycerol (0.20%), and ampicillin (100 μg/ml). The most convenient way of making large quantities of fresh CM medium is to prepare it from stock solutions of mineral salts (M63 basal), a relevant carbon source (20% glycerol stock), vitamins (1% vitamin B1 stock), amino acids (4% stocks), and antibiotics. A kit containing 20 standard L-amino acids can be purchased from Sigma.

Functional Expression of Subtilisin Libraries in E. coli TOP10 F'

The expression of subtilisin E variant libraries is performed in microtiter (MT) plates (Evergreen 96-round bottom wells, clear, Phoenix Research Products, Hayward, CA).

1. Dispense 200 μl of the standard nutrient-rich medium with ampicillin (100 μg/ml) into each well of a MT plate.
2. Transfer one single cell colony from a petri plate (Fig. 3) to the medium in each well using a sterile toothpick. Secure the lid of the MT plate with parafilm to prevent evaporation.
3. Grow overnight at 37° with shaking at 225 rpm.
4. Prepare a corresponding number of new MT plates with 200 μl of CM with ampicillin and store at 4°.
5. Following growth, warm up the MT plates containing CM medium to 37° and inoculate each well with 2 μl of each culture from the overnight (ON).

---

[38] G. E. Sroga and J. D. Dordick, *Protein Eng.* **14**, 929 (2001).

6. Grow the cultures for approximately 8 h.

7. Prepare a corresponding number of new MT plates with 200 μl of CM with ampicillin.

8. Reinoculate the cultures as described earlier and grow them ON at 37° with shaking at 225 rpm.

9. On the following day, depending on the number of different assays to be performed (e.g., amidase, protease, or various esterase assays), transfer 65–100 μl of each ON culture to two or three new MT plates, centrifuge (Marathon 21000 R, MT plate rotor, Fisher Scientific, Pittsburgh, PA), and resuspend in 200 μl (OD$_{590}$ = ca. 0.40) of the CM medium.

10. Incubate the clones for 2 h at 37° with shaking at 225 rpm.

11. Induce the cells by adding arabinose to a final concentration of 0.15% (w/v).

12. Incubate the cultures for 6 h at 20° with shaking at 225 rpm.

13. Harvest the cells for the assays.

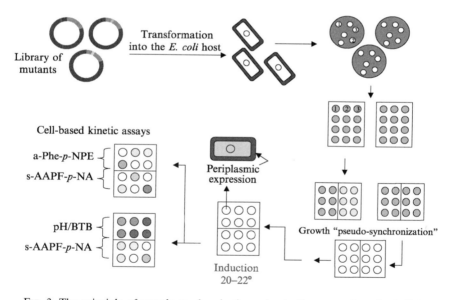

FIG. 3. The principle of pseudo-synchronization prior to the expression of subtilisin E libraries in *E. coli* for cell-based kinetic assays. A library of subtilisin E mutants in plasmid DNA form is transformed into *E. coli*. The resulting single cell colonies are transferred to nutrient-rich medium with an antibiotic. After growth pseudo-synchronization, the synthesis of subtilisin variants is induced at 20–22°. Whole cells are used as biocatalysts in the amidase and esterase kinetic screens. (See color insert.)

Troubleshooting

Consistent and efficient expression of subtilisin variants depends strongly on the growth conditions. Any fluctuation in shaking, speed, or temperature during cell growth and induction negatively influences the process of pseudo-synchronization (from Nos. 1–8) and the resulting efficiency of the recombinant enzyme expression.

Cell-Based Kinetic Screenings

Amidase Activity. The amidase assay on whole cell biocatalysts is performed in osmotic solution I [OS I; 20 mM Tris–HCl, pH 8.00, 2.5 mM EDTA, 2 mM CaCl$_2$, 20% (w/v) sucrose] using AAPF-NA (Sigma). The release of *p*-nitroaniline (*p*-NA) is measured spectrophotometrically at 405 nm.

1. Transfer 100 μl of each induced cell culture to a new MT plate.
2. Centrifuge MT plates with the respective cultures at 22° for 10 min at 1000g.
3. Resuspend bacterial pellets in 200 μl of OS I with 300 μM of AAPF-NA.
4. Immediately place the MT plate into the MT plate reader (HTS 7000 Plus bioassay reader; Perkin Elmer).
5. Collect initial reaction rate data for 40–60 min at 25–30° at 405 nm.
6. In calculations of the initial reaction rates, use the extinction coefficient of *p*-NA ($\varepsilon_{405} = 9.33 \times 10^{-3}\ \mu M^{-1}\ cm^{-1}$).

Esterase Activity. Preparation of bacterial cells and measurement of esterolytic activity of subtilisin E on Phe-NPE (Sigma) is performed similarly to the procedure described previously for amidase activity. The final concentration of Phe-NPE in OS I is 300 μM, and the respective extinction coefficient of the released *p*-nitrophenol is $\varepsilon_{405} = 1.57 \times 10^{-2}\ \mu M^{-1}\ cm^{-1}$.

The esterase assay on sucrose esters, S1′A and S6A,[39] using whole cell biocatalysts is performed in osmotic solution III (OS III; 5 mM Tris–HCl, pH 7.80, 1 mM EDTA, 2 mM CaCl$_2$) with 1 mM of S1′A and 0.5 mM BTB. Formation of the basic form of BTB ($\varepsilon_{405} = 7.51 \times 10^{-3}\ \mu M^{-1}$ cm^{-1}) is monitored at 405 nm.

1. Transfer 100 μl of each induced cell culture to a new MT plate.
2. Centrifuge MT plates with the respective cultures at 22° for 10 min at 1000g.

[39] O. J. Park, D. Y. Kim, and J. S. Dordick, *Biotechnol. Bioeng.* **70,** 208 (2000).

3. Resuspend bacterial pellets in 200 μl of OS III with 1 mM S1'A (or S6A) and 0.5 mM BTB, mix well, but do not cause bubble formation.
4. Immediately place the MT plate into the MT plate reader.
5. Collect initial reaction rate data for 3–4 h at 24–25° at 405 nm.
6. In calculations of the initial reaction rates, use the extinction coefficient ($\varepsilon_{405} = 7.51 \times 10^{-3} \ \mu M^{-1} \ cm^{-1}$) of the basic form of BTB.

Commentary

The progress of enzymatic reactions for all the assays was monitored at 405 nm. While this corresponds to the maximal absorbance for *p*-NA and *p*-NP, BTB has two absorbance maxima (410 and 620 nm). Lower reading sensitivity for BTB helped avoid the selection of false positives during screens of altered esterolytic activity on sugar esters. Also, use of the 405-nm filter assured *simultaneous* measurement of all tested reactivities in a single MT plate.

Troubleshooting

It is obligatory to prepare OS III buffer with fresh BTB and each sucrose ester. The pH of BTB solution must be checked and adjusted before beginning the kinetic assays.

Conclusions and Future Directions

The methodology developed for the kinetic screening of unnatural enzyme reactivities, where whole cells expressing libraries of enzyme variants in *E. coli* periplasm were used as biocatalysts, led to the successful identification of broad esterolytic subtilisins with significantly improved reactivities on several structurally different esters.[38] Sucrose esters have found application as monomers in the synthesis of biodegradable and biocompatible sugar-based hydrogels[34–37] and polymers.[39] The evolved mutants may also be useful in regioselective ester *synthesis* in nonaqueous media[40–43] and thus find applications[44] ranging from the selective synthesis of sugar-based polymers to synthetic nucleoside derivatives. We expect

[40] D. R. Patil, J. S. Dordick, and D. G. Retchwisch, *Biotechnol. Bioeng.* **37,** 639 (1991).
[41] D. R. Patil, D. G. Retchwisch, and J. S. Dordick, *Macromolecules* **24,** 3462 (1991).
[42] A. M. Kilbanov, *Nature* **409,** 241 (2001).
[43] M.-Y. Lee and J. S. Dordick, *Curr. Opin. Biotechnol.* **13,** 376 (2002).
[44] J. H. Zhang, G. Dawes, and W. P. C. Stemmer, *Proc. Natl. Acad. Sci. USA* **94,** 4504 (1997).

that these protocols may open a new avenue to study extensively enzyme regioselectivity on a wide range of ester substrates; an opportunity that has been difficult to achieve thus far using DME due to the lack of relevant experimental tools.

Acknowledgments

This work was supported by the Biotechnology Research and Development Corporation.

[14] Engineering the Thermotolerance and pH Optimum of Family 11 Xylanases by Site-Directed Mutagenesis

By Ossi Turunen, Janne Jänis, Fred Fenel, and Matti Leisola

Introduction

The potential of using xylanases in industrial applications generated an intensive academic and industrial activity to hunt xylanases with desired properties from all kinds of natural sources. As a consequence, the gene banks contain presently over 100 sequences for family 11 xylanases. Widely used commercial enzymes for biomass modification in animal feeding, baking, and pulp bleaching have come from mesophilic filamentous fungi *Trichoderma reesei* and *Aspergillus niger*.[1,2] The mesophilic xylanases are useful but not optimal in all industrial applications. These enzymes are easily inactivated in the preparation of animal feed; furthermore, they are not very active in hot alkaline conditions of pulp bleaching. The screening activities produced numerous thermophilic xylanases, of which only few have entered the markets. An alternative for screening new xylanases from nature is to optimize by rational design or directed evolution an already known xylanase that has several advantages (e.g., good production system) but is missing an acceptable thermostability, alkalitolerance, or some other key properties.

[1] L. Viikari, A. Kantelinen, J. Sundquist, and M. Linko, *FEMS Microbiol. Rev.* 13, 335 (1994).
[2] R. A. Prade, *Biotechnol. Genet. Eng. Rev.* 13, 101 (1996).

Structural Features of Family 11 Xylanases

The sequence identities among family 11 xylanases range from 40 to 98%. A large number of the conserved sites allow some variation, and the number of highly conserved sites is in the range of 20% of all the sites. The size of family 11 xylanases is quite small, containing approximately 190 amino acid residues. The enzyme has been described to form a right-hand-like structure.[3] About 120 of all the residues belong to β strands.[4] The basic fold is formed by a double-layered β sheet that is twisted to form a large active site cleft. A conserved single α helix is located on the outer surface. Crystallographic analysis and molecular dynamics simulations revealed an open–close movement of the active site.[5] Two hinge regions are involved in this movement. The substrate apparently promotes the closing of the active site.[5] It is evident that the engineering of stabilizing mutations should not disturb the open–close movement of the active site.

Stabilization Strategies

The thermostability of mesophilic xylanases (especially *T. reesei* and *Bacillus circulans* xylanases) has been increased considerably by designed mutations. *T. reesei* xylanase II (XYNII) is stable at 40–45°, whereas temperatures above 50° cause conformational changes and the enzyme activity is lost quite quickly.[6–8] *B. circulans* xylanase is completely inactivated in 30 min at 57°.[9] There are several strategies how to design mutations that could improve the thermostability.

Disulfide Bridges at Protein N Terminus

Xylanases contain naturally disulfide bridges at the α helix and cord region (see structures with PDB-codes 1yna, 1PVX, 1BK1, and 1ukr). Disulfide bridges can also be introduced into different sites that do not naturally contain a disulfide bridge in any of the family 11 xylanases. Such a

[3] A. Törrönen and J. Rouvinen, *J. Biotechnol.* **57**, 137 (1997).

[4] N. Hakulinen, O. Turunen, J. Jänis, M. Leisola, and J. Rouvinen, *Eur. J. Biochem.* **270**, 1399 (2003).

[5] J. Muilu, A. Törrönen, M. Peräkylä, and J. Rouvinen, *Proteins* **31**, 434 (1998).

[6] M. Tenkanen, J. Puls, and K. Poutanen, *Enzyme Microb. Technol.* **14**, 566 (1992).

[7] J. Jänis, J. Rouvinen, M. Leisola, O. Turunen, and P. Vainiotalo, *Biochem. J.* **356**, 453 (2001).

[8] O. Turunen, K. Etuaho, F. Fenel, J. Vehmaanperä, X. Wu, J. Rouvinen, and M. Leisola, *J. Biotechnol.* **88**, 37 (2001).

[9] W. W. Wakarchuk, W. L. Sung, R. L. Campbell, A. Cunningham, D. C. Watson, and M. Yaguchi, *Protein Eng.* **7**, 1379 (1994).

place is the N-terminal region.[9,10] A disulfide bridge between positions 2 and 28 in *T. reesei* XYNII increased the thermostability about 15°, both in the absence and in the presence of the substrate.[10] A disulfide bridge cross-linking protein N and C termini also has a significant stabilizing effect.[9]

Disulfide Bridges at α Helix

The single α helix is another region important for the thermostability of family 11 xylanases. Disulfide bridges have been introduced into several positions of the α helix, with the disulfide bridge in the N terminus of the α helix being most effective.[9,11] These disulfide bridges increased the resistance to heat-induced denaturation, but not the activity at high temperatures (apparent temperature optimum).

Single Amino Acid Mutations

A large number of stabilizing amino acid changes (other than disulfide bridges) have been reported for family 11 xylanases.[8,12–18] Single amino acid mutations have been done at the N-terminal region, α helix, and other regions. Furthermore, an N-terminal extension has been reported to have a stabilizing effect; correspondingly, an N-terminal deletion can be destabilizing.[13,19] Comparison of crystal structures from thermophilic and mesophilic organisms indicated that Thr/Ser and Arg/Lys ratios are higher in thermostable family 11 xylanases.[4]

A combination of various mutations has often a cumulative effect on the thermostability of family 11 xylanases. When mutations with only a small stabilizing effect were combined with the disulfide bridge engineered into the α helix, a considerable increase in thermostability was achieved.[8]

[10] F. Fenel, M. Leisola, J. Jänis, and O. Turunen, *J. Biotechnol.* **108**, 137 (2004).

[11] H. Xiong, F. Fenel, M. Leisola, and O. Turunen, submitted for publication.

[12] A. Arase, T. Yomo, I. Urabe, Y. Hata, Y. Katsube, and H. Okada, *FEBS Lett.* **316**, 123 (1993).

[13] W. L. Sung, M. Yaguchi, and K.Ishikawa, U. S. Patent Number 5,866,408 (1998).

[14] J. Georis, F. de Lemos Esteves, J. Lamotte-Brasseur, V. Bougnet, B. Devreese, F. Giannotta, B. Granier, and J.-M. Frère, *Protein Sci.* **9**, 466 (2000).

[15] W. L. Sung and J. S. Tolan, Patent WO00/29587 (2000).

[16] W. L. Sung, Patent WO0192487 (2001).

[17] O. Turunen, M. Vuorio, F. Fenel, and M. Leisola, *Protein Eng.* **15**, 141 (2002).

[18] J.-M. G. Daran, H. H. Menke, J. P. van den Hombergh, and J. M. van den Laar, Patent EP1184460 (2002).

[19] D. D. Morris, M. D. Gibbs, C. W. Chin, M. H. Koh, K. K. Y. Wong, R. W. Allison, P. J. Nelson, and P. L. Bergquist, *Appl. Environ. Microb.* **64**, 1759 (1998).

However, many thermostable family 11 xylanases achieve their high thermostability without disulfide bridges.

Thermostabilization may increase the rigidity of the enzymes. It is often important for industrial applications that the specific activity does not decrease. The determination of kinetic parameters revealed that an extensive stabilization of *T. reesei* XYNII simultaneously at the protein N terminus and the α helix did not affect enzyme activity negatively, as indicated by unchanged K_m and V_{max} values when compared to the wild type.[11]

Modification of pH-Dependent Properties

Several factors have an influence on the pH-dependent properties of xylanases. The ionization states of the nucleophile and acid/base glutamates control the pH-dependent activity profile.[20] The activity at low pH can be increased by modifying the environment of the acid/base catalyst (Glu-177 in *T. reesei* XYNII and Glu-172 in *B. circulans xylanase*). The mutation of nearby Asn to Asp (N35D in *B. circulans* xylanase) lowered the pH optimum considerably and explains the low pH optimum of acidic xylanases.[20] In addition, acidic xylanases appear to have a higher number of acidic and a lower number of basic amino acids on the protein surface.

The reduced catalytic activity of xylanases at alkaline conditions appears to involve pH-induced unfolding and ionization or deprotonation of key amino acid residues.[21] In principle, the pH optimum can be shifted toward alkaline pH by increasing the pK_a value of the acid/base glutamate. The short distance effects on the pK_a value of a residue are important in determining the pH activity profile of xylanases.[22] A histidine residue located near the acid/base catalyst in the active site cleft is involved in maintaining the high pH optimum in family 10 xylanases.[23] The introduction of arginines into the Ser/Thr surface[17] and modification of the protein N terminus[13] have improved the activity of *T. reesei* XYNII at alkaline pH. Furthermore, amino acid substitutions that do not involve charged amino acid residues have increased the activity of xylanase at alkaline pH.[24] *Bacillus agaradhaerens* is a bacterium that grows at very alkaline sources,

[20] M. D. Joshi, G. Sidhu, I. Pot, G. D. Brayer, S. G. Withers, and L. P. McIntosh, *J. Mol. Biol.* **299**, 255 (2000).

[21] D. Nath and M. Rao, *Enzyme Microb. Technol.* **28**, 397 (2001).

[22] M. D. Joshi, G. Sidhu, J. E. Nielsen, G. D. Brayer, S. G. Withers, and L. P. McIntosh, *Biochemistry* **40**, 10115 (2001).

[23] M. Roberge, F. Shareck, R. Morosoli, D. Kluepfel, and C. Dupont, *Protein Eng.* **11**, 399 (2001).

[24] Y.-L. Chen, T.-Y. Tang, and K.-J. Cheng, *Can. J. Microbiol.* **47**, 1088 (2001).

such as soda lakes. The *B. agaradhaerens* xylanase shows a high alkalistability, but curiously a low pH optimum (pH 5.6) and at pH 9 the enzyme activity is even close to zero.[25] In conclusion, factors controlling the pH-dependent activity of xylanases are only partially known and require much further research. This is a relevant challenge in engineering the pH activity profile of xylanases for industrial purposes.

Engineering of Arginines

We tested how the engineering of additional arginines affects the thermostability of *T. reesei* XYNII. The increase of arginines on different sides of the enzyme did not increase the thermostability. Instead, the introduction of five arginines into the Ser/Thr surface had a considerable thermostabilizing effect.[17] The Ser/Thr surface is part of the outer β sheet layer on the protein surface and contains a large number of serines and threonines. Figures 1 and 2 show results obtained for an arginine mutant of *T. reesei* XYNII.[17] The five engineered arginines increased the apparent temperature optimum by 5° (Fig. 1) and the pH-dependent activity profile shifted clearly to alkaline pH (Fig. 2). The shift was seen in both the acidic and the alkaline side of the bell-shaped pH-dependent activity profile. However, the stabilizing effect was enigmatic because the half-life measured in the absence of the substrate decreased as the number of arginines increased on the Ser/Thr surface. A reason for this could be that the electrostatic repulsion caused by a high positive net charge was destabilizing in the absence of the substrate. Thus, the stabilizing effect of arginines was seen only in the presence of the substrate. The half-life increased four-fold in the presence of the substrate due to five arginines on the Ser/Thr surface. In conclusion, the efficient use of arginines as protein stabilizers requires further research: the location on the protein surface and the local environment need special attention.

Experimental Procedures

Planning of Mutations

The protein databank contains a large number of resolved three-dimensional structures of family 11 xylanases.[4] The available structural information is essential in understanding the structure–function relationships. Structural and sequence comparisons in the large xylanase family are

[25] D. K. Y. Poon, P. Webster, S. G. Withers, and L. P. McIntosh, *Carbohydr. Res.* **338,** 415 (2003).

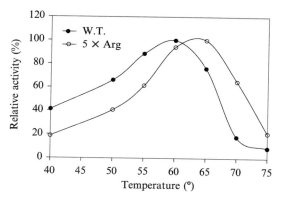

FIG. 1. Effect of engineered arginines on the temperature-dependent activity of *T. reesei* XYNII. *E. coli* culture broth was used as the source for enzymes. Enzyme activity was measured by the DNS assay using a 10-min incubation time in 50 m*M* citrate-phosphate buffer (pH 5). Relative activity values were calculated from absorbance (A_{540}) values obtained from the enzyme assay. The arginine mutant (ST5) contains five arginines on the Ser/ Thr surface. W.T., wild-type XYNII. Reproduced in modified form with permission from Turunen *et al.*,[17] © Oxford University Press.

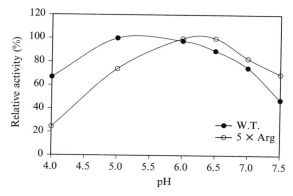

FIG. 2. Effect of engineered arginines on the pH-dependent activity profile of *T. reesei* XYNII. Xylanase activity was measured at 50° in each pH. Otherwise the enzyme assay was done in the same way as in Fig. 1. The arginine mutant was ST5 as in Fig. 1. Reproduced in modified form with permission from Turunen *et al.*,[17] © Oxford University Press.

used to plan mutations that would change the properties of the enzyme in a desired manner. In planning mutations, we have used mainly Swiss-PdbViewer (http://us.expasy.org/spdbv)[26] as a molecular graphical tool to

[26] N. Guex and M. C. Peitsch, *Electrophoresis* **18**, 2714 (1997).

examine the xylanase structure (e.g., lxyp). The program allows the introduction of mutations into the known or modeled xylanase structures. The
effects of the mutations can be evaluated first by the score values calculated
by the program and by inspecting the local interactions of the substituted
residues and also by using the torsion command of the program. Modeling
by other programs may be useful in estimating the conformation of the
mutated residue. Crystallographic B values and molecular dynamics simulations may give useful information about the mobility of different protein
regions. Optimization of the hydrogen-bond network and calculation
of theoretical pK_a values can be done by WHATIF.[27] The alignment of
xylanase protein sequences was done mainly by ClustalX (or ClustalW).

Site-Directed Mutagenesis

Mutations are generated by polymerase chain reaction (PCR) in which
the mutations are introduced into the oligonucleotide primers.[8] The two
(forward and reverse) primers overlap fully and contain the mutated codon
in a middle position, and the primers are designed so that the T_m of the
primers is at least 78°. The PCR conditions are basically those of
the QuikChange mutagenesis system (Stratagene, La Jolla, CA). The
standard PCR reaction mixture contains 5–50 ng template DNA,
0.08 mM dNTP, 0.2 mM of each primer, and 10× PFU reaction buffer
(Stratagene). After a 5-min heating at 95°, 1.0 μl of PFU Turbo polymerase
(Stratagene) is added to the hot PCR reaction mixture. One cycle
is typically as follows: 50 s at 95°/50 s at 60°/1.0 min/kb of plasmid at 68°.
The number of cycles is between 12 and 18, and after the last cycle there
is a 7-min extension at 68°. After the PCR, the parent DNA is digested
by adding 1.0 μl of the DpnI restriction enzyme (New England Biolabs,
Beverly, MA) and incubating for 1–2 h at 37°. Competent *Escherichia
coli* XL1-Blue cells are used in the transformation. *E. coli* clones containing mutant xylanases are grown at 37° overnight on LB agar plates
containing 0.1–0.2% birchwood xylan (Sigma, Steinheim, Germany), which
is coupled to Remazol brilliant blue (Sigma) according to Biely *et al.*[28]
Xylanase activity is indicated by white halos around the positive colonies.

Expression Vectors

In *E. coli* cells, xylanases are secreted into the periplasmic place. In our
work, we have used either α-amylase or pectate lyase secretion signal
sequence.[8] Enough xylanase leaks from the periplasmic place into the

[27] J. E. Nielsen and G. Vriend, *Proteins* **43,** 403 (2001).
[28] P. Biely, D. Mislovicova, and R. Toman, *Anal. Biochem.* **144,** 142 (1985).

culture medium for purification and characterization of the enzyme. *T. reesei* XYNII is produced in *E. coli* XL1-Blue (Strategene) or *E. coli* Rv308 (ATCC 31608) strains using the pALK143 expression construct (ROAL, Rajamäki, Finland) or pKKtac vector (VTT, Espoo, Finland). The low-copy plasmid pWSK29 has been used to construct pALK143, which contains the *T. reesei* xylanase II cDNA, including 11 amino acids from the prosequence, and the *Bacillus amyloliquefaciens* α-amylase signal sequence under the *B. amyloliquefaciens* α-amylase promoter. The xylanase is produced constitutively from pALK143. pKKtac has been constructed from the pKK233 vector (Amersham-Pharmacia Biotech, Uppsala, Sweden) and contains the tac promoter for induction by 1 mM IPTG and lacI^q repressor. This construct has an *Erwinia carotovora* pectate lyase signal sequence, which we fused with the mature XYNII protein without prosequence. Higher quantities of xylanases are usually produced from the pKKtac vector. The prosequence present in pALK143 is not fully removed in some of the mutants.[29]

Purification of Xylanases

Recombinant xylanases are produced in *E. coli* grown in Luria broth supplemented with ampicillin using shake flasks at 30 or 37° (shaking 200–250 rpm). When xylanases are produced from the pKKtac vector, *E. coli* cells are grown for overnight, diluted in ~2 × 10^{-2} dilution, and grown for 3–4 h at 30° ($OD_{600} = 0.5$); 1 mM IPTG is then added overnight to induce xylanase production (growth at 30°). The xylanase is precipitated from the clarified culture broth by 65% saturated ammonium sulfate. The resulting pellet is dissolved in 0.01 M citrate-phosphate buffer (pH 4) and desalted by a PD10 column (Amersham-Pharmacia Biotech) or dialysis (molecular weight cutoff 6000–8000).

Protein purification is carried out by the BioPilot system (Amersham-Pharmacia Biotech).[8] The desalted solution containing xylanase is loaded onto a CM-Sepharose fast flow (Amersham-Pharmacia Biotech) column (1.6 × 11 cm) equilibrated with 0.01 M citrate buffer (pH 4). Elution is done with a linear sodium chloride gradient (0–0.25 M) at a flow rate of 5 ml min^{-1}. Further purification is carried out by hydrophobic interaction chromatography using a phenyl-Sepharose fast flow (Amersham-Pharmacia Biotech) column (1.6 × 10 cm) equilibrated with 1.5 M ammonium sulfate in 0.01 M citrate buffer (pH 4). The enzyme is washed out from the column by decreasing the concentration of ammonium sulfate.

---

[29] J. Jänis, O. Turunen, M. Leisola, P. J. Derrick, J. Rouvinen, and P. Vainiotalo, submitted for publication.

The protein concentration is measured spectrophotometrically by absorbance at 280 nm (A_{280}) using the molar extinction coefficient 54,050 determined for *T. reesei* XYNII.[8]

Enzyme Assay and Characterization of Thermostability and Other Properties

Xylanase activity is determined using the DNS assay to measure the amount of reducing sugars liberated from 1% birchwood xylan.[30] The activity determination in standard conditions for *T. reesei* XYNII is carried out at pH 5–6 and 50°, with a reaction time of 10 min. Citrate-phosphate buffer (50 mM) is used in the xylanase assays at pH 4–7.5 and 50 mM Tris–HCl at pH 7.5–9. Bovine serum albumin (BSA; 0.1 mg/ml) or even inactivated *E. coli* culture broth can be used as a stabilizer in activity and thermostability measurements. When purified enzymes are used in the thermostability assays, a stabilizer is recommended to be included in the diluted enzyme solutions because diluted pure xylanase may disappear quite quickly from the solution (considerable amounts may be lost in 30 min). Several properties of xylanases can even be tested using non-purified enzymes; these include apparent temperature optimum, pH-dependent activity profile, stability against thermal inactivation, half-life, and pH-dependent stability. Determination of kinetic (K_m and V_{max}) parameters, differential scanning calorimetry (DSC), and mass spectrometry are done with purified proteins.

Thermostabilities measured for xylanases are approximations and are dependent on conditions. However, the obtained values can be used to make comparisons of stability and other properties. The stability against thermal inactivation is assayed usually so that the enzyme is heated at a series of different temperatures for a chosen time (e.g., 10 or 30 min), and after rapid cooling the remaining activity is measured at a low temperature (e.g., 50° for *T. reesei* XYNII). The temperature at which 50% of the activity is left (T_{50}) can be compared to the melting point values (T_m) obtained by other methods, such as DSC and hydrogen/deuterium (H/D) exchange analysis with mass spectrometry. However, there can be differences in these values that are dependent on experimental conditions on one side, and on the other side, the enzyme activity can be lost at a lower temperature than where the total unfolding of the protein occurs. Another approach to assay thermostability is that the enzyme is incubated at a given temperature, samples are removed at various time intervals, and then the remaining activity is measured at the standard assay temperature. Half-live

---

[30] M. J. Bailey, P. Biely, and K. Poutanen, *J. Biotech.* **23,** 257 (1992).

values are calculated from the latter graphs obtained for time-dependent inactivation of the enzyme activity. These methods can be used to assess the irreversible inactivation at given conditions, but they do not detect reversible inactivation. Mutations that increase the apparent temperature optimum may reveal a site for reversible unfolding in the enzyme structure.

In xylanases, the substrate appears to increase the thermostability. The measurement of the enzyme activity as a function of time (productivity assay) can be used to calculate the half life of a rough enzyme in the presence of substrate.[17] Arrhenius activation energy can be calculated from the slope of the temperature-dependent activity profile using the equation

$$\ln k = \ln A - (E_a/RT) \tag{1}$$

A plot of $\ln k$ versus $1/T$ gives a straight line with slope $-E_a/R$, from which the Arrhenius activation energy (E_a) can be calculated.[31] k is the rate constant for the reaction, A is the Arrhenius constant, R is the ideal gas constant, and T is the absolute temperature. Our most stable *T. reesei* XYNII disulfide bridge mutants did not show any significant differences in the activation energy to the wild-type value ($\sim$50 kJ gmol^{-1}), whereas the introduction of five arginines into the Ser/Thr surface increased the activation energy by 40%.[11,17]

Use of Mass Spectrometry to Assay Thermostability

Hydrogen/deuterium exchange is a sensitive mass spectrometric application for conformational analyses of proteins.[7,32,33] H/D exchange has been used to probe thermally induced structural changes in proteins. The rate at which hydrogens are replaced with deuterons in solution is a combination of intrinsic exchange rates, hydrogen bonding, and solvent accessibility. Whereas H/D exchange rates of the amino acid side chain hydrogens are typically too fast in a timescale for mass spectrometric detection, the exchange of amide hydrogens is detectable when solution conditions are adjusted carefully. However, in a tightly folded protein structure, H/D exchange rates of hydrogens buried into a hydrophobic protein core or ones involved in hydrogen bonding can be a few magnitudes lower than the corresponding rates of the surface hydrogens that are exposed readily to the solvent. Conformational changes and unfolding of the protein structure expose buried hydrogens to solvent, which can be

[31] P. M. Doran, "Bioprocess Engineering Principles," p. 262. Academic Press, New York, 2000.
[32] V. Katta and B. T. Chait, *J. Am. Chem. Soc.* **115**, 6317 (1993).
[33] A. N. Hoofnagle, K. A. Resing, and N. G. Ahn, *Annu. Rev. Biophys. Biomol. Struct.* **32**, 1 (2003).

detected from the change of the H/D exchange rate. H/D exchange can provide information about the melting point, protein rigidity, degree of exchangeable hydrogens, and stability of protein variants present in the protein sample.

We used H/D exchange for measuring conformational stability on the heat-induced unfolding of *T. reesei* XYNII and its mutants having disulfide bridges and other stabilizing mutations.[7,29] We were also able to measure the thermostability of protein variants caused by the incomplete removal of prosequence, resulting in a difference of even 5° in the T_m values.[29] A typical experimental procedure is as follows. Purified protein in water or buffered solution is diluted with a fully deuterated solvent (typically deuterium oxide, D_2O). A sample tube is placed into a heating block for a predefined time, after which the exchange reaction is quenched either by acid or by placing the tube in an ice bath (0°). Deuterium incorporation versus incubation temperature is measured immediately using mass spectrometry. If a heat-induced conformational change takes place at a certain temperature, this can be monitored by the increase in the H/D exchange rate. If bimodal distribution in deuterium incorporation is detected, i.e., two conformers with distinct exchange rates, relative abundances can be calculated from the spectral peak intensities. Therefore, both qualitative and quantitative aspects of the heat-induced conformational changes can be analyzed. When H/D exchange ESI MS spectra was determined for *T. reesei* XYNII exposed to heat in D_2O solvent in the temperature region of 54–60°, two distinct protein conformers were detected based on their

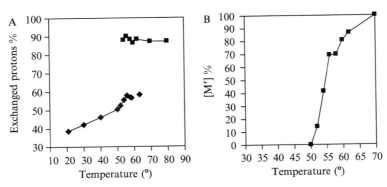

FIG. 3. H/D exchange results for *T. reesei* XYNII. Deuterium incorporation (A) of folded (♦) and partly unfolded (■) conformers and relative abundance (B) of the partly unfolded conformer (■) as a function of incubation temperature. Data are plotted based on ESI MS spectra of XYNII after 20-min incubations in D_2O. Reproduced with permission from Jänis et al.,[7] © Biochemical Society.

different exchange rates. One represents the folded (M), whereas the other represents the partly unfolded (M′) conformer. Average deuterium incorporations of M and M′, as well as the relative abundance of the M′ as a function of the incubation temperature, are presented in Fig. 3.

Conclusions

Because of commercial interests, the stabilization of mesophilic xylanases has been studied quite thoroughly by mutagenesis. The thermostability of these xylanases has been improved by over 20°. However, the stabilizing mutations found thus far may not be applicable to all family 11 xylanases. A few studies have also focused on the alkalitolerance and some progress has been achieved in this area, but we do not know yet how far in alkaline pH the pH optimum of family 11 xylanases can be shifted. The basis for the stability of most thermostable family 11 xylanases is not fully understood yet. There may also be a need to engineer the functional properties (e.g., pH optimum) of highly thermostable xylanases, but a problem may appear that a well-optimized protein structure does not easily tolerate amino acid substitutions.

[15] Screening for Oxidative Resistance

By Joel R. Cherry and Michael H. Lamsa

Introduction

The oxidation of amino acid side chains in proteins has long been recognized as a primary pathway for functional inactivation. Whether a protein drug targeted for intracellular action or an enzyme formulated into laundry detergent, activated oxygen species are generated that can modify protein side chains, significantly altering their hydrophobicity, charge, and size. These modifications in turn render many proteins inactive and more susceptible to aggregation or proteolysis. In addition to the well-known targets of oxidation that include methionine, cysteine, tryptophan, tyrosine, and histidine, other side chains, including phenylalanine, glutamate, leucine, valine, proline, lysine, and glycine, have also been shown to be oxidized by a variety of mechanisms.[1] While removal of all oxidizable amino acids from a protein is impractical, only a small fraction

---

[1] R. T. Dean, S. Fu, R. Stocker, and M. J. Davies, *Biochem. J.* **324**, 1 (1997).

of these residues in a given protein are typically oxidized, making their identification and replacement a reasonable strategy for improving protein oxidative stability.

Creating proteins with improved performance under defined, nonnatural conditions can utilize a wide array of molecular approaches covering a spectrum from rational design to random mutagenesis. Often the most successful approaches combine multiple techniques for the introduction of diversity within a protein, as represented in Fig. 1. Random mutagenesis is perhaps the simplest approach in that it requires nothing beyond a target gene, but because the majority of mutations are either neutral or deleterious to protein function, many variants must be screened to identify one with improved function amidst the large background of wild type, dead, or unimproved mutants. For a target gene with identified and obtainable homologues, family shuffling can be employed. In this technique, genes with sequence identity above 60–70% can be recombined *in vitro* and screened for improvement. Because recombination is occurring between genes that encode active proteins, the likelihood of producing functional variants is higher. Previous work has also clearly demonstrated that family shuffling has the potential to produce variants with phenotypes differing from the

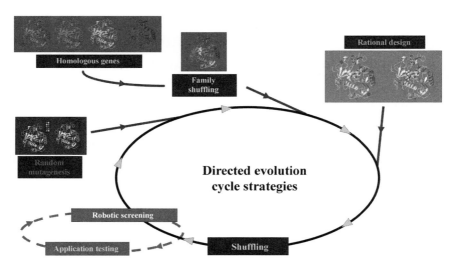

Fig. 1. Directed evolution is best viewed as an iterative cycle in which gene diversity is generated by one or more methods, variants are screened for improvement in a selected phenotype, and the process is repeated with the best variants resulting from the screening. Here, multiple methods of introducing diversity are selected based on the knowledge of structure/function relationships in the protein, as described in the text. A critical component is the validation of the screening system by testing selected variants ("screening winners") in the intended application. (See color insert.)

parents, meaning that the parental genes may not need to have measurable differences in oxidative stability to produce more stable progeny variants.[2] With structural information provided by either nuclear magnetic resonance or X-ray crystallographic data, rational design can be used to substitute amino acid side chains in exposed positions with similarly sized amino acids more resistant to oxidation. In addition, region-specific randomization can be performed to target solvent-exposed residues for substitution with a subset of amino acids. As rational design moves from single substitutions toward randomization, the number of variants produced that must be screened increases dramatically. Complete randomization of five amino acid positions produces 3.2 million possible variants!

The key to the directed evolution process is establishment of a screening system that accommodates the predicted diversity generated by the mutagenesis technique and does not return a high rate of false positives. While few screening systems satisfy these requirements entirely, evaluation of the throughput and assay reproducibility are critical to success. The systematic error in a screening assay sets the limits for detectable improvement. For example, it is unlikely that single mutations will increase the stability of a protein more than 15 or 20% in a residual activity assay, so a screening assay with an error of 30% is unlikely to identify meaningful improved variants.

Methods to Grow and Screen Microbial Cultures in Microtiter Plate Formats

It is the intention of this section to give general guidelines rather than specific methods for growing microbial cultures in microtiter plate formats to screen common laboratory microorganisms. A wide variety of methods to set up a screening system for enzymes secreted by microbial hosts is illustrated in Fig. 2.[3,4] In general, bacterial and yeast hosts work well in this scheme; special consideration will be needed for fungal hosts that grow in a mat-like morphology. In these cases it may be possible to achieve a suitable way to obtain broth samples in a miniaturized format simply by reducing the nutrients (primarily the carbon and nitrogen) in the media, even by simply diluting the media (by a straight dilution with sterile water or buffer or by maintaining the salts, trace elements, and a suitable buffer system for

---

[2] J. E. Ness, M. Welch, L. Giver, M. Bueno, J. R. Cherry, T. V. Borchert, W. P. C. Stemmer, and J. Minshull, *Nature Biotech.* **17**, 893 (1999).

[3] M. Lamsa, N. B. Jensen, and S. Krogsgaard, *in* "Enzyme Functionality: Design, Engineering, and Screening" (A. Svendsen, ed.), p. 527. Dekker, New York, 2003.

[4] H. Pedersen, M. Lamsa, P. K. Hansen, H. Frisner, J. Vind, S. Ernst, L. Kongsbak, B. R. Joergensen, T. C. Beck, T. L. Husum, and I. Von Ossowski, Microtiter Plate Based High Throughput Screening (HTS) Assays, WO01/32844.

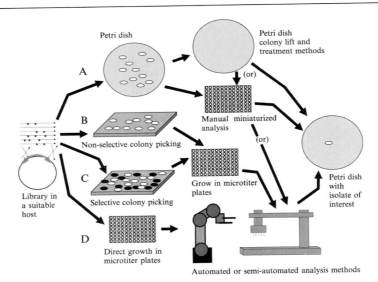

FIG. 2. An illustration of the many ways a screening program may be approached, starting with a library expressing variants in a suitable host. At least four paths may be taken, and variations of these paths can result. In scheme A, a library is plated on a petri dish, which can be picked manually (toothpicks) into a suitable format microtiter plate for growth followed by treatment and analysis in a manual or automated fashion. The alternative is to have a more sophisticated colony replication, lift, and treatment method that allows one to screen very large numbers of variants very rapidly, resulting in a few variants of interest or in a population of variants that will need to go through further and more detailed analysis. Schemes B and C are very similar; the difference is in one instance (C) an analytical method to detect active colonies is incorporated into the agar. In either case, an automated colony picking device (or, alternatively, manual with toothpicks) is used to load a microtiter plate to be grown up for automated or semiautomated analysis. Scheme D depicts a method used to directly inoculate a library without intermediate manipulation into a suitable microtiter plate format. Although this has been done in low-density plates, methods now available allow this to be done in very high-density formats of tens of thousands of wells. In all of these cases, intermediate steps of isolations and retesting will need to be devised to get to the final objective of one or a few candidates that are improved in the application.

the organism). Fungal cultures will generally grow on the bottom of the well in dilute media; more toward the top if the medium is richer; for dilute media, a clear sample can generally be obtained by pipetting from the top of the well. In the case where a fungal mat has formed, it is possible to get clear supernatant samples simply by pushing the mat out of the way with a pipette tip.

Sophisticated devices, such as the HyGrow manufactured by GeneMachines (Redwood City, CA), or simple devices with or without agitation (200 to 350 rpm is a good general range to test on most standard laboratory shakers) and supplied aeration can be used to grow cultures of libraries. In many cases, agitation in a shaking incubator, coupled with some method to minimize evaporation, is sufficient to miniaturize the growth condition from a tube or shake-flask scale. A few general media and their applications are listed.

URA (-) medium is used for expression (yeast with the *URA3* gene for selection and the expressed gene under the control of the *GAL1* promoter). Per liter: 6.7 g of yeast nitrogen base without amino acids [+ $(NH_4)_2SO_4$] (Difco), 20 g NZ amine A (Sigma), 10 g succinic acid, 1 g glucose, 20 g galactose, 7.3 mg $CaCl_2 \cdot 2H_2O$, 8.6 mg/liter $CuSO_4$, and 1000 mg/liter of $MgSO_4$. Generally pH adjust to 6.0 with NaOH.

LBG medium is used primarily for bacterial growth and expression (although it is not optimized for expression). Per liter: 10 g tryptone, 5 g yeast extract, 5 g NaCl, 10 g glucose. Generally pH adjust to 6.0 with NaOH. For plates, add agar at 15 g/liter Bacto agar (Difco).

MY25 medium is used primarily for fungal cultures; however, it also works well for bacterial cultures. Per liter: 25 g maltose, 10 g yeast extract, 2 g urea, 2 g $MgSO_4 \cdot 7H_2O$, 2 g K_2SO_4, 10 g KH_2PO_4, 2 g anhydrous citric acid, 0.2 g $CaCl_2 \cdot 2H_2O$, 1 ml trace metals (per liter: 13.9 g $FeSO_4 \cdot 7H_2O$, 8.5 g $MnSO_4 \cdot 7H_2O$, 14.28 g $ZnSO_4 \cdot 7H_2O$, 1.63 g $CuSO_4$, 0.24 g $NiCl_2 \cdot 6H_2O$, 3 g anhydrous citric acid). Generally pH adjust to 4.0 to 7.0 with NaOH.

TBAB agar plates (per liter: 10 g Bacto tryptone, 3 g Bacto beef extract, 5 g NaCl, 15 g Bacto agar (Difco), 1 ml trace metals (per liter: 13.9 g $FeSO_4 \cdot 7H_2O$, 8.5 g $MnSO_4 \cdot 7H_2O$, 14.28 g $ZnSO_4 \cdot 7H_2O$, 1.63 g $CuSO_4$, 0.24 g $NiCl_2 \cdot 6H_2O$, 3 g anhydrous citric acid).

Antimicrobial compounds such as penicillin (50 mg/ml in water), ampicillin (10 mg/ml in water), tetracycline (10 mg/ml in ethanol), or chloramphenicol (50 mg/ml in ethanol) can be added singly or in combination to inhibit unwanted bacterial growth in all yeast and fungal cultures and, depending on antibiotic markers, in bacterial strains.

In general, media can always be optimized further to improve expression, mostly by manipulation of either the carbon or nitrogen levels or of both simultaneously. Dilution or improvement of buffering capacity can be achieved by supplementing with buffers such as the biological buffers available from Sigma, which have a pH range from 6.0 to 10.7.

Screening Considerations

The analytical methods chosen depend on the size of the library you are screening and the scale you wish to achieve, illustrated in Fig. 3. It is best to keep in mind that screening is a series of reductions of the numbers

Methods consideration for screening approach based on projected total screened										
# Per screen event	Manual	Work-station	Robotic system	Selection methods		During program life	Manual	Work-station	Robotic system	Selection methods
10	▓					10	▓			
100		▓				100	▓			
1000		▓				1000	▓			
10,000		▓	▓			10,000		▓	▓	
100,000			▓	▓		100,000			▓	▓
1,000,000				▓		1,000,000			▓	▓

FIG. 3. This is a very general guideline on methods to consider depending on the scale of the screening program and what is available in the laboratory. Manual methods can involve tubes, petri plates, or microtiter plates and any available instrumentation, such as plate readers, dispensing devices, or multichannel pipettors. A workstation has a fairly broad definition in this example due to the ever-changing landscape of automation.[5] A basic workstation is a single or multichannel pipetting device that has positions for a reasonable number of types of labware in a microtiter plate format. A very sophisticated workstation could have directly attached carousels and other devices plus a gripping device on the workstation to move labware around. Current workstations are nearly equivalent and, in many cases, much better than early robotic systems. Robotic systems can encompass one or more workstations or dedicated devices and instrumentation coupled to a transportation device, usually one or more robotic arms or a track system (potentially with robotic arms or grippers located around the system). With the miniaturization available today, it is possible to easily run automated methods completely on workstations or with a workstation and occasional user intervention.

screened, depending on the desired properties of the enzyme and the particular application. It is almost never possible to model (a complicated) application in a primary screen and achieve reasonable throughput. If the goal is fairly plastic and a large proportion of the variant population will satisfy the requirement, then a screen that is essentially a selection process may produce many suitable variants. Very stringent conditions will eliminate most variants, but could also eliminate interesting intermediate variants that, in combination, could produce a superior variant. It is up to the scientist and project goals to ascertain what is appropriate. In other cases, a very minor change may make a big difference in enzyme performance in the application, but it could be very difficult to detect unless a less stringent screening routine is employed. Less stringency usually means looking at more marginal candidates, and sorting through these will usually involve one or more secondary screens that expand on the characteristics screened for in the primary screen.

It is best to start out manually and scale up incrementally, no matter what equipment is available in your laboratory. Methods should be developed to eliminate a majority of the undesired variants (such as inactive variants) prior to the bulk of the screening. In general, a library to be screened should contain a minimum of 10% active clones and you should have a manual or automated way to pick active clones from agar or liquid culture in microtiter plates. At least 50% active clones is a good point at which it may be feasible to screen all variants, whether active or not, depending on the methods available to pick the active clones and how easily and quickly it can be performed. Poisson statistics can be used to estimate the theoretical distribution of viable colonies in microtiter plates. This statistical method is especially useful if a library is diluted to contain a certain number of viable colonies per milliliter, to be dispensed directly into microtiter plate formats. For example, if 96 total viable colonies are in a volume sufficient to fill one 96-well plate, the distribution will be as follows: 35 empty wells, 35 wells with a single colony, 18 wells with 2 colonies per well, 6 wells with 3 colonies per well, and 2 wells with 4 colonies per well.[6]

Analytical Methods for Screening for Resistance to Oxidizing Conditions in Microtiter Plates

High pH and Thermal Stability Screening of a Peroxidase Enzyme under Oxidizing Conditions

Growth and screening of mutants are performed using automation in 96-well microtiter plates with methods similar to those described earlier. The 96-well plate screen is performed by first growing yeast transformants of a pool of mutants in 50-μl volumes of URA (-) medium, pH 6.0, in 96-well microtiter plates. Cultures are inoculated by dilution into medium and pipetting into 96-well plates. These are placed in a specially designed, humidity-controlled box that is attached to an Innova 4000 (New Brunswick Scientific) incubator set at 30°, 300 rpm and shaken for approximately 5 days. The primary (96-well plate) screening assay is a test of 2,2′-azino-bis(3-ethylbenzthiazoline-6-sulfonate) (ABTS) activity before and after a 20-min incubation (37–53°) while treating a broth dilution with varied concentrations of hydrogen peroxide (200 μM to 15 mM) in 100 mM phosphate/borate buffer, pH 10.0. As the mutants improve from wild type to the most stable mutant, the conditions progress from 37°, 200 μM hydrogen peroxide to 53°, 15 mM hydrogen peroxide. The best mutants selected

[5] D. Leahy, *Nature* **421,** 661 (2003).
[6] J. H. Zar, *in* "Biostatistical Analysis," p. 384. Prentice Hall, Englewood Cliffs, NJ, 1984.

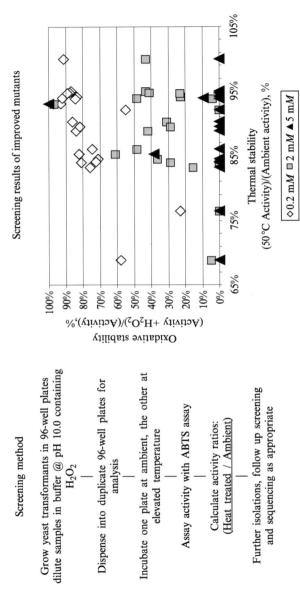

FIG. 4. A simple representation of an oxidative screening method and typical results after several rounds of screening. In this example, thermal stability is defined as the residual activity of each mutant after treatment at 50° for 20 minutes. Oxidative stability is defined as the ratio of activity of each mutant after treatment at 50° for 20 minutes for each concentration of hydrogen peroxide listed in the legend as compared to its activity after treatment at the same levels of hydrogen peroxide listed in the legend for 20 minutes at ambient temperature.

under these conditions are isolated as single colonies by plating diluted cultures. Single colonies are then picked to new growth plates for follow-up testing.

Mutants are grown in 24-well plates in 1 ml of URA (-) medium, pH 6.0, for follow-up screening and are incubated 2–3 days using the same conditions and device as for 96-well plates. Culture broths are assayed for "initial activity" by the ABTS assay at pH 7.0 (20 μl enzyme, 200 μl of 500 μM ABTS in 100 mM phosphate buffer, pH 7.0, OD 405 nM read using a microtiter plate reader). The activity is estimated by comparison to a purified peroxidase standard. Broths are then diluted to a standard peroxidase concentration and incubated in various concentrations of hydrogen peroxide (200 μM to 15 mM), 100 mM phosphate/borate buffer, and pH 10.0 at 50°. After 0, 10, and 20 min, samples are removed and activity is measured using the standard ABTS assay, pH 7.0. Improved mutants are those showing higher residual activity than wild type and are expressed as percentage residual activity relative to the time 0 assay result.[7,8] A simple version of the above screen and results is illustrated in Fig. 4.

Assays used to screen must be validated to assure proper operation under the proposed sample conditions. This is especially true for screening in the presence of oxidizing conditions, as the chemicals may have a negative effect on assay performance. An assay system must be adequately set up to investigate this effect, and, in general, sufficient dilution must be performed after enzyme treatment to eliminate or minimize the effect of the chemical on the assay. Any effect must be taken into account when setting up a control comparison in calculating the percentage remaining activity of the enzyme.

[7] A. H. Pedersen, J. Vind, A. Svendsen, J. R. Cherry, M. Lamsa, P. Schneider, and B. R. Jensen, H_2O_2-Stable Peroxidase Variants, U. S. Patent Number 5,817,495 (1998).
[8] J. R. Cherry, M. H. Lamsa, P. Schneider, J. Vind, A. Svendsen, A. H. Jones, and A. H. Pedersen, Nature Biotechnol. 17, 379 (1999).

[16]   Consensus-Based Engineering of Protein Stability: From Intrabodies to Thermostable Enzymes

By BORIS STEIPE

Stability Engineering

Protein stability* is not only of interest for biotechnology. Because the function of a protein depends on its structure, and the native structure is presumed to be acquired as the nascent peptide traverses a funnel-like energy landscape to descend into a unique global minimum, we can argue that stability—the goal and driving force of this process—is an experimental metric of the information encoded in a protein sequence. In 1994 we proposed an approach to predict stabilizing mutations in proteins, based on the analysis of sequence profiles.[1] This approach has allowed us to engineer hyperstable immunoglobulin domains that can be expressed as intrabodies. The procedure is entirely general, requiring neither knowledge of protein structure nor elaborate computational procedures; only a number of closely related homologous sequences are needed. Its application by a number of groups to a diverse set of proteins has demonstrated this procedure to be the most reliable strategy for the rational generation of stabilizing mutations. Importantly, the method emphasizes the stochastic nature of residue interactions in natural proteins and thus contributes to our understanding of the principles that govern sequence-folding relationships.

Canonical Sequence Approximation

It is not obvious why it should be possible to improve protein function by consensus mutations, one could equally well argue that every protein might have evolved to acquire a unique set of cooperative, favorable residue interactions and thus consensus mutations must be destabilizing as they would not take this specific context into account. Antibody domains are nature's workbench for the evolutionary engineering of affinity and an excellent model system with which to study the question of consensus versus uniqueness of the encoding of folding information in protein sequence. Clusters of variable residues on a conserved framework of

---

* Protein stability in this manuscript refers to thermodynamic stability: the free energy difference between the ideally unfolded state and the native state, determined at equilibrium.

[1] B. Steipe, B. Schiller, A. Plückthun, and S. Steinbacher, *J. Mol. Biol.* **240,** 188 (1994).

                    0076-6879/04 $35.00

structure[2] arise from a process of domain hypermutation and selection of successful variants in the B cell.[3,4] This process of affinity maturation ultimately selects unique, high-affinity, high-specificity variants, but the measure of fitness in this process must be a composite of factors that include antigen affinity, domain stability, assembly and interaction of the heavy-and light-chain, variable, and constant domains, protease resistance, and competence for export and secretion. In order to bind its antigen, the antibody must fold first. The relative importance of any residue for these factors depends on its position in the sequence; only the stability requirement has to be fulfilled in every position, as disruptive mutations are eliminated from the pool of functional proteins.

One can model this situation by imagining a large pool of evolved immunoglobulin sequences, each one optimized individually for antigen binding in a stochastic process of mutation and selection to conserve a minimal folding stability and each one folding into approximately the same global conformation. This model suggests an analogy to a thermodynamic canonical ensemble. The individual sequence positions are independent components of a system and can occur in any number of microstates, corresponding to the nature of the individual residue, but the whole system must satisfy global constraints. Accordingly, the distribution of microstates can be analyzed individually in terms of the constraints. Note that we are not explicitly considering covariation in this analysis but treating residues independently. This makes the canonical sequence approximation technically a mean-field approach. Applying Boltzmann's law to such an ensemble predicts that the states that occur most frequently possess the highest fitness of all alternatives. In the case of immunoglobulins, such an averaged ensemble would not include antigen affinity as a component of the fitness function, as we are averaging over all possible antigens. Accordingly, factors that determine the structure, specifically thermodynamic stability, play a dominant role. In the case of enzymes, particularly when using sets of orthologous sequences, functional requirements are represented as well.

The key assumptions made are residues evolve randomly and independently; the collection of sequences analyzed is an unbiased sample of the sequence space of the protein in which each domain is at an equilibrium of random stabilizing and destabilizing mutations; and a requirement for minimum thermodynamic stability is the dominant factor that constrains variability at each position.

[2] T. T. Wu and E. A. Kabat, *J. Exp. Med.* **132,** 211 (1970).
[3] S. Tonegawa, *Nature* **302,** 575 (1983).
[4] C. Berek and M. Ziegner, *Immunol. Today* **8,** 400 (1993).

To the degree that these assumptions hold, the prediction of stabilizing mutations follows a simple rule: a residue that replaces another that is observed significantly less frequently in a set of related sequences is expected to stabilize the protein. Furthermore, individual replacements can be combined, as independent effects are expected to be additive.

Immunoglobulin Domain Engineering

As a proof of principle, we have predicted point mutations in the well-studied V_κ[1] domain of the antibody McPC603[1] using a manually edited alignment of immunoglobulin sequences. We have shown that predictions of stabilizing mutations that replace native residues with the V_κ consensus residue were correct in 6 of 10 cases, three replacements were neutral and only one (Q79E) was found to be somewhat destabilizing. Random mutations are usually estimated to produce a large excess of destabilizing mutations, on the order of 100:1.[5] We have noted that stability effects are equally well predicted in framework regions and in complementarity-determining regions (CDR). Predictions failed predominantly when they were made for residues that would participate in interdomain interactions in the heterodimeric antibody, such as P43S,[†] Q79E, and A100G. Presumably these would be conserved for interaction requirements, rather than global domain stability in the context of the isolated V_κ domain. Had we used our knowledge of antibody structure to exclude such residues from the prediction, the success rate for mutations that stabilize the isolated domain would increase to 7 out of 8. However, the strategy of simply choosing the most prevalent amino acid in every position was found to carry a penalty of less than 2 kJ mol^{-1} at worst. Moreover, an improvement of Fv fragment association might offset the slight loss of folding stability we have observed, while for practical reasons it also might be a goal for optimization in its own right.

Interestingly, loop lengths of the first complementarity determining region of the V_κ domain are also variable and observed with widely differing frequencies in the database. Changing the native loop length to the consensus length further stabilizes the domain.[6] Thus it is not only residue identity that can be analyzed in terms of consensus predictions, but other features as well, as long as they can be quantified in terms of their frequency of occurrence in a set of related sequences.

[5] F. C. Christians, L. Scapozza, A. Crameri, G. Folkers, and W. P. C. Stemmer, *Nature Biotechnol.* **17**, 259 (1999).

[†] To specify mutations, we use the one-letter code for the native residue, followed by the sequence position, followed by the one-letter code for the new residue.

[6] E. C. Ohage and B. Steipe, *J. Mol. Biol.* **291**, 1119 (1999).

Construction of Intrabodies

One of the hallmarks of the immunoglobulin fold is a strictly conserved intramolecular disulfide bridge. Such disulfide bridges normally cannot be formed in the reducing environment of the cytoplasm.[7] Estimating the contribution of the disulfide bond to stability from its effect on the reduction of entropy in the unfolded state[8] and comparing it to typical stabilities of native immunoglobulin domains show that this disulfide bridge alone would contribute more than the net free energy of folding to the stability of each domain. The consequences have been observed experimentally many times over: attempts to express functional antibody domains in a reducing environment fail in general[9] and chemical reduction denatures the domain *in vitro*. While exceptions have been reported[10–12] and, for many practical applications, the problem has been overcome through the clever application of randomization and screening strategies,[13–15] by and large low stability, low expression rates, and unpredictable behavior have limited the application of functional intrabodies.[16–18]

In a study aimed to overcome this problem by rational engineering, we have combined individually stabilizing mutations and showed that this achieves additive stabilization. The hyperstable V_κ we have obtained with this strategy can be expressed in the cytoplasm, solubly and with good yield.[6] As well, predicted point mutations in a V_H domain were similarly stabilizing,[19] allowing us to construct a framework for loop grafting of novel specificities. This strategy led to the successful expression of a

[7] S. Biocca, F. Ruberti, M. Tafani, A. P. Pierandrei, and A. Cattaneo, *Biotechnology* **13**, 1110 (1995).

[8] C. N. Pace, G. R. Grimsley, J. A. Thomson, and B. J. Barnett, *J. Biol. Chem.* **263**, 11820 (1988).

[9] R. Glockshuber, T. Schmidt, and A. Plückthun, *Biochemistry* **31**, 1270 (1992).

[10] J. P. Maciejewski, F. F. Weichold, N. S. Young, A. Cara, D. Zella, M. J. Reitz, and R. C. Gallo, *Nature Medicine* **1**, 667 (1995).

[11] Y. Wu, L. Duan, M. Zhu, B. Hu, S. Kubota, O. Bagasra, and R. J. Pomerantz, *J. Virology* **70**, 3290 (1996).

[12] I. J. Rondon and W. A. Marasco, *Annu. Rev. Microbiol.* **51**, 257 (1997).

[13] A. Auf der Maur, C. Zahnd, F. Fischer, S. Spinelli, A. Honegger, C. Cambillau, D. Escher, A. Plückthun, and A. Barberis, *J. Biol. Chem.* **277**, 45075 (2002).

[14] T. Tanaka and T. H. Rabbitts, *EMBO J.* **22**, 1025 (2003).

[15] F. Gennari, S. Mehta, Y. Wang, A. St. Clair Tallarico, G. Palu, and W. A. Marasco, *J. Mol. Biol.* **335**, 193 (2004).

[16] W. A. Marasco, *Gene Ther.* **4**, 11 (1997).

[17] N. Gargano and A. Cattaneo, *FEBS Lett.* **414**, 537 (1997).

[18] A. Cattaneo and S. Biocca, *Trends Biotech.* **17**, 115 (1999).

[19] P. Wirtz and B. Steipe, *Protein Sci.* **8**, 2245 (1999).

heterodimeric, catalytic intrabody Fv fragment by loop grafting of the CDR regions of the esterolytic antibody 17E8 into our stabilized framework.[20]

Other Examples

A number of similarly successful studies applying this principle have underscored the generality of our approach for the rational engineering of immunoglobulins for a wide range of objectives, from biotechnology to immunotherapy and ranging from individual domains to whole antibodies. These studies include the construction of frameworks for the display of randomized CDR libraries of scFvs,[21] improved production of a Diels–Alder catalytic antibody,[22] stabilization of an internalizing anti-CD22 scFv, with potential application for non-Hodgkin lymphoma therapy,[23] engineering for improved expression yield of an scFv-β-lactamase fusion protein for antibody-directed prodrug activation cancer therapy,[24] improved secretion of whole antibodies in mammalian cells,[25] and thermostabilization of immunoglobulin C_{H3} constant domains.[26]

Due to the generality of the approach, application of the canonical sequence approximation to other proteins is equally straightforward in principle. Successful examples include the stabilization of GroEL minichaperones,[27] stabilization of p53,[28] and SH3 domains.[29]

Perhaps the most convincing example overall has been contributed from the engineering of thermostable phytases for animal feed technology. Lehmann et al.[30] compiled a multiple sequence alignment of 13 homologous fungal phytases as a starting point for consensus engineering. Instead of introducing individual point mutations, they generated the entire

[20] E. C. Ohage, P. Wirtz, J. Barnikow, and B. Steipe, J. Mol. Biol. 291, 1129 (1999).
[21] A. Knappik, L. M. Ge, A. Honegger, P. Pack, M. Fischer, G. Wellnhofer, A. Hoess, J. Wolle, A. Plückthun, and B. Virnekäs, J. Mol. Biol. 296, 57 (2000).
[22] A. Piatesi and D. Hilvert, Can. J. Chem. 80, 657 (2002).
[23] M. A. E. Arndt, J. Krauss, R. Schwarzenbacher, B. K. Vu, S. Greene, and S. M. Rybak, Intl. J. Cancer 107, 822 (2003).
[24] C. F. McDonagh, K. S. Beam, G. J. Wu, J. H. Chen, D. F. Chace, P. D. Senter, and J. A. Francisco, Bioconjug. Chem. 14, 860 (2003).
[25] E. A. Whitcomb, T. M. Martin, and M. B. Rittenberg, J. Immunol. 170, 1903 (2003).
[26] S. J. Demarest, J. Rogers, and G. Hansen, J. Mol. Biol. 335, 41 (2004).
[27] Q. H. Wang, A. M. Buckle, N. W. Foster, C. M. Johnson, and A. R. Fersht, Protein Sci. 8, 2186 (1999).
[28] P. V. Nikolova, J. Henckel, D. P. Lane, and A. R. Fersht, Proc. Natl. Acad. Sci. USA 95, 14675 (1998).
[29] K. L. Maxwell and A. R. Davidson, Biochemistry 37, 16172 (1998).
[30] M. Lehmann, L. Pasamontes, S. F. Lassen, and M. Wyss, Biochim. Biophys. Acta 1543, 408 (2000).

consensus sequence by gene synthesis. This synthetic protein was not only 15 to 26° more thermostable than any of the progenitor sequences, it was also enzymatically as active. In a follow-up study, the group included further sequence changes that were suggested by the inclusion of additional sequences into the alignment, characterized them individually with respect to the earlier consensus sequence, removed a small number of destabilizing mutations, and finally arrived at a phytase with an unfolding temperature of 90.4°—an increase of almost 30° from the most stable natural phytase of *Aspergillus fumigatus*.[31] The effects of individual mutations are small. Large, global stabilization is achieved by the additive contributions of a number of independent improvements. This makes the strategy complementary to evolutionary engineering methods, which rely on large individual effects to create a screenable signal and which have problems generating saturating libraries with multiple substitutions due to the combinatorially large size of the sequence spaces involved.

The ability to generate synthetic, enzymatically active proteins from consensus sequences emphasizes the statistical nature of natural protein sequences as products of neutral drift and selective pressure. Stabilization is not the only objective for which the approach is useful: others could include the selection of residues for nondisruptive destabilizing mutations to reduce the lifetime of proteins or to generate proteins that can be used to screen for second-site revertants in evolutionary engineering strategies. They could also include guiding the choice of preferred replacements of residues that need to be changed for other purposes, such as engineering of electrostatic properties or alkaline stability.[32]

Work Flow

Consensus engineering is simple and successful predictions do not require elaborate tools. The work flow is represented in Fig. 1 and is discussed in this section.

Sequence Sets

A prerequisite for consensus engineering is the availability of homologous[‡] sequences at moderate to high degrees of similarity, and obviously the very large sequence collections of the immunoglobulins have been an excellent resource in this respect. However, given the rapid growth of sequence databases, it has become more and more likely that a sufficient

[31] M. Lehmann, C. Loch, A. Middendorf, D. Studer, S. F. Lassen, L. Pasamontes, A. P. G. M. van Loon, and M. Wyss, *Protein Eng.* **15**, 403 (2002).
[32] S. Gulich, M. Linhult, S. Stahl, and S. Hober, *Protein Eng.* **15**, 835 (2002).

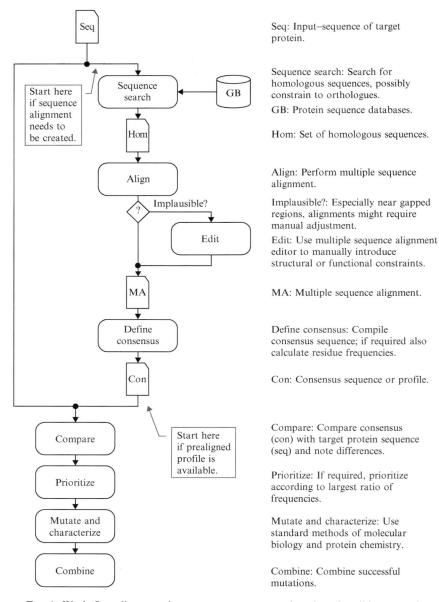

Seq: Input–sequence of target protein.

Sequence search: Search for homologous sequences, possibly constrain to orthologues.

GB: Protein sequence databases.

Hom: Set of homologous sequences.

Align: Perform multiple sequence alignment.

Implausible?: Especially near gapped regions, alignments might require manual adjustment.

Edit: Use multiple sequence alignment editor to manually introduce structural or functional constraints.

MA: Multiple sequence alignment.

Define consensus: Compile consensus sequence; if required also calculate residue frequencies.

Con: Consensus sequence or profile.

Compare: Compare consensus (con) with target protein sequence (seq) and note differences.

Prioritize: If required, prioritize according to largest ratio of frequencies.

Mutate and characterize: Use standard methods of molecular biology and protein chemistry.

Combine: Combine successful mutations.

Fig. 1. Work flow diagram of consensus sequence engineering, describing procedures, principal intermediary results or data sets, and their sequential connection. Details of the individual steps are discussed in the text.

number of suitable sequences can be retrieved through homology searches for any protein of interest. Even where this is not possible, the option of randomly collecting homologous sequences from DNA pools by PCR and sequencing can be considered, as identification of the progenitor organism is not required for the procedure. The data sets need not be large, as long as a reasonably unbiased selection can be obtained. It is not the number of sequences, but the ratio of frequencies that is important.

Typically, a protein–protein BLAST search would be the starting point for the collection of homologous sequences. A standard BLAST search (http://www.ncbi.nlm.nih.gov/BLAST/) with default parameters should be sufficient to retrieve the sequences of interest. We would preferably not use very divergent sequences retrieved from highly sensitive search strategies, such as PSI-BLAST, as these sequences require introducing additional assumptions about the conserved structural context of individual positions and conserved function.

Sequences can be selected on the output Web form, retrieved and saved to disk, to be used later as input for multiple alignment.

If orthologous sequences are required, e.g., for the engineering of enzymes, three main options are available to select appropriate sequences.

i. Annotations of function can be retrieved from the file headers on the BLAST results page. This is fast, but carries the risk of false positives since function may have been annotated from sequence similarity alone, as well as false negatives, since annotations may be missing, wrong, or using unfamiliar synonyms.

ii. One can attempt to demonstrate that the homologous sequences of interest are reciprocal best matches in their respective genomes. This is tedious, but would give the most convincing support. Obviously, this requires that complete genomes are available that include all sequences that are to be analyzed.

iii. If available, sequences can be obtained from the Clusters of Orthologous Groups (COGs) database (http://www.ncbi.nlm.nih.gov/COG/), which stores several thousand precompiled clusters of reciprocal best matches of genes in bacterial genomes.[33] Note that the source

---

‡ Homologous sequences have diverged from a common ancestral sequence. Typically homology is inferred from observing high degrees of sequence similarity, but structural alignments of distantly related proteins may serve as well to define alignments. Orthologous sequences have the same function in different organisms, while paralogous sequences have divergent functions, usually as the result of gene duplications.

[33] R. L. Tatusov, D. A. Natale, I. V. Garkavtsev, T. A. Tatusova, U. T. Shankavaram, B. S. Rao, B. Kiryutin, M. Y. Galperin, N. D. Fedorova, and E. V. Koonin, *Nucleic Acids Res.* **29,** 22 (2001).

sequences, as well as complete CLUSTAL multiple alignments, can be retrieved from this site.

Alignment

As a next step, multiple sequence alignments are required. A number of options are available; the curated bioinformatics links directory of the Canadian Bioinformatics Workshops provides a good overview (http://www.bioinformatics.ca/). Excellent resources are offered through Web interfaces and do not require local installation of alignment tools. CLUSTAL W has long been the method of first choice (http://www.ebi.ac.uk/clustalw/) but it has become apparent that the T-Coffee algorithm (http://www.ch.embnet.org/software/TCoffee.html) produces alignments that appear biologically more meaningful, especially for sequences that are not closely related.

Multiple alignment algorithms optimize a global score across an alignment. This score is typically based on sequence similarity alone and does not take structural or functional information into account. CLUSTAL can be an exception, as it allows the input of templates that guide alignments to be compatible with knowledge about secondary structure. Accordingly, in many cases, visual inspection of the generated alignment will suggest the need for manual edits to better represent biochemical knowledge about the target protein. CINEMA (http://bioinf.man.ac.uk/dbbrowser/CINEMA2.1/) and JALVIEW (http://www2.ebi.ac.uk/~michele/jalview/) are multiple alignment editors that have been in use for quite some time. An interesting newer alternative in terms of simplicity of access may be GoCore (http:// www.helsinki.fi/project/ritvos/ GoCore/), which has been programmed as an Excel plug-in.

Prealigned Profiles

Where available, prealigned sequence collections may be the easiest source to guide consensus engineering projects.

For immunoglobulin variable domains, we have compiled sequence profiles for the major immunoglobulin subtypes based on the machine-readable version of the fifth edition of the Kabat database of immunoglobulin sequences.[34] The sequence alignments have been edited manually to align insertions into the positions where they have been determined crystallographically. The frequency profiles have been posted on the Web (http://biochemistry.utoronto.ca/steipe/research/canonical.html) and allow one to determine the consensus residue itself as well as obtain an

[34] E. A. Kabat, T. T. Wu, H. M. Perry, K. S. Gottesman, and C. Foeller, "Distribution Files of the Fifth Edition of Sequences of Proteins of Immunological Interest," 1992.

overview of how strongly constrained a position is; they can be used to calculate frequency ratios.

Many protein domains have been defined by bioinformatics analysis of sequence databases; these are being made available via the conserved domain database (CDD) at the NCBI (http://www.ncbi.nlm.nih.gov/Structure/cdd./cdd.shtml). The display of CDD profiles can be customized through a Web interface. Consensus sequences are calculated and variability can be estimated visually from the alignments or tabulated manually. As a *caveat*, the domain alignments are built primarily for the purpose of database searches and frequently do not extend all the way to the domain termini. In many cases it will thus be desirable to construct a custom multiple alignment. Even for this purpose, the CDD is an efficient way to retrieve source sequences, as the corresponding GenPept records are linked to the sequences displayed on the page. The use of COGs for this purpose was discussed previously.

Generating Consensus Sequences

Consensus sequences are included in our immunoglobulin sequence alignment, with CDDs, or can be created from CINEMA or GoCore output. When the consensus sequence is not obvious (e.g., for very large alignments), the EMBOSS package of molecular biology tools provides a number of options to process multiple alignments from locally installed tools or online (http://bioinfo.pbi.nrc.ca/EMBOSS/). To compile specific frequency distributions, a Perl program can be obtained from the author of this article.

Sequence Prioritizations

The strongest assumption of our approach is that thermodynamic stability is the dominating component of selective pressure in the positions we would like to mutate. We have noted earlier that this is not necessarily the case for residues that contribute to interdomain interfaces. Generating a homology model (e.g., with the automated SwissModel Web server http://www.expasy.org/swissmod/SWISS-MODEL.html) can help distinguish such residues and prioritize mutations, if necessary. However, there is a second class of residues for which the consensus rule explicitly will not hold: residues involved in the active sites of enzymes. Studies on a number of enzymes suggest a partition of amino acid roles into structural and functional residues, as was demonstrated by Beadle and Shoichet[35] (see also references contained therein). Most residues appear to contribute to structural frameworks through favorable interactions, but a few provide

---

[35] B. M. Beadle and B. K. Schoichet, *J. Mol. Biol.* **321**, 285 (2002).

specific functions in precisely controlled conformations, frequently at the cost of steric or electrostatic strain, exposed hydrophobic surface, or unsatisfied hydrogen-bonding donors or acceptors. These functional residues are highly conserved, yet significant increases of stability can be gained by replacing them. While this is important to recognize in principle, it is a nonissue for the optimization of enzymes as the example of phytase engineering has shown, since presumably the source sequence has the same functional residues conserved as its homologues, especially if these are orthologues, and the enzymatic activity likely needs to be conserved in the engineering process. Where this is not the case, and stabilization without concern for activity is the objective, other methods than consensus engineering are required. We have had some success in this respect with an extension of the canonical sequence approximation to the convergent evolution of nonhomologous structure motif consensus sequences.[36,37]

Possible mutations may be prioritized. In keeping with the *canonical sequence approximation*, we would not only consider the absolute frequency of the consensus residue but also the ratio of frequencies of the consensus residue and the original residue. If this frequency ratio is used for quantitative predictions as $\Delta\Delta G = -RT\ (\ln f_{consensus} - \ln f_{original})$, we typically achieve coefficients of correlation of better than 0.6 (B. Steipe, unpublished), although, given the assumptions that go into the model, such a quantitative treatment has to be treated with caution and, for practical purposes, qualitative predictions are sufficient. The final steps of mutagenesis, characterization and combination of successful residues, are specific for each experimental system at hand and should be straightforward.

Summary

The excellent track record of consensus sequence engineering, together with the ease of obtaining predictions, makes this strategy a method of first choice for biotechnological purposes, as well as for the study of sequence–structure relationships in proteins. The method is general, complements strategies of evolutionary engineering well, and can be combined with alternative methods of stability prediction. It is applicable if nothing but sequence is known for a protein and it can be refined to consider structural and functional knowledge.

---

[36] E. C. Ohage, W. Graml, M. M. Walter, S. Steinbacher, and B. Steipe, *Protein Sci.* **6,** 233 (1997).
[37] M. Niggemann and B. Steipe, *J. Mol. Biol.* **296,** 181 (2000).

[17] Improving the Functional Expression of *N*-Carbamoylase by Directed Evolution using the Green Fluorescent Protein Fusion Reporter System

By Hee-Sung Park, Ki-Hoon Oh, and Hak-Sung Kim

Introduction

It is generally known that bacterial expression systems are simple, efficient, high-yield, and low-cost methods for the production of heterologous proteins in large amounts. However, many proteins often do not fold correctly and aggregate in insoluble form when overexpressed. They are accumulated in the cytoplasm of bacteria in the form of inclusion bodies and it is very difficult to refold the aggregates correctly even with a treatment of strong chaotropic reagents such as urea and guanidium HCl. In order to increase the functional expression of proteins, a number of strategies have been employed: optimization of growth conditions such as culture temperature and medium; coexpression of folding accessory proteins such as foldases and chaperons; and an extracellular secretion system. Recent approaches have focused more on protein fusions and tags such as glutathione *S*-transferase (GST), maltose-binding protein (MBP), protein A, FLAG, and His$_6$ because the high folding tendency of fusion partners usually leads to an increased functional expression of target protein and easy purification.[1–3]

Nevertheless, most of the conventional methods do not alter the inherent insoluble nature of the target protein, resulting in the formation of an inclusion body, and alternative approaches to improve the intrinsic folding stability and functional expression of protein are required. Directed evolution, which offers an efficient way of creating genetic diversity by mimicking the natural evolution,[4] has been implemented successfully to improve the expression of foreign proteins. The folding of green fluorescent protein (GFP) was enhanced greatly after several rounds of DNA shuffling, and the resulting GFPuv showed a 45-fold increased fluorescent emission compared to the wild-type counterpart.[5] Expression of a single chain fragment

[1] R. C. Hockney, *Trends Biotechnol.* **12,** 456 (1994).
[2] G. Hannig and S. C. Makrides, *Trends Biotechnol.* **16,** 54 (1998).
[3] G. S. Waldo, *Curr. Opin. Chem. Biol.* **7,** 33 (2003).
[4] W. P. C. Stemmer, *Nature* **370,** 389 (1994).
[5] A. Crameri, E. A. Whitehorn, E. Tate, and W. P. C. Stemmer, *Nature Biotechnol.* **14,** 315 (1996).

(scFv) of antibody in *Escherichia coli* was reported to result in an unsatisfied level of expression. Proba *et al.*[6] demonstrated that directed evolution of scFv, after mutating the cystein residues responsible for disulfide bond formation, cannot only make up for the loss of the disulfide bond, but also produce a more soluble form of the single chain fragment. Similarly, the solubilities and expression levels of horseradish peroxidase and fungal galactose oxidase were improved considerably through the accumulation of beneficial mutation by directed evolution.[7,8]

N-Carbamoyl-D-amino acid amidohydrolase (*N*-carbamoylase), which catalyzes the hydrolysis of the *N*-carbamoyl group from *N*-carbamyol-D-amino acid, is currently used in the commercial production of unnatural D-amino acid in conjunction with D-hydantoinase.[9,10] However, undesirable properties of *N*-carbamoylase, such as relatively low oxidative and thermostabilities and a high tendency to form aggregates under overexpression conditions, are considered to be drawbacks for industrial purposes. In our previous work,[11] the oxidative and thermostabilities of *N*-carbamoylase were improved successfully by directed evolution. In an attempt to develop an efficient enzymatic process, a bifunctional fusion enzyme composed of *N*-carbamoylase/D-hydantoinase was constructed, and the functional expression and stability of the resulting fusion protein were increased significantly using a similar approach.[12] Other groups revealed that coexpression with GroELS has enabled *N*-carbamoylase from *Agrobacterium tumefaciens* NRRLB11291 and AM10 to fold correctly to some extent.[13,14]

This article presents a simple and efficient method used to improve the functional expression of *N*-carbamoylase in *E. coli* using directed evolution. GFP was used as a reporter protein for screening of the *N*-carbamoylase variant, showing a higher level of functional expression or correct folding.

[6] K. Proba, A. Worn, A. Honegger, and A. Pluckthun, *J. Mol. Biol.* **275**, 245 (1998).

[7] Z. Lin, T. Thorsen, and F. H. Arnold, *Biotechnol. Prog.* **15**, 467 (1999).

[8] L. Sun, I. Petrounia, M. Tagasaki, G. Bandara, and F. A. Arnold, *Protein Eng.* **14**, 699 (2001).

[9] G. J. Kim, D. E. Lee, and H. S. Kim, *Appl. Environ. Microbiol.* **66**, 2133 (2000).

[10] J. H. Park, G. J. Kim, and H. S. Kim, *Biotechnol. Prog.* **16**, 564 (2000).

[11] K. H. Oh, S. H. Nam, and H. S. Kim, *Protein Eng.* **15**, 689 (2002).

[12] G. J. Kim, Y. H. Cheon, and H. S. Kim, *Biotechnol. Bioeng.* **68**, 21 (2000).

[13] Y. Chao, C. Chiang, T. Lo, and H. Fu, *Appl. Microbiol. Biotechnol.* **54**, 348 (2000).

[14] D. Sareen, R. Sharma, and R. Vohra, *Protein Express. Puri.* **23**, 374 (2001).

Materials and Methods

Materials

Imidazole is from Sigma. *N*-Carbamyl-D-hydroxyphenylglycine (NC-HPG) is synthesized from D,L-HPH by using D-hydantoinase as described in previous work.[15] All other reagents are of analytical grade.

Construction of Fusion Protein and Screening of a Mutant Library

For construction of the fusion protein with GFP, the *N*-carbamoylase gene from *A. tumefaciens* NRRL B11291 is amplified and digested with *Nco*I and *Eco*RI and is then ligated into pTrc99A containing the GFP gene. To avoid steric hindrance during folding, 10 amino acids are incorporated between *N*-carbamoylase and GFP as a linker as shown in Fig. 1.

For random mutation using error-prone polymerase chain reaction (PCR), the reaction mixture (100 μl) contains 0.2 mM dATP, 0.2 mM dGTP, 1 mM dTTP, 1 mM dCTP, 2.5 mM MgCl$_2$, 0.3 mM MnCl$_2$, 50 mM KCl, 10 mM Tris–HCl (pH 8.0), 0.1% Triton X-100, 2 unit *Taq* polymerase, and 100 pmol primers. The cloning host used is *E. coli* JM109, and transformation is conducted through electroporation. Transformants are plated on a Luria–Bertani (LB) plate with 100 μg/ml ampicillin, and 0.1 mM isopropylthiogalactoside (IPTG) is added for the induction of fusion proteins.

Positive clones are screened by exciting the library plates directly with a hand-held type UV lamp at 395 nm and by picking clones showing an improved fluorescence emission compared to the wild-type counterpart. Selected clones are cultured and induced for further analysis in terms of fluorescence emission of GFP, and expression and activity of mutant *N*-carbamoylases. Site-directed mutagenesis is performed by the overlap extension method using PCR.

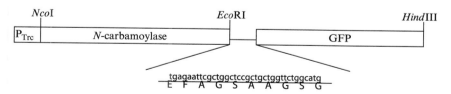

Fig. 1. Structural organization of the gene encoding *N*-carbamoylase/GFP fusion enzyme. A linker with 10 amino acids is shown at the bottom.

[15] G. J. Kim and H. S. Kim, *Enzyme Microb. Technol.* **17**, 63 (1995).

The fluorescence intensity of the cell-free extract or purified protein is measured with a spectrofluorimeter (DMX1100, SLM-AMINCO, MD).

Purification of N-Carbamoylase

For purification of the nonfusion form of *N*-carbamoylase, genes of mutant *N*-carbamoylases are amplified with primers containing the C-terminal His tag, and the resulting fragments are cloned into pTrc99A after digestion with *Nco*I and *Hind*III. For expression of mutant *N*-carbamoylases, cells are cultivated in 400 ml LB with ampicillin and induced with 0.5 m*M* IPTG. After a 3-h induction, cells are harvested by centrifugation and resuspended in 40 ml of 20 m*M* potassium phosphate buffer (pH 7.0) with 1 m*M* phenylmethylsulfonyl fluoride suspended cells are disrupted by sonication. After centrifugation, the supernatant is loaded onto the column packed with Ni-NTA resin (Amersham Pharmacia Biotech, Uppsala, Sweden). The column is washed with 20 m*M* potassium phosphate buffer (pH 7.4) containing 10 m*M* imidazole and eluted with the same buffer with 50 m*M* imidazole. For gel filtration chromatography using the fast performance liquid chromatography (FPLC) system (Amersham Pharmacia Biotech, Uppsala, Sweden), the concentrated fraction is loaded onto a Superdex 200HR 10/30 equilibrated with 20 m*M* potassium phosphate buffer (pH 7.0). Fractions exhibiting *N*-carbamoylase activity are pooled and analyzed further.

Enzyme Assay and Analysis

N-Carbamoylase activity is assayed by incubating cell-free extracts or purified *N*-carbamoylases with 25 m*M* NC-HPG in 100 m*M* potassium phosphate buffer (pH 7.0) at 40° for 30 min. Produced D-HPG is analyzed by HPLC (Shimazu Co. Kyoto, Japan) as described previously.[15] The protein concentration is determined using the Bradford method.

Improvement of Functional Expression of N-Carbamoylase

Screening System Using Fusion Proteins

Several promising methods used to screen protein variants with higher expression or correct folding have been developed. These strategies usually comprise two distinctive steps: (1) generation of a mutant library by various methods and (2) a screening system based on the reporter tags by which soluble protein variants with improved property can be selected easily either *in vivo* or *in vitro*. In these cases, the protein of interest is expressed as an amino-terminal fusion with reporter tags, such as GFP, chloramphenicol aminotransferase (CAT), and lacZα. Each reporter protein has the following characteristics for screening and selection of protein variants with higher functional expression: GFP is an excellent indicator of protein folding when considering that the fluorescent intensity of the downstream GFP protein domain is directly proportional to the folding ability of the upstream target protein.[16–18] CAT enables the host to survive on media containing antibiotic chloramphenicol; therefore, cells expressing a higher level of soluble target protein might produce more functional CAT and are screened on higher selective conditions.[19] lacZα (α fragment of β-galactosidase) activates inactive β-galactosidase, thereby endowing the host with the ability to grow on medium containing lactose as the sole carbon source.[20]

Also, enzymes that induce a color change on selective agar plates can be excellent fusion partners. Of them, D-hydantoinase from *Bacillus stearothermophilus* SD1 is an industrial enzyme and is used currently in the enzymatic production of D-amino acid, together with *N*-carbamoylase from *A. tumefaciens* NRRL B11291. In the process, D-hydantoin is hydrolyzed by D-hydantoinase, and the resulting *N*-carbamyl-D-amino acid is further converted into D-amino acid by *N*-carbamoylase. D-Hydantoinase can be useful for the detection of the improved functional expression of proteins based on the acid-producing catalytic property; the selective agar plate, containing 1% hydantoin and 0.005% phenol red, exhibits background red color at pH 7.7. D-Hydantoinase converts hydantoin into hydantoic acid, lowering the pH, and the plate turns yellow.

Improvement of Functional Expression of N-Carbamoylase Using Reporter Proteins

As a fusion protein, we first attempted D-hydantoinase as described earlier. When D-hydantoinase was fused into the C-terminal of *N*-carbamoylase, expression of the *N*-carbamoylase/D-hydantoinase fusion protein was unstable and was subject to serious proteolysis *in vivo* probably due to the low structural stability of the N-terminal fusion partner, *N*-carbamoylase. Colonies exhibiting the color change were screened from the selective plate.

[16] G. S. Waldo, B. M. Standish, J. Berendzen, and T. C. Terwilliger, *Nature Biotechnol.* **17**, 691 (1999).

[17] C. A. Kim, M. L. Phillips, W. Kim, M. Gingery, H. H. Tran, M. A. Robinson, S. Faham, and J. U. Bowie, *EMBO J.* **20**, 4173 (2001).

[18] J.-D. Pedelacq, E. Piltch, E. C. Liong, J. Berendzen, C.-Y. Kim, B.-S. Rho, M. S. Park, T. C. Terwilliger, and G. S. Waldo, *Nature Biotechnol.* **20**, 927 (2002).

[19] V. Sieber, C. A. Martinez, and F. H. Arnold, *Nature Biotechnol.* **19**, 456 (2001).

[20] W. C. Wigley, R. D. Stidham, N. M. Smith, J. F. Hunt, and P. J. Thomas, *Nature Biotechnol.* **19**, 131 (2001).

After three rounds of directed evolution using DNA shuffling, a mutant F11 containing two mutations (S213C and F394S) in the N-carbamoylase region was selected.[12] The evolved fusion enzyme F11 was found to be resistant to proteolysis, maintaining its structural integrity *in vivo*, and displayed a considerably improved functional expression and stability compared to the parent enzyme.

In order to utilize GFP as a reporter, we constructed N-carbamoylase/GFP fusion protein. GFP was fused into the C-terminal of N-carbamoylase with 10 amino acid linker as in the case of N-carbamoylase/D-hydantoinase fusion (Fig. 1). It was found that the parent N-carbamoylase/GFP fusion construct did not show any fluorescent emission. It seems that the C-terminal partner (GFP) could not form correct conformation and did not show any biological activity, as the N-terminal partner (N-carbamoylase) was structurally unstable and highly prone to form an insoluble aggregate. As the background fluorescence was clear, we could readily identify the positive colonies emitting fluorescence directly from the screening plate. From the first round of error-prone PCR, five fluorescent colonies were selected from a library of 1.0×10^4 mutants.

To evaluate the correlation between the fluorescence intensity of GFP and the functional expression of N-carbamoylase among wild-type and mutant N-carbamoylase/GFP fusions, selected clones were cultivated and induced under identical conditions. Although the intensities of fluorescence emitted from GFP were not directly proportional to the activities of mutant N-carbamoylases, all positive clones with improved fluorescence intensities were found to display elevated N-carbamoylase activities at the same time when compared with the wild type (Fig. 2). Mutant Ex1 was found to have a 30-fold increased fluorescence emission and about 4-fold enhanced N-carbamoylases activity simulataneously. To compare the expression levels among these evolved N-carbamoylases, the C-terminal fusion partner (GFP) was removed and the 6x His tag was fused with wild-type and mutant N-carbamoylases. Under overexpression conditions, expression levels of the selected mutant N-carbamoylases were shown to be increased significantly when compared with the parent enzyme (data not shown). The expression level of Ex1 was consistent with enhanced fluorescence emission and N-carbamoylase activity.

For further improvement of evolved N-carbamoylases, an additional round of *in vitro* recombination through DNA shuffling was attempted, but we could not screen improved mutants. It is likely that the increment in fluorescence of the first-round positives was so large that the background fluorescence of the plate was high and screening of clones with improved fluorescent emission against the background became difficult by visual inspection.

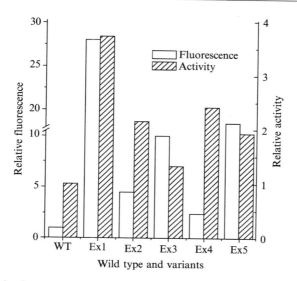

Fig. 2. Relative fluorescence and activity of selected *N*-carbamoylase/GFP fusion variants. Fluorescence was verified using 100 μg/ml of cell lysate at 25°. Activity was determined by measuring the conversion rate of NC-HPG to D-HPG by 100 μg/ml of the cell lysate at 40° for 30 min.

Characterization of Evolved Mutants

Mutant Ex1 was found to have two amino acid substitutions: V40A and T262A. In order to investigate the mutations that cause a major impact on the elevated expression of evolved *N*-carbamoylase Ex1, we constructed two variants with a single point mutation by site-directed mutagenesis and analyzed the specific activities and expression levels for the wild-type and mutant N-carbamoylases. As shown in Table I, V40A and T262A mutants have 2- and 1.5-fold increased activities, respectively, whereas Ex1 has about a 4-fold increase when compared with wild-type N-carbamoylase. In addition, expression profiles of the soluble fraction revealed that each single mutant V40A and T262A exhibits a similar improvement in the expression levels and Ex1 possesses a distinct expression level, as shown in Fig. 3. These results indicate that two single mutations, V40A and T262A, seem to have a similar effect on the expression of *N*-carbamoylase and also make an additive contribution to the functional expression of Ex1 as a whole. Activity analysis with purified wild type and mutants demonstrates that V40A and T262A mutations do not affect the catalytic properties of *N*-carbamoylase (Table I), which means that an increase in activities

TABLE I
SPECIFIC ACTIVITIES OF CELL LYSATE AND PURIFIED ENZYME

	Specific activities (U/mg protein)			
	WT	V40A	T262A	Ex1
Cell lysate	0.36	0.73	0.50	1.32
Purified enzyme	5.0	4.8	4.1	4.5

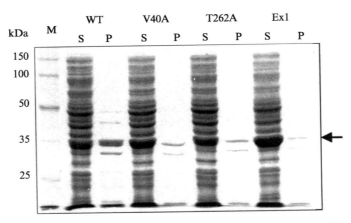

FIG. 3. Overexpression of N-carbamoylase mutants. WT, wild type; V40A, V40A mutant; T262A, T262A mutant. S and P indicate soluble and insoluble fractions after sonication, respectively. Arrow indicates the N-carbamoylase.

of the whole cell lysate resulted mainly from the elevated expression of mutants. It should also be noted that the insoluble fraction of wild-type N-carbamoylase, which is largely caused by inclusion body formation, was reduced remarkably by single mutations and only a trace amount of insoluble fraction was detected in Ex1 crude lysate (Fig. 3). It seems that cooperative action of the mutations leads to the functional expression of N-carbamoylase by improving the innate folding ability of the protein. Interestingly, these two mutations were also found in thermostable N-carbamoylase variant 2S3 of our previous work, which possesses four additional mutations in addition to V40A and T262A.[12] From the stability analysis of single point mutants of 2S3, it was revealed that while V40A mainly gave rise to the increase in oxidative stability rather than thermostability, T262A resulted in a significant improvement in both oxidative stability and thermostability. Generally, it is well known that when heterologous polypeptides fail to fold into their native state, they are either degraded by the cellular proteolytic machinery or accumulated as an inclusion body. Therefore, it is suggested that V40A and T262A play a critical role in enhancing the folding efficiency of *N*-carbamoylase in the first place and, at the same time, T262A additionally increases the stability of *N*-carbamoylase *in vivo*, thereby preventing it from degradation by cellular proteases.

Concluding Remarks

This article showed that the functional expression of *N*-carbamoylase in *E. coli* can be improved significantly by directed evolution. A screening system using a fusion reporter protein such as D-hydantoinase or GFP can be used effectively for selecting proteins with higher expression or correct folding in directed molecular evolution.

Section III

Applications: Directed Evolution of Enzymatic Function

[18]   Directed Evolution of Lipases and Esterases

By Marlen Schmidt, Markus Baumann, Erik Henke,
Monika Konarzycka-Bessler, and Uwe T. Bornscheuer

Introduction

Directed evolution has emerged as a powerful tool to improve biocatalysts as well as to broaden our understanding of the underlying principles for substrate specificity, stereoselectivity, and mechanism.[1–4] In contrast to rational protein design, directed evolution does not require knowledge of the three-dimensional structure of a given enzyme and about the relationship among structure, sequence, and mechanism.

In general, directed evolution is composed of two steps: (1) the generation of mutant libraries and (2) the identification of desired variants using a suitable screening or selection system. Two different strategies for the generation of mutant libraries are described: asexual (nonrecombinant) and sexual (recombinant) evolution. Sexual methods start from homologous parent genes. This can be genes encoding for homologous enzymes or genes obtained from an asexual evolution or from rational protein design.

The earliest method for sexual evolution is DNA (or gene) shuffling, which was developed by Stemmer.[5,6] The gene is digested randomly with DNase I in many fragments, which are then recombined without and with primers in a subsequent polymerase chain reaction (PCR), yielding full-length chimeric genes. In the *random priming recombination* approach, gene fragments are produced by short PCR cycles using random primers. The advantage in comparison to DNA shuffling is the need of much smaller amounts of DNA. Other methods include the staggered extension process (StEP), incremental truncation for the creation of hybrid enzymes (ITCHY), SCRATCHY (a combination of ITCHY and DNA shuffling), random chimeragenesis on transient templates (RACHITT), and sequence homology-independent protein recombination (SHIPREC).

The most widely used asexual method is error-prone PCR (epPCR). Here, nonoptimal reaction conditions are used to create a mutant library.

[1] F. H. Arnold and A. A. Volkov, *Curr. Opin. Chem. Biol.* **3,** 54 (1999).
[2] C. Schmidt-Dannert and F. H. Arnold, *Trends Biotechnol.* **17,** 135 (1999).
[3] U. T. Bornscheuer and M. Pohl, *Curr. Opin. Chem. Biol.* **5,** 137 (2001).
[4] U. T. Bornscheuer, *Biocat. Biotrans.* **19,** 84 (2001).
[5] W. P. C. Stemmer, *Nature* **370,** 389 (1994).
[6] W. P. C. Stemmer, *Proc. Natl. Acad. Sci. USA* **91,** 10747 (1994).

For example, the error rate of the commonly used polymerase, from *Thermus aquaticus* (*Taq*), can be increased using $MnCl_2$ instead of $MgCl_2$ salts, nonbalanced nucleotide (dNTP) concentrations, and alterations of cycle number and duration. However, some disadvantages, such as non-biased distribution of amino acid mutations and unsatisfactory ligation efficiency of the PCR products, have been reported.

Another very simple way to get a mutant library is to use mutator strains, such as the commercially available strain *Epicurian coli* XL1-Red (Stratagene).[7] This has defects in its DNA repair mechanisms so that mutations are maintained during replication. The only disadvantages compared to epPCR are that the mutation rate cannot be adjusted and that the entire plasmid is mutated and not only the gene of interest.

Lipases and Esterases

These two enzymes from the class of hydrolases are important biocatalysts and are especially suitable for industrial applications as they are very stable and also active in organic solvents.[8] Moreover, they very often exhibit high stereoselectivity and are therefore suitable for the synthesis of optically active compounds, for which >1000 examples can be found in the literature. A considerable number of lipases and esterases are available commercially.

However, not all enzymes are perfectly designed by nature for a certain application. Thus, in addition to other approaches,[9] directed evolution was also applied to lipases and esterases.

Examples for the Directed Evolution of Lipases and Esterases

Examples investigated at our group are presented in detail later on. The groups of Jaeger and Reetz used directed evolution to enhance the enantioselectivity of a lipase from *Pseudomonas aeruginosa* for the resolution of 2-methyl decanoate. The wild-type enzyme shows almost no selectivity ($E = 1.1$), but several rounds of mutation using error-prone PCR, shuffling, saturation, and cassette mutagenesis led to the creation of variants with selectivities of $E = 51$. Moreover, mutants were also identified with good selectivity for the opposite enantiomer.[10]

[7] U. T. Bornscheuer, M. M. Enzelberger, J. Altenbuchner, and H. H. Meyer, *Strategies* **11**, 16 (1998).

[8] U. T. Bornscheuer and R. J. Kazlauskas, "Hydrolases in Organic Synthesis: Regio- and Stereoselective Biotransformations." Wiley-VCH, Weinheim, 1999.

[9] U. T. Bornscheuer, C. Bessler, R. Srinivas, and S. H. Krishna, *Trends Biotechnol.* **20**, 433 (2002).

Another example is the enhancement of phospholipase (PLA$_1$) activity for lipases from *Staphylococcus aureus* (SAL) and from *Bacillus thermocatenulatus* (BTL2). In case of SAL, four cycles of epPCR created two new enzyme variants with 8–9 amino acid substitutions exhibiting a 7.8- to 9.2-fold increased phospholipase activity and a 5.9- to 6.9-fold increase in phospholipase/lipase activity ratio. Further improvement was achieved by DNA shuffling.[11] Similar results were reported for BTL2.[12]

Moore and Arnold[13] evolved a *Bacillus subtilis* esterase by a combination of epPCR and DNA shuffling, yielding a variant with 150 times higher activity in 15% dimethylformamide (DMF) in hydrolysis of the *p*-nitrobenzyl ester of Loracarbef.[13] Also, thermostability of an esterase was increased by 14° within six generations of random mutagenesis without compromising its catalytic activity at lower temperatures.[14]

In our earliest report using directed evolution, an esterase from *Pseudomonas fluorescens* (PFE), cloned into *Escherichia coli*,[15] was used to alter substrate specificity, as the wild-type esterase was not able to hydrolyze a sterically hindered 3-hydroxy ester. Mutant libraries were created using the mutator strain *Epicurian coli* XL1-Red and active variants were identified by an agar plate assay based on pH indicators and enhanced growth. This led to the identification of a double mutant (Ala209Asp and Leu181Val), which was able to act on the substrate. However, enantioselectivity was rather low.[16]

Next, directed evolution was used to increase the enantioselectivity of PFE, as this was also rather low for other substrates.[15] This time, the gene encoding PFE was subjected to rounds of epPCR or using the mutator strain described in Materials and Methods. To identify esterase variants with enhanced enantioselectivity in mutant libraries, a chromophoric assay using optically pure resorufin esters of the model substrate [(*R*)- or (*S*)-3-phenylbutyric acid] was used and employed for a screening in microtiter plates (MTP) (Fig. 1 and Scheme 1) using a high-throughput screening system (HTS) as described later.[17] Mutants thus identified were confirmed in preparative scale biocatalysis using the corresponding methylester and

[10] M. T. Reetz, S. Wilensek, D. Zha, and K. E. Jaeger, *Angew. Chem. Int. Ed. Engl.* **40**, 3589 (2001).

[11] M. D. van Kampen and M. R. Egmond, *Eur. J. Lipid Sci. Technol.* **102**, 717 (2000).

[12] I. Kauffmann and C. Schmidt-Dannert, *Protein Eng.* **14**, 919 (2001).

[13] J. C. Moore and F. H. Arnold, *Nature Biotechnol.* **14**, 458 (1996).

[14] L. Giver, A. Gershenson, P.-O. Freskgard, and F. H. Arnold, *Proc. Natl. Acad. Sci. USA* **95**, 12809 (1998).

[15] N. Krebsfänger, K. Schierholz, and U. T. Bornscheuer, *J. Biotechnol.* **60**, 105 (1998).

[16] U. T. Bornscheuer, J. Altenbuchner, and H. H. Meyer, *Biotechnol. Bioeng.* **58**, 554 (1998); U. T. Bornscheuer, J. Altenbuchner, and H. H. Meyer, *Bioorg. Med. Chem.* **7**, 2169 (1999).

[17] E. Henke and U. T. Bornscheuer, *Biol. Chem.* **380**, 1029 (1999).

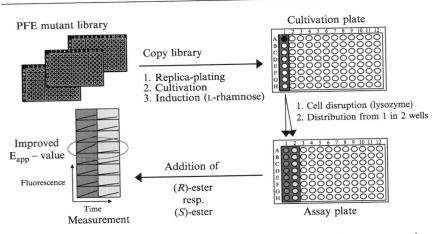

FIG. 1. Principle of the screening of mutant libraries as exemplified for an esterase from *Pseudomonas fluorescens* (PFE). First, clones are transferred from a master plate to a new MTP. After cultivation of *E. coli* harboring the esterase gene and production of the enzyme by induction with L-rhamnose, esterases are isolated by cell disruption using lysozyme and freeze/thaw cycles. The enzyme from the supernatant is then split into two wells of another MTP and optically pure (*R*)- or (*S*)-substrate is added. The hydrolytic activity is quantified by measurement of fluorescence. From the initial rate, apparent enantioselectivity (E_{app}) can be calculated. (See color insert.)

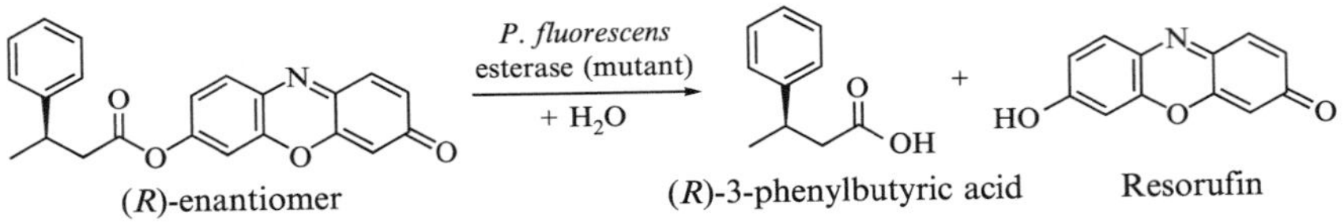

SCHEME 1. Chromophoric assay based on resorufin esters of chiral carboxylic acids.[17]

determination of enantiomeric excess by gas chromatography using a chiral column. Thus, enantioselectivity was enhanced from $E = 3.4$ (wild type) to $E = 5.2$ (Ala209Gly), 5.8 (Asp158Asn), and 6.6 (Leu181Gln).[17] Next, saturation mutagenesis at positions 158 and 181 led to a further increase in enantioselectivity to $E = 12$ (unpublished).

Assay Systems

As outlined earlier, a range of methods are available for the creation of mutant libraries. Most crucial for a successful directed evolution is the availability of suitable assay systems, which allow for a rapid, reliable, and fast identification of desired variants. In literature, a considerable range of HTS systems has been described and readers are referred elsewhere for further information.[18]

As mentioned earlier, a resorufin-based fluorescence assay was developed to identify mutants of PFE with enhanced enantioselectivity. Colonies obtained from epPCR were transferred from agar plates to microtiter plates, cultivated, and, by the addition of dimethyl sulfoxide (DMSO), master plates were prepared. After replica plating, esterase variants were produced via induction with L-rhamnose. An aliquot of each well was then split into two wells, diluted with buffer, and the assay solution consisting of optically pure (R)- and (S)-3-phenylbutyric acid resorufin ester was added. From the ratio of initial rates determined in the PFE-catalyzed hydrolysis of these esters by measurement of the fluorescence of released resorufin (excitation 544 nm, emission 590 nm), the apparent enantioselectivity E_{app} was calculated (Fig. 1 and Scheme 1).

Although this assay is very sensitive and allowed the identification of mutants with enhanced enantioselectivity with good agreement between E_{app} and E_{true} (determined by gas chromatography from the kinetic resolution of racemates), it also has some disadvantages when generalized for other substrates. First, it is only applicable to (chiral) carboxylic acids as substrates. Second, it might turn out that mutants are identified with high selectivity for the bulky resorufin ester, but reactions with the true substrate (i.e., a methyl ester) can provide a different result.

In order to overcome this problem and especially to allow a screening of enzymes with altered substrate specificity or enhanced enantioselectivity toward secondary alcohols, an alternative hydrolytic assay was developed (Scheme 2). This method also has the major advantage that no surrogate substrates are used, as acetates are the preferred esters for the resolution of secondary alcohols.

Hydrolysis of acetates using an esterase (or lipase) releases acetic acid. This is then "converted" stoichiometrically in an enzyme cascade reaction into NADH. The increase in NADH concentration is finally quantified spectrophotometrically at 410 nm[19] (Scheme 2). It turned out that this HTS is very reliable and fast, as the exact determination of activity and enantioselectivity is possible within minutes. In addition, the acetic acid kit is available commercially. Screening of mutant libraries for enantioselectivity can follow the same principle as described earlier, i.e., use of optically pure (R)- or (S)-acetates of a secondary alcohol in separate wells of a microtiter plate.

[18] D. Wahler and J.-L. Reymond, *Curr. Opin. Chem. Biol.* **12,** 535 (2001); M. T. Reetz, *Angew. Chem. Int. Ed.* **41,** 1335 (2002).

[19] M. Baumann, R. Stürmer, and U. T. Bornscheuer, *Angew. Chem. Int. Ed.* **40,** 4201 (2001).

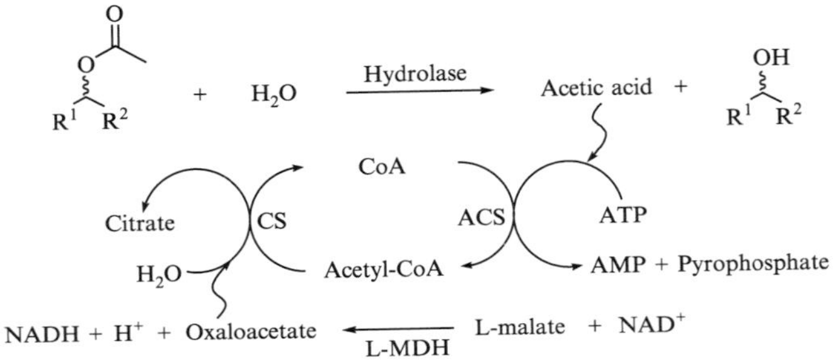

SCHEME 2. Assay based on the conversion of acetic acid released in the hydrolase-catalyzed reaction in a subsequent enzyme cascade yielding an increase in NADH.[19]

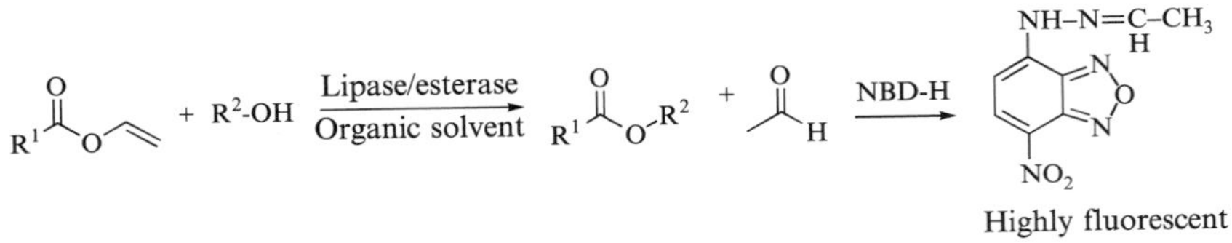

Highly fluorescent

SCHEME 3. Assay used to determine the synthetic activity of lipases and esterases by *in situ* derivatization of acetaldehyde with a hydrazine yielding a fluorescent hydrazone.[20]

As lipases and esterases are very active in organic solvents, many biocatalytic reactions are performed as transesterifications in the absence of water. This has the advantage that the reaction is shifted from hydrolysis to synthesis, conditions can be widely influenced by the choice of organic solvent, enantioselectivity is often altered, and chemists are more used to organic solvents rather than water. However, not all enzymes show satisfying performance in organic media, even if they are highly active in aqueous systems. Thus, a HTS system was developed that allows the direct determination of synthetic activity in a microtiter plate format.[20] The method is based on transesterification between an alcohol and a vinyl ester of a carboxylic acid. Acetaldehyde generated from the vinyl alcohol by keto-enol tautomerization is reacted with a hydrazine (NBD-H) to produce the corresponding hydrazone, which is then quantified by fluorimetric measurement (Scheme 3). As demonstrated for a range of lipases and esterases, this principle allows the rapid identification of active enzymes. Moreover, enzyme-catalyzed transesterification and *in situ* conversion into the highly

[20] M. Konarzycka-Bessler and U. T. Bornscheuer, *Angew. Chem. Int. Ed.* **42,** 1418 (2003).

fluorescent hydrazone can be performed in one well of a microtiter plate using a broad range of solvents (e.g., toluene, hexane, ether).

Materials and Methods

Creation of Error-Prone PCR Libraries

For this the wild-type gene of the esterase from *P. fluorescens* is amplified from pJOE2792.1[21] by a mutagenic PCR protocol essentially as described by Leung *et al.*[22] The plasmid, isolated using the QIAprep kit (Qiagen, Hilden, Germany), is used in the epPCR at a final concentration of 0.1 ng/μl. To the plasmid preparation are added 10 μl 10× *Taq* polymerase buffer, 1 μl 1 M β-mercaptoethanol, 10 μl DMSO, 10 μl 5 M MnCl$_2$, 1.5 μl of each appropriate primer, 10 μl 10 mM dGTP, 10 μl 2 mM dATP, 10 μl 10 mM dTTP, 10 μl 10 mM dCTP, 12 μl H$_2$O, and 100 μl mineral oil. DNA is denatured by 94° for 7 min, the primer is annealed at 50° for 1 min, and 2 μl 2 U/μl *Taq* DNA polymerase (New England Biolabs) are added. The chain elongation reaction proceeds at 70° and the next denaturation step is followed at 94°. Cycles of 94° for 1 min, 50° for 1 min, and 70° for 4 min are repeated 24 times. The PCR products are purified by gel extraction, digested with *Eco*RI and *Bam*HI to generate cohesive ends, recloned into plasmid pJOE2792.1, and transformed into *E. coli* DH5α.

To achieve a mutation frequency of about one substitution per 130 amino acids, the protocol can be modified as follows[11]: 100 μl end volume, 0.2 mM of each dNTP, 1.75 mM MgCl$_2$, 0.25 mM MnCl$_2$, 100 pmol of each primer, 0.3 μg template DNA, 2.5 U *Taq* polymerase, 10 mM Tris–HCl, pH 9.0, 50 mM KCl, and 0.1% Triton X-100), for 5 min at 94°, a subsequent 35 cycles of 1 min at 94°, 1.5 min at 48°, and 2.5 min at 72°, with a final step of 5 min at 72°.

Esterase Production in Microtiter Plates (Fig. 1)

Plasmids obtained from random mutagenesis are transformed in freshly prepared competent *E. coli* DH5α using a literature method,[23] and cells are plated on LB agar plates supplemented with ampicillin (LB-Amp). After growing overnight in an incubator at 37°, single colonies are picked into 96-well microtiter plates containing 150 μl LB-Amp media per well. These

[21] N. Krebsfänger, F. Zocher, J. Altenbuchner, and U. T. Bornscheuer, *Enzyme Microb. Technol.* **22,** 641 (1998).

[22] D. W. Leung, E. Chen, and D. V. Goeddel, *Technique* **1,** 11 (1989).

[23] C. T. Chung, S. L. Niemela, and R. H. Miller, *Proc. Natl. Acad. Sci. USA* **86,** 2172 (1989).

plates serve as master plates. After cell growth for 24 h at 37° and 220 rpm, the master plates are duplicated by transferring a 2-μl aliquot into a new microtiter plate (containing 200 μl LB-Amp media per well) used for the subsequent production of esterase (production plates). The master plates are supplemented with DMSO (final concentration 10%, v/v) and stored at −80°. The production plates are incubated for 3 h at 37° and 220 rpm and esterase production is then induced by the addition of 2 μl per well of a sterile L-rhamnose solution (20%, w/v). After cultivation for approximately 6 h at 37° and 220 rpm, cells are disrupted by three freezing and thawing cycles or by the addition of lysozyme before transferring them to the assay plate to determine enantioselectivity.

High-Throughput Assays

Resorufin Assay (Scheme 1). Production plates containing esterase variants produced as described earlier are diluted 1:10 with sodium phosphate buffer (pH 6.5, 50 mM) using a microtiter block with a capacity of 2.2 ml per well. Two hundred microliters of the diluted crude extract from one well is then split into two wells of a new microtiter plate (screening plate). Then 20 μl of either (R)- or (S)-3-phenylbutyric acid resorufin ester (0.1 mM dissolved in DMSO) is added, and the rate of hydrolysis is determined in parallel for both enantiomers by fluorescence measurement (excitation 544 nm; emission 590 nm). Each reaction is monitored for 30 min taking one data point every 90 s. Enantioselectivity is then calculated by comparing the rates of hydrolysis for both enantiomers. The resulting value is named apparent enantioselectivity (E$_{app}$) because true enantioselectivity (E$_{true}$) must be calculated in a competing experiment in which both enantiomers are present. Mutants that show improved enantioselectivity are picked and cultivated in 500 ml LB-Amp media. After the expression of esterase variants, cells are harvested by centrifugation, washed, and resuspended in sodium phosphate buffer (pH 6.5, 50 mM) before being disrupted by sonication. Cell debris is removed by centrifugation, and the crude extract is lyophilized to obtain the mutated enzyme in a stable form. Enantioselectivity is then verified by repeating the hydrolysis of optically pure resorufin esters as well as by preparative hydrolysis of the corresponding (R,S)-methyl ester.[17]

Acetic Acid Assay (Scheme 2). The test kit for the determination of acetic acid released (initially produced and distributed by Roche Diagnostics, Penzberg, Germany) is from R-Biopharm GmbH (Darmstadt, Germany) and applied according to the manufacturers protocol.

To a mixture of the test kit components (150 μl), PFE (20 μl, 2 mg ml^{-1}) is given. Reactions are started by adding a solution of 20 μl acetates of secondary alcohols (obtained by standard chemical synthesis) as sodium phosphate buffer (10 mM, pH 7.3), and the increase of NADH is monitored at 340 nm. Mixtures of the test kit with buffer or cell lysates of noninduced *E. coli* harboring the gene encoding recombinant PFE serve as controls. In a similar manner, reaction rates are determined using optically pure (*R*)- or (*S*)-acetates. For reactions with crude cell extract, PFE is produced in microtiter plates as described earlier.

Synthesis Assay (Scheme 3). A known amount of enzyme is dissolved in phosphate buffer (pH 7.5, 50 mM), transferred to a MTP, and freeze dried. A mixture of vinyl laurate (15 μmol), 1-propanol (150 μmol) (1:10 ratio), and NBD-H (50 nmol) in organic solvent (e.g., 1-propanol, *n*-hexane, isooctane, petrol ether, DMSO) is then added, the plates (total volume: 150 μl per well) are tape sealed and inserted into a fluorimeter (shaken at 45° at 170 rpm), and the increase of fluorescence is determined at 45°, excitation 485 nm, and emission 520 nm for 15 min. As controls, the same format is used without enzyme or NBD-H is incubated with enzyme or solvent only. To avoid nondesired hydrolysis by the presence of water, it is recommended to dry all reactants and solvent before the reaction over activated molecular sieves.

Conclusions

The examples shown in this article demonstrated that directed evolution presents a versatile method to improve biocatalysts. The most important key to success is the availability of suitable HTS systems, and the methods developed at our laboratory allow for the rapid and reliable identification of active lipase and esterase variants, but also with improved enantioselectivity toward chiral carboxylic acids and alcohols. Moreover, the determination of synthetic activity is also now feasible.

Acknowledgments

The authors are grateful to Degussa AG and BASF AG for financial support.

[19] Protein Engineering of the Cytochrome P450 Monooxygenase from *Bacillus megaterium*

By Vlada B. Urlacher and Rolf D. Schmid

Introduction

The role and importance of cytochrome P450 enzymes (CYP) in drug development, biodegradation processes, and biocatalysis have been widely acknowledged. P450 monooxygenases exhibit an extremely wide substrate spectrum, which is the basis of their ability to activate or detoxify a large variety of target molecules. P450 monooxygenases have been isolated from bacteria, yeasts, and insects, as well as from mammalian and plant tissues. Currently, the enzyme family is one of the best known gene subfamilies with over 3000 characterized members (http://drnelson.utmem.edu/CytochromeP450.html).

Many studies have been dedicated to structural models of cytochrome P450 in order to improve our understanding of the mechanistic details of the enzymes,[1,2] their substrate specificity, and their pronounced stereo- and regiospecificity.[3] In addition, homology modeling of mammalian P450s[4–6] and quantitative structure activity relationship (QSAR) analyses using chemicals, which are metabolized by P450s,[7] have added considerably to our understanding of the metabolic variations and functions of the enzyme.

Cytochrome P450 enzymes are of considerable interest to pharmaceutical and chemical industry and have thus become targets for protein engineering approaches. Protein engineering is generally defined as the modification of an enzyme by site-directed or random mutagenesis with the aim of altering its properties. Site-directed mutagenesis requires a solid structural basis and profound knowledge of the catalytic mechanism of the enzyme, which was provided by determining the structures of CYPs using

---

[1] I. Schlichting, J. Berendzen, K. Chu, A. M. Stock, S. A. Maves, D. E. Benson, R. M. Sweet, D. Ringe, G. A. Petsko, and S. G. Sligar, *Science* **287**, 1615 (2000).

[2] M. Hata, Y. Hirano, T. Hoshino, and M. Tsuda, *J. Am. Chem. Soc.* **123**, 6410 (2001).

[3] D. A. Rock, B. N. Perkins, J. Wahlstrom, and J. P. Jones, *Arch. Biochem. Biophys.* **416**, 9 (2003).

[4] R. Dai, M. R. Pincus, and F. K. Friedman, *J. Protein Chem.* **17**, 121 (1998).

[5] N. V. Belkina, M. Lisurek, A. S. Ivanov, and R. Bernhardt, *J. Inorg. Biochem.* **87**, 197 (2001).

[6] D. F. Lewis, *Xenobiotica* **32**, 305 (2002).

[7] D. F. Lewis and M. Dickins, *J. Enzyme Inhib.* **16**, 321 (2001).

X-ray crystallography at high resolution. Ten of the 12[8-17] crystallized cytochrome P450s are of prokaryotic origin and water soluble. From a technical point of view, microbial P450s are easier to handle than P450 enzymes from plants and animals. They are not membrane associated and exhibit a relatively high stability. Eukaryotic cytochrome P450 enzymes are membrane-associated proteins and are hence more difficult to crystallize. Currently, only the X-ray structures of two membrane-bound mammalian P450s, rabbit CYP2C5[18] and human CYP2C9,[19] are known. Models of other mammalian P450s were built based on the structure of CYP2C5[20] and its bacterial analogues.[21]

P450cam, the cytochrome P450 monooxygenase from *Pseudomonas putida*, is the best characterized microbial P450 enzyme.[1,22-25] A large number of other soluble prokaryotic P450 enzymes have been identified, isolated, subcloned in *Escherichia coli*, overexpressed, and characterized since the late 1980s.[10-17]

[8] T. L. Poulos, B. C. Finzel, and A. J. Howard, *Biochemistry* **25**, 5314 (1986).

[9] K. G. Ravichandran, S. S. Boddupalli, C. A. Hasermann, J. A. Peterson, and J. Deisenhofer, *Science* **261**, 731 (1993).

[10] C. A. Hasemann, K. G. Ravichandran, J. A. Peterson, and J. Deisenhofer, *J. Mol. Biol.* **236**, 1169 (1994).

[11] K. Nakahara, H. Shoun, S. Adachi, T. Iizuka, and Y. Shiro, *J. Mol. Biol.* **239**, 158 (1994).

[12] J. R. Cupp-Vickery and T. L. Poulos, *Nature Struct. Biol.* **2**, 144 (1995).

[13] S. Y. Park, K. Yamane, S. Adachi, Y. Shiro, K. E. Weiss, and S. G. Sligar, *Acta Crystallogr. D Biol. Crystallogr.* **56**, 1173 (2000).

[14] L. M. Podust, Y. Kim, M. Arase, B. A. Neely, B. J. Beck, H. Bach, D. H. Sherman, D. C. Lamb, S. L. Kelly, and M. R. Waterman, *J. Biol. Chem.* **278**, 12214 (2003).

[15] L. M. Podust, T. L. Poulos, and M. R. Waterman, *Proc. Natl. Acad. Sci. USA* **98**, 3068 (2001).

[16] J. K. Yano, F. Blasco, H. Li, R. D. Schmid, A. Henne, and T. L. Poulos, *J. Biol. Chem.* **278**, 608 (2003).

[17] S. Nagano, H. Li, H. Shimizu, C. Nishida, H. Ogura, P. R. Ortiz De Montellano, and T. L. Poulos, *J. Biol. Chem.* **278**, 44886 (2003).

[18] P. A. Williams, J. Cosme, V. Sridhar, E. F. Johnson, and D. E. Mcree, *J. Inorg. Biochem.* **81**, 183 (2000).

[19] P. A. Williams, J. Cosme, V. Sridhar, E. F. Johnson, and D. E. Mcree, *Mol. Cell* **5**, 121 (2000).

[20] D. F. Lewis, C. Sams, and G. D. Loizou, *J. Biochem. Mol. Toxicol.* **17**, 47 (2003).

[21] G. D. Szklarz and J. R. Halpert, *Life Sci.* **61**, 2507 (1997).

[22] R. Davydov, T. M. Makris, V. Kofman, D. E. Werst, S. G. Sligar, and B. M. Hoffman, *J. Am. Chem. Soc.* **123**, 1403 (2001).

[23] C. F. Harford-Cross, A. B. Carmichael, F. K. Allan, P. A. England, D. A. Rouch, and L.-L. Wong, *Protein Eng.* **13**, 121 (2000).

[24] K. J. French, D. A. Rock, J. I. Manchester, B. M. Goldstein, and J. P. Jones, *Arch. Biochem. Biophys.* **398**, 188 (2002).

[25] O. Sibbesen, J. J. De Voss, and P. R. Montellano, *J. Biol. Chem.* **271**, 22462 (1996).

Cytochrome P450 BM-3 from *Bacillus megaterium* is catalytically self-sufficient. It contains a P450 heme domain of 54 kDa and an FAD/FMN reductase domain of 64 kDa on a single polypeptide chain.[26] The enzyme catalyzes the subterminal oxidation of saturated and unsaturated fatty acids with a chain length of 12 to 20 carbons. High-resolution X-ray crystal structures are available for substrate-free,[9] palmitic acid-bound,[27] and *N*-palmitoylglycine-bound[28] wild-type and mutant[29] P450 BM-3 enzymes. The structure resolved by nuclear magnetic resonance (NMR) is also available.[30] The well-known structure, the availability of the CYP102A1 gene that encodes the protein, and the possibility of expressing the protein in *E. coli* have encouraged a number of research groups to undertake site-directed mutagenesis studies in order to identify key amino acids. Insights into the mechanisms of P450 BM-3 have been gained and transferred to eukaryotic P450 enzymes, because P450 BM-3 is an excellent model for addressing questions on the wide substrate specificity of P450s in general. Techniques involving the mutagenesis of P450 BM-3 have led to a variety of biocatalysts with features of industrial interests. This article summarizes the recent research on this particular P450 enzyme.

Isolation and Expression of P450 BM-3

The CYP102A1 gene from *B. megaterium*, coding for a heme domain and an FAD/FMN reductase, was cloned using Southern blot technique.[31] Initially, protein expression was performed in *E. coli* cells harboring the recombinant plasmid (pUC13 derivative) under the control of the original *B. megaterium* promoter.[31] A 5-kb DNA fragment containing the CYP102A1 gene was sequenced. One open reading frame was found whose sequence could be matched to the amino acid sequences of P450 and reductase domains generated by trypsin digestion.[32]

As the DNA sequence of CYP102A1 is known, it can be usually isolated directly from genomic DNA of *B. megaterium* by a polymerase chain reaction (PCR) using primers that introduce appropriate restriction

[26] H. Y. Li, K. Darwish, and T. L. Poulos, *J. Biol. Chem.* **266,** 11909 (1991).
[27] H. Li and T. L. Poulos, *Nature Struct. Biol.* **4,** 140 (1997).
[28] D. C. Haines, D. R. Tomchick, M. Machius, and J. A. Peterson, *Biochemistry* **40,** 13456 (2001).
[29] H. Yeom, S. G. Sligar, H. Li, T. L. Poulos, and A. J. Fulco, *Biochemistry* **34,** 14733 (1995).
[30] C. F. Oliver, S. Modi, M. J. Sutcliffe, W. U. Primrose, L. Y. Lian, and G. C. Roberts, *Biochemistry* **36,** 1567 (1997).
[31] L. P. Wen and A. J. Fulco, *J. Biol. Chem.* **262,** 6676 (1987).
[32] R. T. Ruettinger, L. P. Wen, and A. J. Fulco, *J. Biol. Chem.* **264,** 10987 (1989).

sites upstream of the ATG (methionine) and downstream of the stop codon.

The overexpression of heterologous proteins in *E. coli* requires a strong promoter and an efficient ribosome-binding site at an optimal distance from the first methionine codon. CYP102A1 has been subcloned into pUC derivatives, utilizing the *lac* promoter.[31] *In vivo* promoter activity is controlled efficiently by the *lac* repressor protein, which is encoded by the *lacI* gene located inside the *lac* operon. Transcription can be induced by isopropyl thio-β-D-galactosidase (IPTG). Many mammalian cytochrome P450 enzymes could be expressed in *E. coli* using the pCWORI(+) vector (a derivative of plasmid pHSe5).[33–35] Therefore, the use of this vector was also tested with P450 BM-3.[36] The transcription/translation region of pCWORI(+) contains a *lacUV5* promoter and two copies of a *tac* promoter. The pGLW11 vector (a derivative of pK223), utilizing a *tac* promoter, was also applied successfully for the expression of P450 BM-3 in *E. coli*.[37]

In our laboratory, two *E. coli* expression vectors were used for the recombinant expression of the P450 enzyme from *B. megaterium*. Standard methods for the manipulation of DNA were employed.[38] The CYP102A1 gene was isolated by PCR using genomic DNA of *B. megaterium* (DSM 32T) as a template and primers introducing a *Bam*HI site at the N terminus and an *Eco*RI site at the C terminus of the gene. This fragment was used for the subsequent subcloning into the pCYTEXP1 vector, resulting in the plasmid pT-USC1BM3.[39] The pCYTEXP1 vector contains the bacteriophage tandem promoters P_R and P_L, which are preceded by the *clts857* repressor gene, and the transcription terminator from the *fd* bacteriophage. Plasmid DNA was transformed using $CaCl_2$-treated *E. coli* DH5α cells.[38] Expression was induced by a temperature shift from 37 to 42° for 5 h, yielding 300–350 nmol of CO-reactive P450 per liter of cell culture. When σ-aminolevulinic acid (1 m*M* final concentration) was added to the growing cells after induction, a P450 yield of up to 500 nmol per liter broth was

[33] H. J. Barnes, *Methods Enzymol.* **272**, 3 (1990).

[34] C. W. Fisher, D. L. Caudle, C. Martin-Wixtrom, L. C. Quattrochi, R. H. Tukey, M. R. Waterman, and R. W. Estabrook, *FASEB J.* **6**, 759 (1992).

[35] C. W. Fisher, M. S. Shet, and R. W. Estabrook, *Methods Enzymol.* **272**, 15 (1990).

[36] E. T. Farinas, U. Schwaneberg, A. Glieder, and F. H. Arnold, *Adv. Synth. Catal.* **343**(6 + 7), 601 (2001).

[37] A. B. Carmichael and L. L. Wong, *Eur. J. Biochem.* **268**, 3117 (2001).

[38] J. Sambook and D. W. Russell, "Molecular Cloning." Cold Spring Harbor Laboratory Press, Cold Spring Harbor, NY, 2001.

[39] U. Schwaneberg, C. Schmidt-Dannert, J. Schmitt, and R. D. Schmid, *Anal. Biochem.* **269**, 359 (1999).

reached. P450 BM-3 mutants were expressed in the same manner as a wild-type enzyme.

P450 BM-3 and its mutants could also be expressed fused to a His6 tag in a pET system. The pET system has a strong bacteriophage T7 promoter and is one of the most popular expression systems. Under the control of a T7 promoter the target gene can only be transcribed by the phage T7-DNA-polymerase, which is only present in specially engineered *E. coli* strains such as BL21 (DE3) or HMS174 (DE3).

The gene encoding P450 BM-3 was amplified from pT-USC1BM-3[39] by PCR using primers designed specifically to facilitate the cloning of the gene into pET28a(+) vector between *Bam*HI and *Eco*RI restriction sites.[40] The initial cloning was performed in DH5α *E. coli* cells, which have a high transformation efficiency and give an excellent plasmid yield. Subsequently, the gene was expressed in strain BL21 (DE3).

For the high-level expression of active P450, the reaction was induced by the addition of IPTG to a final concentration of 0.5 mM, and the cells were grown at 30° and 120 rpm. After 4–5 h the cells were harvested by centrifugation (20 min, 6000 rpm, 4°). After cell disruption and centrifugation, 400–500 nmol of cytochrome oxidase-reactive P450 per liter of cell culture was obtained even without the addition of σ-aminolevulinic acid prior to induction.

The advantage of this expression system is the high expression level of heterelogous protein. Upon optimization of the expression protocol (lower incubation temperature: 25–30°, or lower concentration of the inductor IPTG), the number of inclusion bodies, which are usually observed for mammalian P450s in a pET system, can thus be reduced drastically. One should mention that stirring speed influences the expression level of active P450 BM-3 and hence is a parameter that must not be neglected.

Roles of Key Amino Acids of P450 BM-3

Several residues in the P450 BM-3 are believed to be important for the catalytic reaction of the enzyme. The role of these residues and their manifold effects have been examined by structural analysis and molecular modeling followed by site-directed mutagenesis and functional characterization. The information gained adds to our understanding of the regio-, stereo-, and chemoselectivity of the oxidation, the water accesses to active site, electron transport, and redox state of the enzyme.

---

[40] S. C. Maurer, H. Schulze, R. D. Schmid, and V. Urlacher, *Adv. Synth. Catal.* **345**, 802 (2003).

The active site of P450 BM-3 consists of a long, hydrophobic channel, extending from the heme to the protein surface.[27] Comparison of substrate-free[9] and substrate-bound[27] crystal structures and site-directed mutagenesis studies points to the important role of the amino acid residues R47 and Y51 located at the entrance of the active center. These two positions interact with the carboxylate of the fatty acid and are thus crucial for the proper positioning of the substrates. Mutagenesis experiments at residue R47 confirmed the important contribution the guanidinium group of arginine on enzyme activity.[41–45] Although R47E, R47A,[41,46,47] and R47G[47] mutants retained their activity toward C12–C16 fatty acids, their k_{cat}/K_m values are 5- to 15-fold lower than those of the wild-type enzyme. The combination of R47L/Y51F increases the oxidation activity toward phenanthrene, fluoranthene, and pyrene up to 40-fold.[37] However, the substitution at position 51 has less impact on enzyme activity than that at position 47.[47]

The phenylalanine residue at position 87 is highly conserved. It is located in the active site of the protein and is very important for the correct orientation of the fatty acid hydrocarbon chain.[30] Comparison of substrate-free and substrate-bound crystal structures of P450 BM-3 revealed a substantial conformational difference that is caused by the phenyl ring of phenylalanine.[9,27] Mutations of P450 BM-3 at position 87 can affect its activity and stereo- or regioselectivity.[48] An unfavorable substitution at position F87 can lead to irreversible conformational changes during catalytic turnover, which will then result in a decrease or loss of catalytic competence.[47] In some cases, the combination of a mutation at position 87 with other mutations, in and outside the active site, has revealed a surprisingly strong effect on substrate selectivity[37,49,50] and peroxygenase activity of the enzyme.[36]

[41] T. W. Ost, C. S. Miles, J. Murdoch, Y. Cheung, G. A. Reid, S. K. Chapman, and A. W. Munro, *FEBS Lett.* **486**, 173 (2000).

[42] T. W. Ost, C. S. Miles, A. W. Munro, J. Murdoch, G. A. Reid, and S. K. Chapman, *Biochemistry* **40**, 13421 (2001).

[43] T. A. Kunkel, *Proc. Natl. Acad. Sci. USA* **82**, 488 (1985).

[44] Q. S. Li, U. Schwaneberg, M. Fischer, J. Schmitt, J. Pleiss, S. Lutz-Wahl, and R. D. Schmid, *Biochim. Biophys. Acta* **1545**, 114 (2001).

[45] C. F. Oliver, S. Modi, W. U. Primrose, L. Y. Lian, and G. C. Roberts, *Biochem. J.* **327**(Pt 2), 537 (1997).

[46] L. A. Cowart, J. R. Falck, and J. H. Capdevila, *Arch. Biochem. Biophys.* **387**, 117 (2001).

[47] M. A. Noble, C. S. Miles, S. K. Chapman, D. A. Lysek, A. C. Mackay, G. A. Reid, R. P. Hanzlik, and A. W. Munro, *Biochem. J.* **339**, 371 (1999).

[48] S. Graham-Lorence, G. Truan, J. A. Peterson, J. R. Falck, S. Wei, C. Helvig, and J. H. Capdevila, *J. Biol. Chem.* **272**, 1127 (1997).

[49] Q. S. Li, J. Ogawa, R. D. Schmid, and S. Shimizu, *Appl. Environ. Microbiol.* **67**, 5735 (2001).

It is assumed that the threonine residue at position 268, located in the distal I helix, plays an important role in oxygen binding and activation. Enzymatic properties and the crystal structure of the heme domain of the W268A mutant were determined using sodium laurate as a substrate. The mutant exhibited slower rates of NADPH and oxygen consumption and much higher uncoupling rates of electron transfer and substrate hydroxylation.[29]

The crystal structure of a complex between P450 BM-3 and *N*-palmitoylglycine at a resolution of 1.65 Å and site-directed mutagenesis revealed features of the active site that had not been determined previously.[28] Binding of the substrate leads to a conformational change, resulting in a shift of the A264 carbonyl away from the heme iron. The pivotal water molecule in the active site is thus brought in close vicinity of the heme group and can fill the sixth coordination site of the heme iron.

Interactions of Monooxygenase and Reductase Domains and Electron Transport

Protein–protein interactions are of great importance in the cytochrome P450 system. They provide essential electron functions and control the rate-limiting electron transfer.

Potential roles of specific residues in the heme domain of P450 BM-3, as well as in the reductase domain, relate to specific affinity interactions, electrostatic charge, and the direct electronic coupling of redox centers. Earlier investigations proposed a similar docking of redox partners (ferredoxin or flavin reductase) in the prokaryotic and eukaryotic P450 systems at a proximal part of the heme. The involvement of both electrostatic and hydrophobic protein–protein interactions has been demonstrated.[9,10,12,51] However, no clear evidence exists for the direct influence of certain amino acid residues on the electron transfer between P450 and its redox partner.

The reductase domain of P450 BM-3 is very similar to the microsomal NADPH-dependent cytochrome P450 reductase. However, there is a significant difference in the reduction mechanism of the heme iron.[52] In the mammalian enzyme, the fully reduced flavin ($FMNH_2$) is the electron donor to the heme iron, whereas in the case of the P450 BM-3 reductase

[50] A. Glieder, E. T. Farinas, and F. H. Arnold, *Nature Biotechnol.* **20**, 1135 (2002).
[51] A. Muller, J. J. Muller, Y. A. Muller, H. Uhlmann, R. Bernhardt, and U. Heinemann, *Structure* **6**, 269 (1998).
[52] I. F. Sevrioukova and J. A. Peterson, *Biochimie* **77**, 562 (1995).

domain the 1-electron donor is reduced semiquinone (FMNH). Based on analysis of sequence similarities between the P450 BM-3 reductase and those flavoproteins whose three-dimensional structures were already known, binding sites for FMN, FAD, and NADPH were suggested. The crystal structure of the complex between the heme and the FMN-binding domain (2.3 Å)[53] shows that the flavin domain is located at the proximal end of the heme domain. The region between P382-Q387 is assumed to be involved in transferring electrons from the FMN to the heme iron (electron transfer pathway through the polypeptide chain). Site-directed mutagenesis was applied to validate this observation. Cysteine residues were introduced at positions 104 and 387, which were expected to be responsible for the interactions between the two domains. Because position 372 is located on the other side of the heme domain and so cannot be involved in electron transfer, glutamate at this position was substituted through cysteine and served as a control. The cysteine residues were subsequently modified with a bulky sulfhydryl reagent (DC modification) in order to prevent close protein interactions between the FMN domain and the heme domain. In addition, the cysteine residue at position 156 was substituted by alanine.

Procedure

Mutations E372C and Q387C were introduced in a first PCR with modified 5'-primers, which carried the desired mutations. An oligonucleotide corresponding to the carboxyl terminus of the heme domain was used as a 3' primer. The mutations L104C and C156S were introduced accordingly, but the 3' primers were modified. In a second PCR, 0.2- to 0.5-kb fragments served as 5' or 3' primers. The 1.4-kb fragments were cloned into the pNEB vector. To create double and triple mutants, the corresponding regions of the gene were excised with endonucleases and exchanged.[53]

Spectral analyses, laser flash photolysis experiments, and DC modification of the mutants have shown that C387 has no significant effect on the electron transport from the FMN to the heme iron. In contrast, the C104 residue and surrounding area are most critical for the docking of the redox partner.

The tryptophan residue at position 96 in the heme domain was also assumed to have an effect on the electron transfer from the FMN to the heme.[54] Substitutions of this tryptophan residue by alanine, phenylalanine, or tyrosine caused a lower heme content of P450 BM-3 while the levels of

[53] I. F. Sevrioukova, J. T. Hazzard, G. Tollin, and T. L. Poulos, *J. Biol. Chem.* **274**, 36097.
[54] A. W. Munro, K. Malarkey, J. Mcknight, A. J. Thomson, S. M. Kelly, N. C. Price, J. G. Lindsay, J. R. Coggins, and J. S. Miles, *Biochem. J.* **303**(Pt 2), 423 (1994).

catalytic activity remained unaltered. This indicates that W96 plays a role in the association of the prosthetic group of the heme and is probably required for an efficient redox interaction between heme and flavin domains.

Other residues such as W574, W536, and G570 of the reductase domain were ruled out as being crucial for FNM binding. Only the tryptophane at position 574 has a minor effect on the electron transport.[55]

Redox Potential of the P450 Heme Iron

The analysis of the structure of P450 BM-3 and the characterization of mutants demonstrated the important role of phenylalanine at position 393 in controlling the reduction potential of the P450 heme iron.[42] This position is one of very few highly conserved amino acid residues of P450 enzymes. Changing this residue obviously affects the catalytic properties of the enzyme. Unfortunately, no details are available that would explain this effect. F393Y, F393A, and F393H mutants were expressed in *E. coli*. No effect of F393 on stabilizing the heme was observed. Analysis of the oxidation products of myristic acid generated by the F393H mutant revealed the same product proportions as for the wild-type enzyme. Nevertheless, a lower turnover rate of the mutant was observed. The crystal structure of the mutant and spectroscopic analysis suggested the thermodynamic control of F393 over the heme iron: this position seems to establish the equilibrium between the rate of heme reduction and the rate at which the ferrous heme can bind and, subsequently, reduce molecular oxygen.[56]

Artificial P450 Systems Using P450 BM-3 Domains

As a natural fusion protein, P450 BM-3 is the catalytically most efficient P450 enzyme currently known. P450 BM-3 does not require additional expression and purification of the redox partners. Both P450 BM-3 domains were used to create artificial fusion proteins. The aim of such projects was the creation of highly active and soluble P450 enzymes that are applicable on an industrial scale.

Nonphysiological partners such as flavodoxin from *Desulfovibrio vulgaris* and the reductase domain of P450 BM-3 have been used for

[55] M. L. Klein and A. J. Fulco, *J. Biol. Chem.* **268,** 7553 (1993).
[56] T. W. Ost, A. W. Munro, C. G. Mowat, P. R. Taylor, A. Pesseguiero, A. J. Fulco, A. K. Cho, M. A. Cheesman, M. D. Walkinshaw, and S. K. Chapman, *Biochemistry* **40,** 13430 (2001).

constructing such fusion proteins.[57] The artificial combination of *D. vulgaris* flavodoxin and the heme domain of P450 BM-3 was made possible through the PCR-based introduction of an additional *Nla*III restriction site at the 3′ terminus of the P450 BM-3 heme domain and at the 5′ end of the flavodoxin gene. The artificial genes were expressed in *E. coli*. Electrochemical experiments demonstrated the feasibility of this assembled protein.[57]

The solubility of microsomal P450 enzymes has also been addressed by fusing the human P450 2E1 gene with the P450 BM-3 reductase gene. The membrane-bound N terminus of the human enzyme was replaced by that of the P450 BM-3 heme domain.[57] The P450 2E1 reductase of the artificial BM-3 protein was constructed in two steps. The first 80 N-terminal residues of P450 2E1 were replaced by the first 54 amino acid residues of the heme domain of P450 BM-3. The modified P450 2E1 was then fused with the reductase domain. After cloning into the pT7 vector, the fusion protein was expressed in a soluble form in BL21 (DE3) CL *E. coli* cells using the standard protocol and purified on DEAE-Sepharose. After ultracentrifugation of the cell lysate at 100,000*g*, the protein still remained in the soluble fraction.

Changing the Properties of P450 BM-3 by Mutagenesis

"Rational Evolution"

A combination of computer-assisted protein modeling with methods of directed evolution allows one to improve significantly the efficiency of the search for the enzyme variant with new properties. In a procedure termed "rational evolution" (protein design combined with directed evolution), substrate specificity was shifted from fatty acids with 12 carbons to those with 10 and 8 carbons. In these experiments, the fatty acid pseudosubstrates 10- and 8-*p*-nitrophenoxycarboxylic acid were used.[44]

Substrate docking was examined on the basis of the crystallographically determined structure of the palmitate-bound P450 BM-3, which was obtained from the Protein Data Bank. A model of 8-*p*NCA was used as a substrate molecule. The chemical structure of 8-*p*NCA and the mutations in the binding pocket were deduced using the biopolymer tool of SYBYL. The C1–C4 atoms of the substrate were placed in analogy of the C6–C9 atoms of palmitic acid. The C7 and C8 atoms of 8-*p*NCA were placed at a distance of 4 and 3.6 Å from the heme iron. This distance was determined previously by NMR for P450 BM-3-laurate and 12-bromolaurate complexes. The

[57] G. Gilardi, Y. T. Meharenna, G. E. Tsotsou, S. J. Sadeghi, M. Fairhead, and S. Giannini, *Biosens. Bioelectron.* **17**, 133 (2002).

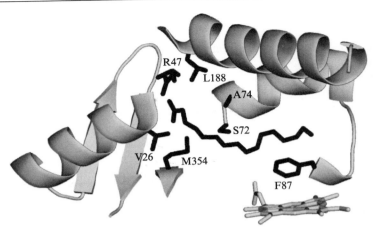

FIG. 1. P450 BM-3–palmitate complex model showing the mutations that can affect substrate specificity.

p-nitrophenoxy group was placed manually into the binding pocket. Eight sites of the binding pocket were selected for further saturation mutagenesis: V26, R47, S72, A74, F87, L188, and M354 (Fig. 1).

Saturation Mutagenesis and Recombination of Mutations

Saturation mutagenesis describes the substitution or insertion of codons encoding all possible amino acids at any predetermined position in a gene. PCR was utilized to produced the first library of P450 BM-3 mutants rapidly. Mutagenic PCR was performed using a slightly modified protocol from the Stratagene Quik-Change kit, which allows the introduction of the nucleotide exchange, which is necessary to obtain all possible codons.

Mutant F87A was used as template DNA. After saturation mutagenesis, subsequent ligation into the pCYTEXP1 vector, and protein expression in *E. coli* strain DH5α, the mutant library was tested with different *p*NCA substrates with a chain length of fewer than 15 carbons. Compared to standard assays, which are based on the consumption of NADPH or oxygen, the *p*NCA assay is substrate specific and allows detection of the reaction product *p*-nitrophenolate. ω-Hydroxylation of *p*NCA by P450 BM-3 first leads to formation of an instable hemiacetal intermediate, which then dissociates into ω-oxocarboxylic acid and yellow *p*-nitrophenolate. A pH of 8.2 is preferable, as 92% of all *p*-nitrophenolate is deprotonated at this pH and thus exhibits a yellow color. In addition, P450 BM-3 remains active. The spectral absorption of *p*-nitrophenolate was measured at

410 nm, with the extinction coefficient being $\varepsilon = 13,200 \ M^{-1} \ cm^{-1}$. As *p*NCA is not dissolved easily in water, small amounts (1%) of organic solvent [dimethyl sulfoxide (DMSO) or acetone] should be added to the reaction mixture.[44]

Synthesis of pNCA. Synthesis of *p*NCA is a multiple-step process. In the first three steps, the corresponding ω-bromocarboxylic acids are esterified. This is followed by an *sn2* reaction with sodium *p*-nitrophenolate. The last step, the hydrolysis of the esters, can be done chemically or enzymatically using a lipase.[39]

p-*NCA Assay in 1-ml Scale.* Eight microliters of 6 nmol 10- or 8-*p*NCA or 12-*p*NCA, dissolved in DMSO (1%), is added to 892 μl Tris–HCl buffer (100 m*M*, pH 8.2) in a 1-ml cuvette. The mixture is incubated for 5 min at room temperature before the reaction starts with the addition of 100 μl of an aqueous solution of 1 m*M* NADPH. The amount of *p*-nitrophenolate formed in the reaction is determined photometrically at 410 nm.[39]

p-*NCA Assay in Microtiter Plates.* Two hundred microliters containing 18 nmol of 10- and 11-*p*NCA, 12 nmol 12-*p*NCA, or 10 nmol of 15-*p*NCA dissolved in 2 μl DMSO (acetone) is placed into a 96-well microtiter plate. The reaction is started by adding 20 μl of 1 m*M* NADPH solution to each well. The procedures are performed by an automated workstation. The amount of *p*-nitrophenolate formed in each well is determined with a microtiter plate reader.[39]

Mutants with the highest activity toward 8-*p*NCA and 10-*p*NCA were cultivated again and their specific activities are calculated in terms of P450 concentrations. The nucleotide sequences of the mutants were analyzed in order to determine the respective mutation. Complete structural and kinetic characterizations of the expressed and purified enzymes provide important information for the subsequent recombination experiments. As a result, the mutations V26T, R47F, S72G, A74G, F87A, L188K, and M354T were selected. Recombination of the corresponding positive mutations was carried out by subsequent site-directed mutagenesis. All mutant enzymes had a full complement of flavin and heme, which indicated that the mutations did not disrupt or change the tertiary structure. After each mutagenic step, a structural model of a new mutant was designed according to which generalized effect on substrate selectivity was studied. F87A is regarded as a key substitution that affects the substrate specificity of P450 BM-3 by altering the contact between substrate and heme irons. The combination of L188K and F87A results in the formation of a new carboxylate-binding site with increased activity toward 12- and 10-*p*NCA. The A74G substitution obviously plays an important role in altering the chain-length specificity of P450 BM-3. Residue A74 is located at the N terminus of the α-helix B, and the side chain of alanine interacts sterically with the neighboring residues. If alanine is replaced by glycine at this particular position and phenylalanine is substituted with alanine at position 87, this increases the size of the binding site and so enables the hydroxylation of shorter chain fatty acids with the bulky p-nitrophenoxy group. It is well known that R47 is important for an efficient catalysis, as it builds a carboxylate-salt bridge with a fatty acid and contributes to its binding. The addition of R47F to the triple mutant increased the k_{cat} for 10-pNCA and 8-pNCA. With this mutant we sought to understand whether preventing the original carboxylate binding and thus an increased hydrophobicity at the entrance of the binding pocket enables the diffusion of shorter chain substrates further into the active site. It was noted that the introduction of M354T, V26T, or S72G mutations to (F87A)LAR decreased k_{cat} values for shorter chain acids. Active mutants also showed activity toward the free C-8 and C-10 fatty acids[58] and also toward the unnatural substrates naphthalene, n-octane, and 8-methylquinoline.[59]

Activity Against Indole

The rational evolution of P450 BM-3 toward a catalyst for the hydroxylation of shorter chain fatty acids also revealed some colonies that produced a blue pigment. High-performance, thin-layer chromatography detected a rapidly moving blue and slower moving red component. Mass spectrometry and NMR analysis confirmed the presence of indigo and indirubin in the mixture.[60]

Sequence analysis revealed that all mutants producing this pigment contained mutations at either one or several of the three positions A74, F87, or L188. A strategy for the subsequent site-specific randomization of each site, starting with the best mutant from the previous mutagenesis step, was developed. The best mutant, which had a 10-fold higher activity toward indole than the wild type, was the A74G, F87V, L188Q mutant.

Activity Assay for Indole Hydroxylation. The mixture of indole solution in DMSO and enzyme in the appropriate amount of 0.1 M Tris–HCl buffer in a final volume of 900 μl is preincubated for 10 min at room temperature. The reaction is started by the addition of 50 μl of 1 mM NADPH solution. The reaction is stopped after 2 min by adding 1.2 mM

[58] O. Lentz, Q.-S. Li, U. Schwaneberg, S. Lutz-Wahl, P. Fischer, and R. D. Schmid, *J. Mol. Catal. B Enzym.* **15**, 123 (2001).

[59] D. Appel, S. Lutz-Wahl, P. Fischer, U. Schwaneberg, and R. D. Schmid, *J. Biotechnol.* **88**, 167 (2001).

[60] Q.-S. Li, U. Schwaneberg, P. Fischer, and R. D. Schmid, *Chem. Eur. J.* **6**, 1481 (2000).

KOH. Indigo formation is determined spectrophotometrically at 670 nm (extinction coefficient $\varepsilon = 3.9$ M^{-1} cm^{-1}). The formation of indirubin has only a weak effect on the absorption at 670 nm.[60]

Activity Against Polycyclic Aromatic Hydrocarbons

Several mutants of P450 BM-3 with activity against several polycyclic aromatic hydrocarbons were designed. Two hydrophobic substitutions—R47L and Y51F—increased the activity of this enzyme against phenanthrene, fluoranthene, and pyrene up to 40-fold. After combination of these mutations with mutation A264G, PAH oxidation increased another 5-fold, with simultaneous enhancement of NADPH oxidation and coupling efficiency between NADPH oxidation and substrate hydroxylation. All of the aforementioned residues are located in the substrate-binding pocket of the enzyme. However, substitution of M354 and L437, also located within the active site, reduced PAH oxidation activity.[37]

Directed Evolution

Directed evolution involves either random mutagenesis (e.g., error-prone PCR) or recombination of gene fragments (e.g., gene shuffling, staggered extension process). It has proved very efficient in improving the enzymatic activity[36,61] and stability of P450 BM-3.[62] Directed evolution does not rely on the structural information about a certain protein. Instead, it can lead to mutants that help us understand the relationship between structure and function.

Error-prone PCR has been applied to alter the features of P450 BM-3. Error-prone PCR procedures are generally modifications of a standard PCR protocol, which are aimed at increasing the natural error rate of the polymerase. A standard error-prone reaction mixture contains a higher concentration of $MgCl_2$ (6–7 mM) compared to basic PCR (1.5 mM), which is required for the stabilization of noncomplementary pairs. To increase the error rates, either $MnCl_2$ (0–0.2 mM) or unbalanced amounts of dNTPs can be added to the reaction mixture. *Taq* polymerase, which is typically applied for error-prone PCR, has a high error rate of its own. However, several newly constructed polymerases allow increasing the number of variations in the mutants.[63]

[61] P. C. Cirino and F. H. Arnold, *Angew. Chem. Int. Ed. Engl.* **42**, 3299 (2003).
[62] O. Salazar, P. C. Cirino, and F. H. Arnold, *Chembiochem.* **4**, 891 (2003).
[63] P. C. Cirino, K. M. Mayer, and D. Umeno, *Methods Mol. Biol.* **231**, 3 (2003).

Procedure

Mutagenic PCR is performed on the heme domain of P450 BM-3, on the whole CYP102A1 gene or its mutant. The standard reaction mixture in 50 μl contains 5 μl 10× PCR buffer (100 mM Tris–HCl, pH 8.8 at 25°; 500 mM KCl, 0.8% Nonidet P40, and 15 mM MgCl$_2$), 4 μl dNTPs mix in unbalanced concentrations (2.5:10:10:10 mM) forward and reverse primers introducing the restriction sites (the same as for gene amplification) (40 pmol each), template DNA (10 ng), and MnCl$_2$ (0–0.2 mM).

The PCR program, subsequent restriction of the PCR products, and transformation of the plasmid are the same as in the standard PCR protocol. To find the appropriate mutagenesis rate, the percentage of inactive clones among mutants should be estimated.

To take advantage of the powerful error-prone PCR, an efficient expression system (see earlier) is required. Therefore, protein induction and isolation are performed in a 96-well microtiter plate scale.

The colonies are picked with a robot and inoculated into a microtiter plate containing 150 μl Luria–Bertani (LB) supplemented with ampicillin (100 μg/ml). The plates are incubated overnight at 37° on an orbital shaker at 200 rpm. These microtiter plates are also used as stock plates. Then, 25 μl of overnight culture is added to a 1.2-ml well plate with square wells containing 375 μl LB medium with ampicillin (100 μg/ml). Cells are grown at 37° and 200 rpm to an OD$_{578}$ $\sim$ 0.8 and are then induced with 500 μM IPTG. After incubation at 30° and 200 rpm for 12–16 h after induction, cells are harvested by centrifugation and the medium is discarded.

The cell pellet is resuspended in lyso buffer [potassium phosphate buffer (50 mM, pH 7.5) containing 1 mg/ml lysozyme, 1 mg/ml DNAse, protease inhibitor, and 40 mM NaCl]. After incubation at 4° for 30 min, the plates are centrifuged at 2500g. The supernatant contains the active P450 BM-3 enzyme.

For activity measurements, a rapid and reproducible assay is necessary that is sensitive to minute changes. In addition, it should be adapted to high-throughput screening. The standard method used to determine P450 activity involves measurement of the rate of NADPH consumption in the presence of substrate. This method was also adapted to a microtiter plate scale.[64] However, the rate of NADPH consumption can be very low in the case of unusual substrates, and the uncoupling between reductase and monooxygenase activity may be significant. To overcome this problem, assays with substrate analogs can be used. These can include colorimetric assays for fatty acid hydroxylation (*p*-NCA assay)[65] (see earlier) or alkane hydroxylation

---

[64] A. Glieder and P. Meinhold, *Methods Mol. Biol.* **230,** 157 (2003).

(*p*-pnpane assay),[36] which are optimized for HTS. Screening for alkane oxidation activity is based on the use of a substrate analog that generates aldehyde and yellow *p*-nitrophenolate after terminal hydroxylation.

Glieder and colleagues studied directed evolution for the conversion of P450 BM-3 into an efficient alkane hydroxylase.[50] After five mutagenesis rounds starting with the wild-type P450 BM-3 and subsequent screening for a better activity, one variant with 11 substitutions was found that could accept not only hexane and octane as a substrate, but also gaseous alkanes such as butane and propane. Surprisingly, some mutants were found that led to either (*R*)- or (*S*)-enantiomer products of alkane hydroxylation (Glieder, personal communication).

Another direction of research was to employ hydrogen peroxide as both an oxidant and an electron donor, thereby substituting NADPH.[61] Active peroxygenase[61] and some thermostable variants[62] were evolved, which could use hydrogen peroxide instead of oxygen for the hydroxylation of various substrates.

Conclusions

Protein engineering approaches are powerful tools with regard to elucidating the crucial role of key residues of the P450 BM-3 monooxygenase. Site-directed and random mutageneses enable the manipulation of catalytic properties of this enzyme. Both methods are also suitable for investigating the electron transfer between the reductase and the heme domain. Although insights into different aspects of mechanism and catalytical functions of P450 enzymes have already been provided, many unresolved issues remain. The chemical activation of the substrate, physical aspects of protein–protein and protein–substrate interactions, reactions without oxygen insertion, or multistep reactions can be solved by combining protein engineering and molecular modeling based on the huge natural variation of P450s in structure and sequence.

Note

Since this article has been submitted, significant progress in the protein engineering of P450 BM-3 has been achieved. The stability and activity of P450 BM-3 in the presence of different polar organic solvents were increased significantly using methods of directed evolution.[66] Using a combination of error-prone PCR and site-directed mutagenesis, different

[65] U. Schwaneberg, C. Otey, P. C. Cirino, E. Farinas, and F. H. Arnold, *J. Biomol. Screen.* **6,** 111 (2001).

[66] T. Seng Wong, F. H. Arnold, and U. Schwaneberg, *Biotechnol. Bioeng.* **85,** 351 (2004).

P450 BM-3 variants were engineered, which are able to hydroxylate linear alkanes from C3 to C10 regio- and enantioselectively.[67]

Acknowledgments

We thank Juergen Pleiss and Stephan Tatzel who provided the model of CYP102A1 shown in Fig. 1. We acknowledge financial support by EU Grant QLRT 200000725 and by BASF AG, Ludwigshafen. R. D. Schmid acknowledges generous support by the Verband der Chemischen Industrie (VCI), Frankfurt.

[67] M. W. Peters, P. Meinhold, A. Glieder, and F. H. Arnold, *J. Am. Chem. Soc.* **125,** 13442 (2003).

[20]   Directed Evolution of Aldolases

By DIRK FRANKE, CHE-CHANG HSU, and CHI-HUEY WONG

Introduction

Asymmetric aldol condensation is one of the most useful reactions in synthetic organic chemistry.[1] It is an attractive reaction wherein a carbon–carbon bond is formed while introducing two new stereogenic centers. Much effort has been directed toward controlling the regio- and stereoselective course of the reaction. Although several chemical methods are available, chemoenzymatic aldol condensation catalyzed by aldolases has emerged as a competitive method that guarantees a product with defined and highly predictable stereochemistry in excellent yields.[2] One key limitation of the chemoenzymatic method is that stereoselectivity is controlled by the enzyme rather than the structure or the stereochemistry of the substrate, thus limiting the range of accessible products to the spectrum of available aldolases. However, newly established methods of *in vitro* screening and directed evolution (combinatorial approach) or directed evolution in combination with elements of enzyme design (rational approach) have paved the way to systematically alter and engineer the catalytic activity of enzymes in order to generate new catalysts for desired processes.

Typical points of interests for the alteration of industrially important aldolases may include, but is not limited to,

[1] T. D. Machajewski and C.-H. Wong, *Angew. Chem. Int. Ed. Engl.* **39,** 1352 (2000).
[2] K. M. Koeller and C.-H. Wong, *Nature (Lond.)* **409,** 232 (2001).

i. Altering the substrate specificity to also accept diastereomeric derivatives of the natural substrate.

ii. Switching the substrate specificity to accept enantiomeric sugars (D- vs L-sugars).

iii. Elimination of phosphate dependence.

iv. Increasing thermal and/or chemical enzyme stability in order to develop higher process stability.

The purpose of this article is to outline detailed protocols for generating aldolases with new catalytic properties. The procedures described originally[3–5] and those presented herein allow for efficient improvement of the catalytic activity of aldolases.

Methods

General Materials and Methods

DNA manipulations can be done according to standard procedures.[6] Restriction enzymes and T4 DNA ligase are available from New England Biolabs (Beverly, MA). *Taq, Pfx,* and *Pfu* polymerases are provided commercially by Stratagene (La Jolla, CA). All chemicals, antibiotics, and other enzymes are obtainable from Sigma (St. Louis, MO).

A high-level protein expression system consisting of pTrcHis vectors (Invitrogen, Carlsbad, CA) and *Escherichia coli* JM109 (wild type) was found to be ideal. Additionally, pKK223-3 (Pfizer, New York) in *E. coli* KM3 (*fda⁻* strains)[7] or pET30 vectors (Novagen, Madison, WI) in combination with *E. coli* XL1-Blue[8] or in combination with *E. coli* BL21 (DE3)[5] have also been used as expression systems. *E. coli* XL1-Blue is used as the cloning strain. If needed, antibiotics are added at the following final concentrations: ampicillin (100 μg/ml), kanamycin (50 μg/ml), and chloramphenicol (30 μg/ml).

---

[3] M. Wada, C.-C. Hsu, D. Franke, M. Mitchell, A. Heine, I. A. Wilson, and C.-H. Wong, *Bioorg. Med. Chem.* **11,** 2091 (2003).

[4] S. Fong, T. D. Machajewski, C. C. Mak, and C.-H. Wong, *Chem. Biol.* **7,** 873 (2000).

[5] G. DeSantis, J. Liu, D. P. Clark, A. Heine, I. A. Wilson, and C.-H. Wong, *Bioorg. Med. Chem.* **11,** 43 (2003).

[6] J. Sambrook, E. F. Fritsch, and T. Maniatis, "Molecular Cloning: A Laboratory Manual," 2nd Ed. Cold Spring Harbor Laboratory Press, Plainview, NY, 1989.

[7] G. J. Williams, S. Domann, A. Nelson, and A. Berry, *Proc. Nat. Acad. Sci. USA* **100,** 3143 (2003).

[8] N. Wymer, L. V. Buchanan, D. Henderson, N. Mehta, C. H. Botting, L. Pocivavsek, C. A. Fierke, E. J. Toone, and J. H. Naismith, *Structure* **9,** 1 (2001).

Plasmid Construction

The gene encoding aldolase can be amplified using the standard poly-merase chain reaction (PCR) from a genomic DNA preparation using the Qiagen kit (Düsseldorf, Germany) and primers flanking the gene with engineered restriction sites. Purify the resulting fragment on an agarose gel, digest with appropriate restriction enzymes, and ligate into the expres-sion vector. The ligation product can then be transformed into the appro-priate *E. coli* cloning strain by electroporation.[9] Plasmids recovered from transformants are sequenced and used subsequently for protein expression and as mutation templates for the construction of gene libraries.

Gene Disruption of Wild-Type Enzyme

Gene disruption of a chromosomally encoded wild-type enzyme should be performed without prior cloning via recombination with linear DNA in *recBCD*-deficient *E. coli* strains, following the procedure of Murphy *et al.*[10] We used hyperrecombinogenic *E. coli* K-12 strains JC8679[11] and KM32[10] for the generation of knockout variants. The following procedure can be implemented for the generation of *E. coli* strains with deficiencies in *N*-acetyl-D-neuraminic acid aldolase (encoded *nanA*) and *N*-acetyl-D-neur-aminic acid transporter (encoded by *nanT*). Primers are designed and used for the amplication of the 500-bp fragment A upstream of the desired knockout region and for the 500-bp fragment B downstream of the knock-out region. We successfully used primers GGCTGAACATGCCACC-GATG and GAGACACAACGTGGCTTTCCCGTTGCCATAAATAC-CTCTGAAGTG for amplification of the fragment A upstream of *nanT* and GGCAGAGCATTACGCTGACTTGAGTTAATCCTGTTGCCC-GGTC and GGTATAGCCAGAAAGCGTAGTGCC for the fragment B downstream of *nanA* (the annealing region on chromosomal DNA is underlined). For the amplification of fragment K harboring the marker gene *kan*[R] for kanamycin resistance from the pUC4K vector (Pfizer), we used primers GGAAAGCCACGTTGTGTCTC and CAAGTCAGCG-TAATGCTCTGCC.

For the primary PCR generating fragments A and B, 50 mM KCl, 1 mM MgSO$_4$, 0.7 μM distal primers, 2.8 μM proximal primers, 125 μM of each dNTP, 0.7 μg/ml chromosomal DNA, and 25 U/ml *Taq* DNA polymerase should be combined in a total volume of 100 μl of 20 mM Tris–HCl (pH 8.4). PCR reactions are found to be optimal using the following

[9] M. T. Dower, J. F. Miller, and W. Ragsdale, *Nucleic Acids Res.* **16,** 6127 (1988).
[10] K. C. Murphy, K. G. Campellone, and A. R. Poteete, *Gene* **246,** 321 (2000).
[11] J. R. Gillen, D. K. Willis, and A. J. Clark, *J. Bacteriol.* **145,** 521 (1981).

thermocycler program: 95° for 30 s, followed by 30 cycles of 95° for 1 min, 58° for 1 min, 72° for 75 s, and a final extension at 72° for 7 min.

For amplification of fragment K, 50 mM KCl, 1 mM MgSO$_4$, 0.7 μM primers, 125 μM of each dNTP, 50 ng/ml plasmidic DNA pUC(kan), and 25 U/ml Taq DNA polymerase should be combined in a total volume of 100 μl of 20 mM Tris–HCl (pH 8.4). The assembly reaction can be done using the following program: 95° for 4 min, followed by 34 cycles of 95° for 30 s, 55° for 90 s, 72° for 90 s, and a final extension at 72° for 90 s. To determine the yield and quality of amplification, run small aliquots of the amplified products A, B, and K on agarose gel.

Crude products of fragments A, B, and K, 3.3 μl of each, 50 mM KCl, 1 mM MgSO$_4$, 125 μM of each dNTP, and 25 U/ml Taq DNA polymerase are combined in a total volume of 100 μl of 20 mM Tris–HCl (pH 8.4). The assembly reaction should be performed using the following program: 95° for 30 s, followed by 26 cycles of 95° for 1 min, 54° for 1 min, 72° for 3 min, and a final extension at 72° of 3 min. The expected 2000-bp fragment in mixture with equal amounts of 500- and 1000-bp fragments should be verified by agarose gel electrophoresis.

For gene knockout ($nanAT$), digest 10 μl of the crude amplification mixture with DpnI, purify on an agarose gel, and electroporate into 100 μl competent cells of $E.\ coli$ strains JC8679 and KM32. About 20% of the Km^R mutants grown overnight are unable to grow on media containing ampicillin due to the chromosomal replacement of $nanAT$ by kan^R. In contrast to the parental strains, the knockout strains JC8679Δ nanAT::Km^R and KM32ΔnanAT::Km^R are not capable of growing in mineral salt medium with D-sialic acid (10 g/liter) as a sole carbon source.

$Note.$ The single-stage PCR product with 50-bp extensions homologous to regions adjacent to the wild-type gene can also be used for gene disruption.[12] However, we found that when using the single-stage method, the recombination events occur with significantly lower frequency, resulting in a higher percentage of false-positive clones (presumably containing plasmidic kan^R).

Preparation of Substrate

Substrates for the retroaldol reaction can be synthesized enzymatically from aldehyde and keto compounds using wild-type aldolase or advantageously evolved mutants (with favorable k_{cat} values). The preparation of substrate for pyruvate-dependent aldolases is done routinely as follows: sodium pyruvate (50 mg, 0.45 mmol), aldehyde (0.5 mmol), and aldolase

[12] K. A. Datsenko and B. L. Wanner, $Proc.\ Nat.\ Acad.\ Sci.\ USA$ **97**, 6640 (2000).

(0.5 U or more) dissolved in 2.5 ml of 50 mM potassium phosphate (KP) buffer (pH 7.5) containing 5 mM 2-mercaptoethanol are incubated at 37°. After 5 days, the pH is adjusted to 2 using 7% perchloric acid, and debris are removed by centrifugation. The supernatant is brought to a total volume of 10 ml with water, and the solution is applied to an anion-exchange column (20 g DOWEX 1 × 8, 100–200 mesh, HCO_2^- form equilibrated with water). The product is eluted by applying a linear gradient of water to 1 M formic acid. Fractions containing product are pooled and concentrated *in vacuo*. The residue is stripped with water and dried *in vacuo* several times to remove residual formic acid. Further purification is obtained using silica gel chromatography with the appropriate eluents. Finally, the product is recrystallized from ethanol–water to remove pyruvate contamination.

Library Construction

Procedure 1: Preparation with epPCR

Mutagenic PCR should be carried out under standard error-prone conditions.[13]

1. Combine 40 pmol of each primer, 10 ng template, 10 mM Tris–HCl (pH 8.3), 50 mM KCl, 1.5 mM $MgCl_2$, 0.5 μl dimethyl sulfoxide (DMSO), 0.2 mM $MnCl_2$, 0.2 mM of each dNTP, and 2.5 U *Taq* polymerase in a total volume of 50 μl.
2. Run 30 amplification cycles of 94° for 1 min, 55° for 1 min, and 72° for 1 min. Extend at 72° for an additional 2 min.
3. Run a small aliquot of the reaction on agarose gel to verify amplification.
4. Purify the reaction product with the highest yield and the lowest amount of nonspecific products on agarose gel, digest with appropriate restriction endonucleases, and ligate into the expression vector.
5. Transform *E. coli* cloning strains with the plasmid library. Purify plasmids recovered from several mutants with the Qiagen miniprep and determine the frequency of gene insertion by running small samples on agarose gel.

Note. If plasmids purified from a *dam* methylation-positive strain (DH5α, XL1-Blue) are used as templates, the amplification product might be incubated with *Dpn*I endonuclease to remove parental DNA and

---

[13] D. W. Leung, E. Chen, and D. V. Goeddel, *Techniques* **1,** 11 (1989).

decrease the background of nonrecombinant clones. Combine 50 μl of the amplification reaction, 5 μl of *Dpn*I reaction buffer, and 1 μl of *Dpn*I restriction endonuclease (5–10 U). Incubate at 37° for 1 h. For insertion of the mutant library into the expression vector, a DNA molar ratio of 4:1 (library:plasmid) yields the best results. More library DNA may be used to increase the insertion rate, if desired.

Procedure 2: Preparation with DNA Shuffling

DNA shuffling can be performed according to the method of Stemmer.[14] Current modifications of the original method have been described elsewhere.[15] Fragments of 50–100 bp are isolated and used for the reassemble PCR. A standard PCR of a mixture of two to four selected plasmids from the first-generation library can be used as substrates for shuffling. The following thermocycler program is optimal: 94° for 60 s, followed by 35 cycles of 30 s at 94°, 30 s at 55°, and 30 s at 72°.

Procedure 3: Preparation with Site-Directed Mutagenesis

Site-directed mutagenesis can be carried out using overlap extension PCR[16] and two complementary mutagenic primers of approximately 20–25 bp in length. Prepare the PCR reaction mixture containing 50 ng of each mutagenic primer, 100 ng template, 0.1 mM of each dNTP, 5 μl 10× *Pfu* buffer, and 2.5 U *Pfu* polymerase in a total volume of 50 μl water. Fifty amplification cycles are run as follows: 94° for 45 s, 55° for 30 s, and 72° for 90 s. For fragment insertion into appropriate plasmids and control experiments, follow procedure 1, starting from step 3.

Screening of Library

General note. For all retroaldol screening assays, substrate concentrations above the K_M value for the enzyme reaction should be chosen in order to select for evolved enzymes with optimized k_{cat} rather than optimized k_{cat}/K_M values. The latter may result in product inhibition in synthetically interesting aldol condensations.

Preparation of Enzyme Library

The DNA library can be translated into a protein library as follows.

[14] W. P. C. Stemmer, *Proc. Natl. Acad. Sci. USA* **91**, 10747 (1994).
[15] A. A. Volkov and F. H. Arnold, *Methods Enzymol.* **328**, 447 (2000).
[16] R. Higuchi, B. Krummel, and R. K. Saiki, *Nucleic Acids Res.* **16**, 7351 (1988).

1. Transform the *E. coli* expression strain with the mutant plasmid library by electroporation. Spread transformed cells on Luria–Bertani (LB)[6] agar plates (antibiotic) and incubate at 37° for 16 h.
2. Pick individual colonies into 96-well plates containing 1 ml LB medium (antibiotic) and 0.2 m*M* isopropyl-β-D-thiogalactopyranoside (IPTG).
3. Seal the plates with gas-permeable membranes (Breathe Easy, Diversified Biotech, Boston, MA) and incubate at 37°, 200 rpm for 18 h. Following incubation, centrifuge the plates (10,000g, 4°, 1 h) and the supernatant, decant and discard the supernatant.
4. Resuspend pellets in 500 μl of 50 m*M* KP buffer (pH 7.5) containing 500 μg/ml lysozyme.
5. Freeze plates in liquid nitrogen and then thaw at room temperature.
6. Collect cell debris by centrifugation (10,000g, 4°, 1 h).
7. Transfer 100 μl of the supernatant from each well into a 96-well plate.

Note. For decreasing the background of NADH consuming enzymatic tests, the 96-well plate should be incubated at 65° for 50 min after step 5. Most aldolases are stable under these conditions.

For reducing the time, we prefer to use a device with 96 sterile tips for transforming colonies from the agar plate to 96-well plates and reference agar plate without picking individual colonies. A colony density of approximately 50 colonies per cm^2 on a 150-mm agar plate is desirable. It is essential to ensure a high degree of homogeneity for spreading the clones.

Pyruvate-Dependent Aldolases

Retroaldol Assay with L-Lactic Dehydrogenase. In order to monitor the formation of pyruvate product, a coupled L-lactic dehydrogenase assay based on fluorometric detection of NADH was implemented (Fig. 1).

After incubating the plate from the section entitled "Preparation of Enzyme Library" at 25° for 10 min, add 25 μl assay solution containing 50 nmol NADH and 0.1 U L-lactic dehydrogenase in 50 m*M* KP buffer

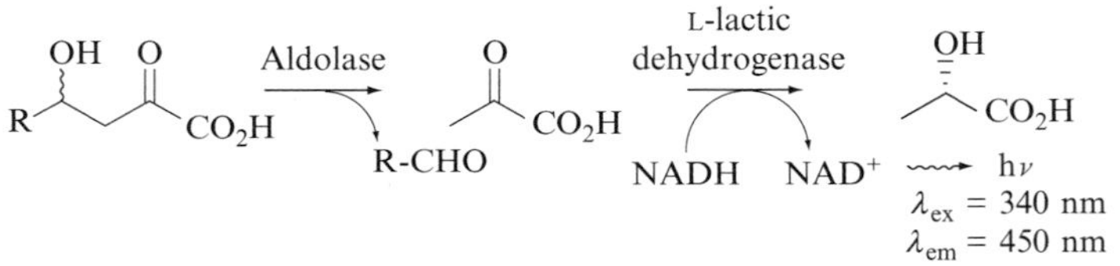

FIG. 1. L-Lactic dehydrogenase assay for the screening of pyruvate-dependent aldolases.

(pH 7.5) to each well using an eight-channel-repeating pipette. Monitor baseline drift (λ_{ex} = 340 nm, λ_{em} = 450 nm) for 2 min. Initiate the reaction by adding 10 μl of substrate at concentrations of 8 to 13 mM in 50 mM KP buffer (pH 7.5). After an initial strong shaking, monitor the fluorescence signal at 1-min intervals for 20 min. The rate of signal decrease can be correlated positively with activity of the mutants.

Aldol Assay with L-Lactic Dehydrogenase. Incubate the plate from the section entitled "Preparation of Enzyme Library" at 25° for 10 min. Upon addition of 25 μl assay solution containing 2 mM sodium pyruvate, 11 mM aldehyde, and 50 mM NADH in 50 mM KP buffer (pH 7.5) into each well, incubate the mixture for 30 min at 25° to allow the aldol reaction to occur. Monitor baseline drift for 5 min. Add 5 U L-lactic dehydrogenase to each well and, after an initial strong shaking, monitor the decrease of absorbance (λ = 340 nm) for 5 min at 20-s intervals. The baselines after the burst phase upon the addition of enzyme can be extrapolated to the zero time point to obtain the y intercepts. Differences between the y intercepts obtained before and after the addition of enzyme should be used to estimate the amount of pyruvate remaining in each sample.

Acetaldehyde-Dependent Aldolases

In Vivo Selection Assay. After transforming *E. coli* SELECT[5] cells with the plasmid library, incubate the cells for 1 h at 37° in 1 ml SOC medium containing 0.1% sodium acetate. Harvest cells by centrifugation (3000 rpm, 4°, 10 min). Wash three times in M9 medium containing 0.2% glucose and resuspend in this medium. Dilute cells to an optical density (λ = 600 nm) of 0.001 in M9 medium (antibiotic) containing 0.2% glucose and 0.01 mM IPTG. Add the appropriate supplementation substrate (sodium acetate, aldol, or aldehyde) to 0.1% (w/w) concentration and incubate the suspension at 37° for 24 to 72 h. Harvest cells by centrifugation (3000 rpm, 4°, 10 min) and isolate plasmids.

Dihydroxyacetone Phosphate (DHAP)-Dependent Aldolases

Aldol Assay with G3P Dehydrogenase. For screening of fructose 1,6-bisphosphate (FDP) aldolase, the generation of glycerol 3-phosphate can be monitored with NAD$^+$-dependent glycerol 3-phosphate (G3P) dehydrogenase and a redox system based on phenazine methosulfate and nitroblue tetrazolium. Add varying concentrations of D-fructose 1,6-bisphosphate to 100 μl of 50 mM Tris–HCl (pH 8.0) containing 0.1 M potassium acetate, 0.5 mg/ml NAD$^+$, 0.3 mg/ml nitroblue tetrazolium, 20 μg/ml phenazine methosulfate, 1.5 mg/ml sodium arsenate, and 20 μg/ml glyceraldehyde

3-phosphate dehydrogenase (Boeringer, Mannheim, Germany). The reactions can be monitored visually (purple dye as formazan) over a time period of 60 min.

Enzyme Expression and Purification

1. Transform plasmid into *E. coli* expression strain.
2. Inoculate individual clone in 5 ml LB medium (antibiotic) and incubate at 37°, 220 rpm, for 14 h.
3. Inoculate 500 ml of LB medium (antibiotic) with preculture from step 2 and incubate at 37°, 220 rpm. Add 0.2 mM IPTG at an optical density ($\lambda = 600$ nm) of 0.4.
4. Harvest cells 6 h after induction by centrifugation (1200g, 4°, 10 min).
5. Resuspend cell pellet in 20 ml of 50 mM KP buffer (pH 7.5) containing 5 mM 2-mercaptoethanol and 300 mM NaCl on ice and lyse by passing through a French press (SLM instruments, Urbana, IL) at 1500 psi three times.
6. Remove cell debris by centrifugation (12,000g, 4°, 1 h) and filter the supernatant through a 0.2-μm cellular acetate membrane filter (Corning, NY).
7. Load a 2.5-ml preequilibrated Ni^{2+}-NTA-agarose column with the supernatant. Wash with 20 ml 50 mM KP buffer (pH 7.5) containing 5 mM 2-mercaptoethanol, 300 mM NaCl, 5% (v/v) glycerol, and 10 mM imidazole and then 20 ml 50 mM KP buffer (pH 7.5) containing 5 mM 2-mercaptoethanol and 10 mM imidazole. Elute with 10 ml 50 mM KP buffer (pH 7.5) containing 5 mM 2-mercaptoethanol and 250 mM imidazole.
8. Dialyze enzyme solution extensively against 50 mM KP buffer (pH 7.5) containing 5 mM 2-mercaptoethanol at 4°.
9. Determine enzyme concentration and purity by the Bradford method[17] and SDS gel electrophoresis.

Note. After step 4, the cell pellet can be stored at −80° for later workup. The enzyme solution can be stored at the same temperature over time periods of months after step 7, as no activity loss is observed upon freezing and thawing. The isolated enzyme is at least 90% pure (SDS–PAGE).

---

[17] M. M. Bradford, *Anal. Biochem.* **72,** 248 (1976).

Enzyme Assays for Isolated Enzymes

Pyruvate-Dependent Aldolases

Retroaldol Assay for L-Lactic Dehydrogenase. Retroaldol activity can be determined easily by a standard coupled assay with L-lactic dehydrogenase (E.C. 1.1.1.27, type II from rabbit muscle) by monitoring the consumption of NADH. For this, 0.8 U L-lactic dehydrogenase, 0.26 μmol NADH, and 10 to 100 μg aldolase mutant are combined in 600 μl of 50 mM KP buffer (pH 7.5) containing 5 mM 2-mercaptoethanol. First, preincubate at 25° for 5 min. Then initiate the assay by addition of an appropriate amount of substrate in 50 mM KP buffer (pH 7.5); the decrease of absorbance (λ = 340 nm) is measured during the first 30 s for initial rate determination.

Aldol Assay for L-Lactic Dehydrogenase. To determine aldol condensation activity, monitor the rate of pyruvate depletion. The pyruvate concentration can be quantified by a method similar to that described previously.[18] For this, 10 mM pyruvate, 250 mM aldehyde, and 30 to 300 μg/ml aldolase mutant are combined in 50 mM KP buffer (pH 7.5). Upon incubating at 37°, take 100-μl aliquots from the reaction mixture at various time points quench the reaction with 30 μl of 7% perchloric acid, and then neutralize with 20 μl of 1 M NaOH. The neutralized samples are diluted with water to 1.15 ml volume. The pyruvate concentration can be determined using the L-lactic dehydrogenase assay.

Aldehyde-Dependent Aldolases

Retroaldol Assay with GPD/TIM. Monitor enzyme activity using the α-glycerophosphate dehydrogenase/triosephosphate isomerase (GPD/TIM) assay. Combine 0.01 to 200 mM substrate, 1.5 U/ml GPD/TIM, and 0.3 mM NADH in 50 mM triethanolamine hydrochloride buffer (pH 7.5). After incubating the mixture at 25° for 10 min, add aldolase mutant (30 to 300 μg/ml) and obtain estimated rates by monitoring the decrease of absorbance (λ = 340 nm) during the first 30 s.

Aldol Assay with Baker's Yeast ADH. An end point assay determining the concentration of acetaldehyde with baker's yeast alcohol dehydrogenase was used to monitor the aldol reaction. Combine 200 mM freshly distilled acetaldehyde, 200 mM glyceraldehyde, and 0.2 mg/ml aldolase

[18] M. C. Shelton, I. C. Cotterill, S. T. A. Novak, R. M. Poonawala, S. Sudarshan, and E. J. Toone, *J. Am. Chem. Soc.* **118**, 2117 (1996).

mutant in 50 mM deoxygenated triethanolamine buffer (pH 7.5). Incubate at 25° under N$_2$ atmosphere and withdraw 50-μl aliquots. Quench the aliquots with 15 μl of 50% perchloric acid and incubate on ice for 5 min. To this add 890 μl of 1 M triethanolamine buffer (pH 7.5) and 45 μl of 4 N NaOH to neutralize the aliquots. Neutralized aliquot are added (20 μl) to a triethanolamine buffer solution (pH 7.5) containing 50 μg/ml ADH and 0.3 mM NADH. Determine NADH consumption by the decrease of absorption ($\lambda = 340$ nm). The total NADH consumed equates to the amount of acetaldehyde remaining in the solution.

Note. The stereochemical outcome of the aldol condensation reaction may also be monitored using ^{31}P and ^{1}H-NMR spectroscopy.[7]

Discussion

One of the crucial challenges in applying directed evolution (DE) approaches to generate novel aldolases is establishing suitable screening systems that work robustly in the environment of crude cell extracts. Unfortunately, the typical product of the aldolase reaction is a highly functionalized and complex carbohydrate that is generally not easy to characterize and quantify, especially when isomeric and anomeric mixtures are involved. Fortunately, enzymatic aldol condensation is found to be highly reversible; therefore the retroaldol condensation can be implemented for screening purposes. The quantification of smaller aldehydes and ketones is generally much easier as the carbonyl group of the precursory structures, as the smaller compounds are ideally activated for either addition or reduction reactions in chemical and enzymatic assays.

By using the retroaldol reaction, the main focus of assay development is reduced to the challenge of establishing preparative access to the aldol condensation product. Depending on the activity of the wild-type enzyme, the substrate demand for the screening process can easily amount to several grams within the first few rounds of evolution. Unfortunately, in all cases where the ultimate goal of DE is to generate aldolases for the preparative production of an aldol of synthetic value, a "chicken-and-egg problem of DE of aldolases" arises. Specifically, an aldolase needs to be developed as a catalyst for the generation of a product, which cannot, or cannot efficiently, be synthesized by other means; however, for the evolution process, this product is a prerequisite on a fairly large scale in the first place. Generally, this problem is addressed by chemically synthesizing enough substrate for the first rounds of DE and then the enzymatic aldol reaction with the best mutants can be used for generating more substrate for further screening.

Aldolases can be divided into three classes, depending on the donor compound (keto compound) of the aldolase reaction: dihydroxyacetone phosphate (DHAP)-dependent, pyruvate-dependent, and acetaldehyde dependent-aldolases.* With regard to mechanism and structure, these classes are further subdivided into two types: lysine-mediated aldolases (type I) and metal-dependent aldolases (type II). For all relevant keto compounds (pyruvate, DHAP, and acetaldehyde), robust enzymatic screening methods for monitoring the cleavage reaction have been developed, paving the way in principle for establishing DE assays for the most synthetically valuable aldolases. Evolution details of some examples, including DE of members of all three classes, are given in Table I.

To our knowledge, type II aldolases have not been used in DE experiments thus far. Compared to DE approaches that quantify natural metabolites (pyruvate, DHAP, or acetaldehyde), the evolution of specificity for substrates consisting of nonnatural aldehyde or keto compounds requires more sophisticated strategies. For example, an enzymatic assay for screening the cleavage of L-fructose by rhamnulose-1-phosphate aldolase (RhaD) to yield L-glyceraldehyde and dihydroxyacetone has not been described so far. This can be attributed to the fact that L-glyceraldehyde is a weak substrate for most natural enzymes. However, assays using dihydroxyacetone are limited by the high intracellular background of dihydroxyacetone. In this case, "caged fluorogenic substrates" derived from methods for DE of transaldolases[19] allow for the highly sensitive detection of aldolase cleavage (Fig. 2).

In these "caged substrates," fluorogenic umbelliferone is released via β elimination from the primary or secondary carbonyl retro-aldol reaction product. This principle has been demonstrated for fluorogenic substrates of a variety of enzymes, including alcohol dehydrogenases, lipases, esterases, amidases, phosphatases, and epoxide hydrolases, as well as for fluorogenic polypropionate fragments that react stereospecifically with aldolase catalytic antibodies.[20] The fluorogenic marker for L-fructose, for example, is easily accessible from L-fructose in five steps,[19] which can be produced in preparative scale on a chemoenzymatic route.[21] The generation of

---

* We are not considering phosphoenolpyruvate class and other two carbon donor aldolases here because the reaction mechanism is significantly different and the reaction is nonreversible for the first class, and for the second class the synthetic applicability is too limited.

[19] E. González-García, V. Helaine, G. Klein, M. Schuermann, G. A. Sprenger, W.-D. Fessner, and J.-L. Reymond, *Chem. Eur. J.* **9,** 893 (2003).

[20] S. V. Taylor, P. Kast, and D. Hilvert, *Angew. Chem. Int. Ed. Engl.* **40,** 3310 (2001).

[21] D. Franke, T. D. Machajewski, C.-C. Hsu, and C.-H. Wong, *J. Org. Chem.* **68,** 6828 (2003).

TABLE I
EXAMPLES OF DE EXPERIMENTS FOR DIFFERENT CLASSES OF ALDOLASES

Enzyme (Ref.)	Substrate[a]		k_{cat} (s^{-1})	K_M (mM)	k_{cat}/K_M (s^{-1} mM^{-1})	$[k_{cat}/K_M](TS)/[k_{cat}/K_M](NS)$ (1)	$[k_{cat}/K_M]$(mutant)/$[k_{cat}/K_M]$(wt) (1)
DRP aldolase (5), acetaldehyde dependent, type I							
	DRP(NS)	Wild type	68	0.64	106	1.9×10^{-5}	1
	DR(TS)		0.107	57	0.002		1
	DRP(NS)	Mutant	41	4.3	9.5	2.7×10^{-4}	0.09
	DR(TS)		0.175	67	0.0026		1.3
KDPG aldolase (4), pyruvate dependent, type I							
	KDPG(NS)	Wild type	284	0.35	811	2.0×10^{-5}	1
	KDG(TS)		4.6	285	0.016		1
	KDPG(NS)	Mutant	11.2	0.46	24	1.8×10^{-2}	0.03
	KDG(TS)		5.2	12.1	0.43		27
NANA aldolase (3), pyruvate dependent, type I							
	NANA(NS)	Wild type	49	2.6	18.7	3.9×10^{-2}	1
	L-KDO(TS)		9.0	13.3	0.72		1
	NANA(NS)	Mutant	16.8	2.6	6.54	8.2×10^{-1}	0.35
	L-KDO(TS)		27	5.0	5.34		7.4
TDP aldolase (7), DHAP dependent, type I							
	TDP(NS)	Wild type	16,800	0.26	64,800	3.0×10^{-3}	1
	FDP(TS)		246	1.3	192		1
	TDP(NS)	Mutant	780	0.016	4800	2.9	0.074
	FDP(TS)		2160	0.16	13,800		72

[a] D-2-Deoxyribose5-phosphate (DRP), D-2-deoxyribose (DR), tagatose1,6-diphosphate (TDP), fructose1,6-diphosphate (FDP), D-2-keto-3-deoxy-6-phosphoguconate (KDPG), D-2-keto-3-deoxyguconate (KDG), N-acetyl-D-neuraminic acid (NANA), 3-deoxy-L-manno-oct-2-ulosonic acid (L-KDO), dihydroxyacetone phosphate (DHAP); in brackets: natural substrate (NS), target substrate (TS).

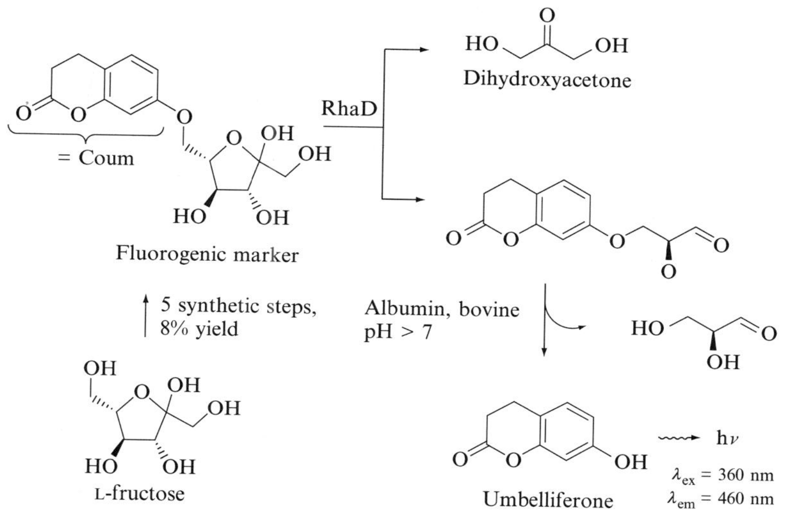

FIG. 2. Use of a fluorogenic marker as a substrate for screening in a directed evolution approach of rhamnulose-1-phosphate aldolase (RhaD) to accept L-fructose.

aldehyde is characteristic for all retro-aldol reactions, making this approach applicable as a general platform for screening and DE of aldolase.

New methods of *in vivo* selection allow the size of the aldolase library to be analyzed to be increased dramatically.[20] We have introduced an *in vivo* selection method for the DE of acetaldehyde-dependent D-2-deoxyribose-5-phosphate (DRP) aldolase that is based on the preparation of acetaldehyde auxotrophic *E. coli* strains by blocking the biosynthetic pathways toward acetyl coenzyme A.[5] Although the evaluation of the applicability of this system is still in progress, preliminary results using preselection in liquid media (see protocols) are promising.[5] In the case of DE experiments of *E. coli* enzymes, the sensitivity and reliability of the screening assay can be increased significantly using mutant strains, deficient in the wild-type enzyme, which can be created by *in vivo* gene replacement.

Using error-prone PCR libraries, amino acid exchanges were found in all areas of the protein structure, with no topological preferences from our experience. However, beneficial amino acid exchanges in, or in close proximity of, the active pocket that would be expected to participate in optimal substrate binding have not been observed in our libraries. Most amino acid exchanges occurred at positions located on the surface rather than in the inner structure of the protein. Such mutations result in slight changes of the $(\beta/\alpha)_8$ barrel structure and changes in catalytic properties, as demonstrated by X-ray structure analysis. We found that further evaluation of amino acid positions where amino acid exchanges were found to be beneficial (putative "hot spots") did not result in further improved variants in our hands.

We are currently addressing the question of how to use DE and rational design for the generation of aldolases that can be used for the efficient synthesis of nonnatural carbohydrates with biological function as carbohydrate mimetics.

Acknowledgments

We thank Professor Valéry de Crécy-Lagard and Dr. David Metzgar (Molecular Biology Department, The Scripps Research Institute) for valuable advice in knockout experiments. We thank Miss Sarah Hanson for helpful discussion. A Feodor-Lynen Fellowship of the Alexander-von-Humboldt Foundation to D. F. is gratefully acknowledged.

[21] Changing the Enantioselectivity of Enzymes by Directed Evolution

By Manfred T. Reetz

Introduction

The development of chiral catalysts for the enantioselective synthesis of optically active organic compounds is of considerable academic and industrial interest, especially in the area of pharmaceuticals, plant protecting agents, and fragrances. Enzymes are becoming increasingly important tools for this purpose.[1,2] In practice, enzymes are used to catalyze the conversion of an unnatural substrate according to the wishes of the practicing organic chemist. In many cases the degree of enantioselectivity is acceptable, specifically when the enantiomeric excess (*ee* value) reaches 95% and/or the selectivity factor E in the case of kinetic resolution of a racemate exceeds 50.[3] In enzyme-catalyzed asymmetric transformations, however,

[1] A. N. Collins, G. N. Sheldrake, and J. Crosby, "Chirality in Industry II: Developments in the Commercial Manufacture and Applications of Optically Active Compounds." Wiley, Chichester, 1997.

[2] K. Drauz and H. Waldmann, "Enzyme Catalysis in Organic Synthesis: A Comprehensive Handbook," 2nd Ed., Vols. I–III. VCH, Weinheim, 2002.

[3] C.-S. Chen, Y. Fujimoto, G. Girdaukas, and C. J. Sih, *J. Am. Chem. Soc.* **104,** 7294 (1982).

substrate specificity remains problematic. Thus the method is far from general and the solution is by no means trivial. Changing the temperature of reaction, varying the solvent, modifying the enzyme chemically, immobilizing, using additives, or performing imprinting techniques are all possible strategies, but their outcome is uncertain.[2,4] The same can be said for protein engineering based on site-specific mutagenesis, where the number of successful examples is limited.[2,4]

In the late 1980s molecular biologists began to develop new and powerful random gene mutagenesis methods, among them the milestone error-prone polymerase chain reaction (epPCR).[5,6] By varying the conditions of epPCR (e.g., $MgCl_2$ concentration), the mutation rate can be adjusted empirically, resulting in an average of one, two, three, or more amino acid exchange events per enzyme molecule. Recombinant methods such as DNA shuffling[7] enriched the repertoire of methods, and molecular biologists, chemists, and chemical engineers began to apply them in the quest to improve the activity as well as the chemical and thermal stability of enzymes.[7-11] The use of multiple cycles of gene mutagenesis, expression, and screening established the process of directed evolution.

Application of the methods of directed evolution to the creation of *enantioselective* enzymes for use as catalysts in synthetic organic chemistry constitutes a special challenge.[12,13] One of the major problems is the need for high-throughput *ee* assays which allow thousands of samples to be screened per day.[14-16] Another difficulty is the optimal choice from the available mutagenesis methods for searching in protein sequence

[4] M. T. Reetz, *Curr. Opin. Chem. Biol.* **6,** 145 (2002).

[5] D. W. Leung, E. Chen, and D. V. Goeddel, *Technique (Philadelphia)* **1,** 11 (1989).

[6] R. C. Cadwell and G. F. Joyce, *PCR Methods Appl.* **2,** 28 (1992).

[7] K. A. Powell, S. W. Ramer, S. B. del Cardayré, W. P. C. Stemmer, M. B. Tobin, P. F. Longchamp, and G. W. Huisman, *Angew. Chem.* **113,** 4068 (2001); *Angew. Chem. Int. Ed. Engl.* **40,** 3948 (2001).

[8] F. H. Arnold, *Nature* **409,** 253 (2001).

[9] S. Brakmann and K. Johnsson, "Directed Molecular Evolution of Proteins (or How to Improve Enzymes for Biocatalysis)." Wiley-VCH, Weinheim, 2002.

[10] F. H. Arnold and G. Georgiou, "Directed Enzyme Evolution (Screening and Selection Methods)." Humana Press, Totowa, NJ, 2003.

[11] M. T. Reetz and K.-E. Jaeger, *Top. Curr. Chem.* **200,** 31 (1999).

[12] M. T. Reetz, A. Zonta, K. Schimossek, K. Liebeton, and K.-E. Jaeger, *Angew. Chem.* **109,** 2961 (1997); *Angew. Chem. Int. Ed. Engl.* **36,** 2830 (1997).

[13] M. T. Reetz and K.-E. Jaeger, *Chem. Eur. J.* **6,** 407 (2000).

[14] M. T. Reetz, *Angew. Chem.* **113,** 292 (2001); *Angew. Chem. Int. Ed. Engl.* **40,** 284 (2001).

[15] M. T. Reetz, *Angew. Chem.* **114,** 1391 (2002); *Angew. Chem. Int. Ed. Engl.* **41,** 1335 (2002).

[16] M. T. Reetz, *in* "Enzyme Functionality—Design, Engineering, and Screening" (A. Svendsen, ed.). p. 559 Dekker, New York, 2004.

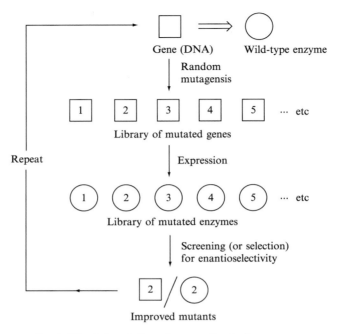

FIG. 1. Directed evolution of an enantioselective enzyme.

space. Figure 1 illustrates the general concept of directed evolution of enantioselective enzymes.[13]

Typically, 2000 to 10,000 clones need to be examined by an appropriate *ee* assay within 1 day. Deconvolution is not relevant because the mutant genes and the encoded enzyme variants are not only linked, but also spatially addressable. Following application of an appropriate mutagenesis method, the mutant genes are inserted into a bacterial host and plated onto agar plates. Individual bacterial colonies are transferred to 96- or 384-well culture plates, either manually using toothpicks or automatically with a colony picker. Because each bacterial colony originates from a single cell, mixtures of mutant enzymes are avoided, provided that inoculation and culture are carried out carefully without cross-contamination. Finally the bacteria or culture supernatants are transferred to the wells of microtiter plates where the reaction of interest and *ee* screening are carried out using the appropriate robotic equipment (Fig. 2).

The focus of this article is on the directed evolution of a lipase from *Pseudomonas aeruginosa* as the catalyst in the hydrolytic kinetic resolution of racemic *p*-nitrophenyl 2-methyldecanoate (**1**) with the

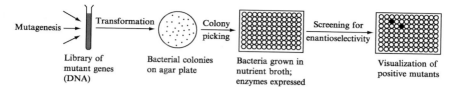

FIG. 2. Experimental stages of the directed evolution of enantioselective enzymes.

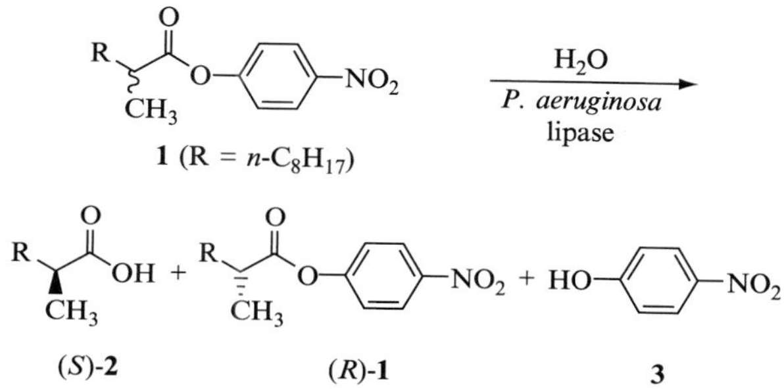

FIG. 3. Scheme illustrating the test reactions.

formation of (S)- and (R)-2-methyldecanoic acid $(\mathbf{2})^{12,13}$ (Fig. 3). The third product, p-nitrophenol $(\mathbf{3})$, is formed concomitantly and can be used to monitor the reaction by UV/VIS spectroscopy. The wild-type enzyme shows a poor enantioselectivity of $E = 1.1$ in slight favor of the (S) acid. This system was chosen here for illustrative purposes not only because it constitutes the first example of the directed evolution of an enantioselective enzyme, but also because it is the most thoroughly studied system to date.

In the present case, as in other studies concerning directed evolution, the problem of protein sequence space needs to be considered. The lipase from $P.\ aeruginosa$ consists of 285 amino acids. Having the algorithm $N = 19^M \times 285!/[(285 - M)! \times M!]$, the size of the library, N, can be calculated as a function of M, the number of amino acid substitutions per enzyme molecule.[12] In the case of $M = 1$, the library would theoretically contain 5415 unique members. However, due to the degeneracy of the genetic code, inter alia, it is essentially impossible to generate an epPCR library that contains all of the 5415 variants. If the mutation rate is increased to yield an average of two amino acid exchanges per enzyme molecule $(M = 2)$, then the number of mutant $P.\ aeruginosa$ lipases predicted by the aforementioned algorithm increases dramatically to about 15 million. When $M = 3$ about 62 billion enzyme variants are possible.

Due to the efforts associated with screening, navigation in protein sequence space should be as efficient as possible, but it is currently difficult to predict the optimal strategy. The *P. aeruginosa* lipase-catalyzed reaction was used to study a variety of different mutagenic approaches, including epPCR at low and high mutation rates, saturation, and cassette mutagenesis, as well as DNA shuffling.

Evolution of an Enantioselective Lipase

Development of an ee Assay

Since conventional chiral gas chromatography or HPLC can handle only a few dozen samples per day, high-throughput *ee* screening systems needed to be developed. The first such assay was based on UV/VIS spectroscopy, designed specifically to monitor the lipase-catalyzed kinetic resolution of the aforementioned chiral ester 1[12,17] (Fig. 3). The *p*-nitrophenol (3), which is formed concomitantly, absorbs at 410 nm, which means that samples in the wells of microtiter plates (96 or 384 format) can be screened in a time-resolved manner with a standard UV/VIS plate reader. However, when using a racemate of substrate 1, only information regarding the overall rate can be obtained. Therefore, the culture supernatant of each lipase mutant was tested in parallel on the enantiomerically pure (*R*)- and (*S*)-substrates 1. Thus, 48 mutants could be screened in a 96-well microtiter plate. The enzyme-catalyzed hydrolysis of each (*R*)/(*S*) pair was monitored by measuring the absorption of the *p*-nitrophenolate anion at 410 nm as a function of time. Figure 4 shows the reaction profile of the original wild-type lipase and that of an improved mutant.[12,14] Although the relative slopes of the two straight lines can serve as a crude basis for calculating the approximate enantioselectivity, this assay was used only to identify hits. About 500–800 samples were evaluated in 1 day, by today's standards medium throughput. In separate experiments the exact enantioselectivity of each mutant lipase "hit" was determined by hydrolyzing the racemate 1 and analyzing the ratio of the (*R*)-2 and (*S*)-2 reaction products by gas chromatography on chirally modified capillary columns.

The main disadvantage of this screening system is the requirement for the *p*-nitrophenol ester instead of the chemically more interesting methyl or ethyl ester. Moreover, the system does not allow the enzyme to interact with the (*R*)- or (*S*) substrate in a competitive manner. Faster, more precise, and more general *ee* assays were developed later (see Table I).

---

[17] K. Liebeton, A. Zonta, K. Schimossek, M. Nardini, D. Lang, B. W. Dijkstra, M. T. Reetz, and K.-E. Jaeger, *Chem. Biol.* **7**, 709 (2000).

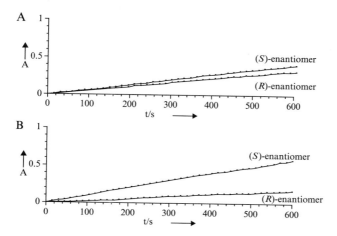

FIG. 4. Course of the lipase-catalyzed hydrolysis of the (R)- and (S)-**1** as a function of time. (A) Wild-type lipase from *Pseudomonas aeruginosa*. (B) Improved mutant in the first generation.

Protocols

Expression of Lipase Genes

Lipase is expressed in *P. aeruginosa* PABST7.1, a lipase-deficient strain containing a chromosomally located copy of the T7-RNA polymerase under control of the *lac*UV5/*lac*I^q element. This system allows the induction of expression by the addition of isopropyl-β-D-thiogalactoside (IPTG). Mutated lipase genes are expressed from plasmid pUCPL6A, derived from pUCPKS, which contains the lipase operon under control of the T7 promoter.[18]

ee Screening

Single colonies of *P. aeruginosa* PABST.1/pUCPL6A are picked with a sterile toothpick and resuspended in wells of a 96-deep-well microtiter plate filled with 1 ml of 2× LB (20 g/liter tryptone, 10 g/liter yeast extract, 10 g/liter NaCl) containing 200 μg/ml carbenicillin and 50 μg/ml tetracycline.[12,17] Cells are grown by shaking the plates overnight at 30°. Lipase expression is induced by the addition of IPTG to 0.4 mM and further shaking at 30° for 4–24 h. Cells are separated from the culture supernatant by centrifugation of

[18] K.-E. Jaeger, B. Schneidinger, F. Rosenau, M. Werner, D. Lang, B. W. Dijkstra, K. Schimossek, A. Zonta, and M. T. Reetz, *J. Mol. Catal. B: Enzym.* **3**, 3 (1997).

the plates at 5000g for 15 min. Aliquots of the 50-μl supernatant are taken from each well and transferred into two wells of a second microtiter plate, each containing 10 μl 100 mM Tris–HCl, pH 8.0, 30 μl Aqua dist. and 10 μl of the (R)- or (S)-enantiomer, respectively, of 2-methyldecanoic acid p-nitrophenyl ester (1) in 10% N, N-dimethylformamide (DMF). Plates are shaken for 10 s, and the reactions are monitored at 410 nm for 6–10 min at 30° using a Spectramax-8 channel photometer (Molecular Devices, Sunnyvale, CA). The E values of the hits are then calculated using the formula of Chen et al.[3] Those mutants showing an improved enantioselectivity as compared to the "parent" of the corresponding generation are analyzed further by reacting a racemic substrate mixture with culture supernatant and analyzing the reaction products by chiral gas chromatography.[12,17]

Random Mutagenesis by epPCR

A 1046-bp BamHI/ApaI fragment from plasmid pUCPL6A containing the complete P. aeruginosa lipase gene lipA is cloned into the corresponding sites of pBluescript II KS (Stratagene) to produce plasmid pMut5. This plasmid serves as the template DNA for error-prone PCR using primers LipH (5'-GGC-GAA-CGC-CAG-TGG-AAT-CAG-CAG-GAG-G-3') and LipB (5'-GCG-TAA-TAC-GAC-TCA-CTA-TAG-GGC-GAA-3'), which aneal downstream and upstream, respectively, of the BamHI/ApaI fragment.[12,17] Error-prone PCR is performed on a RoboCycler gradient 40 (Stratagene, The Netherlands) under the following conditions: a total volume of 50 μl of 75 mM Tris–HCl, pH 8.8, 20 mM $(NH_4)_2SO_4$, 0.01% (v/v) Tween 20, 7 mM $MgCl_2$, 0.25 mM dNTPs, 2% dimethyl sulfoxide (DMSO), 0.4 μM each of primer LipH and LipB, containing 1 ng of template DNA and 2 U Goldstar Taq polymerase (Eurogentec, Belgium).[12,17] Ten parallel samples overlaid with 70 μl paraffin are amplified using the following cycling protocol: 1× (2 min 98°), 25× (1 min 94°; 2 min 64°; 1 min 72°), and 1× (7 min 72°). PCR products are cloned as described later. This protocol results in an average mutation frequency of one to two base substitutions per 1000 bp and an average of one amino acid exchange per enzyme molecule as demonstrated by DNA sequence analyses.[12,17]

Site-Specific Mutagenesis

The "megaprimer" PCR method of Barettino et al.[19] is used to introduce the substitution S155F into wild-type and mutant lipases.[12,17] The first PCR reaction is performed as follows: A 50-μl volume of 75 mM Tris–HCl,

[19] D. Barettino, M. Feigenbutz, R. Valcarcel, and H. G. Stunnenberg, Nucleic Acids Res. 22, 541 (1994).

pH 9.0, 20 mM $(NH_4)_2SO_4$, 0.01% (v/v) Tween 20, 1.5 mM $MgCl_2$, 10% (v/v) DMSO, 0.2 mM dNTPs, 0.2 μM each of the mutagenic primer (5'-GGT-ACC-CAG-AAT-TTT-CTG-GGC-TCG-CTG-3') and the primer LipH containing 1 ng of template DNA (pMut5), and 2 U of Goldstar *Taq* polymerase (Eurogentec, Belgium). The cycling protocol is as follows: $1\times$ (2 min 98°), $30\times$ (1 min 94°; 2 min 64°; 1 min 72°), and $1\times$ (7 min 72°). PCR products are separated from the template DNA by agarose gel electrophoresis, purified with the NucleoSpin Extract kit (Macherey and Nagel), and used as megaprimers in a second PCR, similar to the first PCR but with the following changes: in addition to primer LipH and the mega-primer, LipB is added as a third primer at a concentration of 0.2 μM and the plasmid template is pMut6, derived from pMut5 by digestion with *Kpn*I/*Afl*III, blunting the end with T4 DNA polymerase (Fermentas, Germany) and religation.[12,17] PCR products are cloned as described later.

Saturation Mutagenesis

Saturation mutagenesis is performed as described earlier for site-directed mutagenesis using mutagenic primers in which the codon under investigation has been randomized by mixing equal amounts of nucleo-side phosphoramidates during synthesis.[12,17] Primers are obtained from Eurogentec (Belgium).

DNA Shuffling and Cassette Mutagenesis

The techniques of DNA shuffling and combinatorial multiple cassette mutagenesis (CMCM)[20,21] are combined as follows. The mutants encoding the variants D and E of *P. aeruginosa* lipase A are amplified by PCR using the primers LipH and LipB and the enzyme Pfu (Stratagene, La Jolla, CA) according to the manufacturer's directions. Cycling conditions are $1\times$ (2 min 98°) and $25\times$ (1 min, 94°; 30 s 58°; 90 s 72°). PCR products are purified with the Nucleospin kit and are digested with 3 mU DNase I (Roche, Mannheim, Germany) for 5 min at 0° in a final volume of 96 μl containing 20 mM $MnCl_2$, 100 mM Tris–HCl, pH 7.5, and 2 μg of each variant DNA. Reactions are stopped with 48 μl 100 mM EDTA; fragments are purified using the Qiaex II kit (Qiagen, Hilden, Germany) and eluted in 30 μl H_2O. The extent of digestion is controlled on a 2% agarose gel. Reassembly reactions contain 10 μl purified DNA fragments, 9 nmol each dNTP, and 0.75 U Pfu in 22.5 μl of $1\times$ Pfu buffer. For cassette

[20] M. T. Reetz, S. Wilensek, D. Zha, and K.-E. Jaeger, *Angew. Chem.* **113**, 3701 (2001); *Angew. Chem. Int. Ed.* **40**, 3589 (2001).
[21] A. Crameri and W. P. C. Stemmer, *BioTechniques* **18**, 194 (1995).

mutagenesis, 10 pmol of a single standard oligonucleotide containing the mutations of interest is added to the reaction mixture. Assembly reactions are incubated for a total of 50 cycles using the Pfu amplification profile described earlier. One microliter of the reassembly reaction is used as a template for PCR with the Pfu protocol described previously.

Cloning and Transformation

PCR products are purified with the NucleoSpin kit (Macherey and Nagel, Germany) and digested with *Bam*Hl and *Apa*l according to the manufacturer's directions. The plasmid pUCPLAn, a null mutant derivative of the plasmid pUCPLGA, is digested with the same enzymes. Fragments are separated by gel electrophoresis on a 0.8% agarose gel, and the appropriate bands are excised and purified with the NucleoSpin extract kit (Macherey and Nagel). Ligations are performed with T4 DNMA ligase (Fermentas, Germany) in a volume of 10 μl containing 5 pmol each of insert and vector according to the manufacturer's directions. Ligation mixtures are diluted 1:10 in TMF buffer and are used to transform chemically competent *Escherichia coli* JM 109 cells. Plasmid DNA from the entire JM 109 library is purified using the InViSorp spin plasmid minikit (InViTtek, Germany) and uses *P. aeruginosa* PABST7.1.[12,17] *Pseudomonas aeruginosa* transformants expressing active lipase protein form clear halos on tributyrin agar. Tributyrin plates are prepared as follows: 0.75 g of gum arabic (Sigma, Germany) is dissolved in 7.5 ml of sterile Aqua dist. and mixed with 7.5 ml of tributyrin (Sigma). This mixture is added to 500 ml of LB agar containing 200 μg/ml carbenicillin and 50 μg/ml tetracycline and is emulsified with a Ultraturrax T25 (Ika Labortechnik, Germany) for 1 min at 24,000 U/min prior to pouring the plates.[12,17]

Results and Evaluation

After performing four rounds of epPCR at low mutation rates averaging one amino acid substitution per enzyme molecule and screening only 2000–3000 mutant in each cycle, the selectivity factor in the hydrolytic kinetic resolution of the chiral ester **1** increased from $E = 1.1$ to $E = 11.3$ (variant A).[12] This was accomplished by identifying the most enantioselective mutant (variant A), isolating the corresponding mutant gene, and using it as a template for the next round of epPCR. Figure 5 shows the course of this enantioselective "evolution" and the amino acid exchange determined by DNA sequencing.[12,13,17]

Although it would be theoretically possible to continue to improve enantioselectivity by performing additional rounds of epPCR, such a strategy is not optimal. We reasoned that particular sites of amino acid

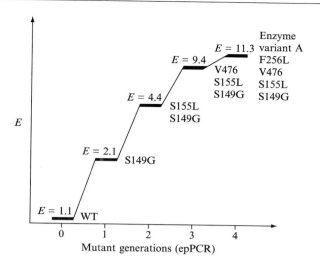

FIG. 5. Increasing E values of the lipase-catalyzed hydrolysis of the chiral ester 1 by cumulative mutations caused by epPCR.

exchange ("hot spots") are important for increasing enantioselectivity, but that the specific amino acids exchanges identified by epPCR experiments are not necessarily optimal. There are two reasons for this conclusion. First, the size of the screened libraries was smaller than the theoretical number of enzyme variants. Second, the amino acid exchanges achieved by epPCR are biased due to the degeneracy of the genetic code.[22] Therefore, several "mutation hot spots" were chosen for saturation experiments.[17] Theoretically, all 20 proteinogenic amino acid substitutions at single position will be contained in a saturation library of $\geq$300 mutants. Indeed, this strategy led to variant B displaying an E value of 20 (Fig. 6). The gene encoding enzyme variant B was then subjected to epPCR, resulting in the improved enzyme variant C ($E = 25$). Thus, the combination of epPCR and appropriate saturation mutagenesis constitutes a viable strategy. Variant C has five amino acid exchanges, most of these at positions remote from the active site.[17,23]

We also applied DNA shuffling, whereby the first question arising is which mutants to recombine.[20] DNA shuffling of the mutant genes produced by low mutation rate epPCR failed to achieve significant improvements,

[22] T. Eggert, M. T. Reetz, and K.-E. Jaeger, in "Enzyme Functionality—Design, Engineering, and Screening" (A. Svendsen, ed.). p. 375, Dekker, New York, 2004.
[23] M. T. Reetz, Tetrahedron 58, 6595 (2002).

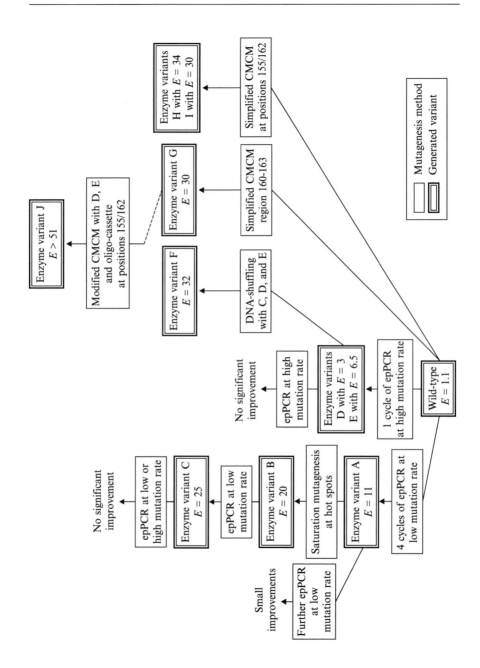

leading us to conclude that higher gene diversity was required. To this end, epPCR of the wild-type gene was repeated, this time at a higher mutation rate corresponding to three amino acid exchange events per enzyme molecule. This process led to identification of the improved enzyme variants D and E ($E = 3$ and $E = 6.5$). The genes encoding the variants D and E and variant C, from the fourth generation of low error rate epPCR, were then subjected to conventional DNA shuffling. This led to variant F displaying the highest enantioselectivity up to that point ($E = 32$).[20] Sequencing results revealed that some of the first weakly positive mutations exert a negative effect on the enzyme enantioselectivity of the mutants evolved in later rounds of epPCR. This observation is not surprising, but fortunately corrections are possible during the processes of DNA shuffling or saturation mutagenesis.

Finally, we extended the region of accessible protein sequence space by developing a modified version of combinatorial multiple cassette mutagenesis (CMCM).[20] The original form of CMCM, as described by Stemmer, is a special type of DNA shuffling, which allows the generation of mutant gene libraries in which the wild-type gene and cassettes composed of defined sequences are randomized.[21] In the modified version, two mutant genes encoding the enzyme variants D and E and a mutagenic oligocassette containing simultaneous saturation mutagenesis at the "hot spots" codons 155 and 162 were subjected to DNA shuffling.[20] This process resulted in an extremely enantioselective variant J, displaying a selectivity factor of $E > 51$. The results of these and other experiments are summarized in Fig. 6.

We have shown that protein sequence space can be explored efficiently with respect to enantioselectivity by applying epPCR at a high mutation rate, using these results to identify hot spots and regions, and then performing saturation mutagenesis and DNA shuffling in the form of CMCM.[13,17,20,23] None of these results were based on prior analysis of the three-dimensional structure of the enzyme or of its mechanism of action. However, it is possible to use these results to improve our understanding of enzymes, specifically because enantioselectivity is a sensitive probe. The most enantioselective variant ($E > 51$) contains six amino acid substitutions.[20] Based on consideration of the X-ray structure of the wild-type enzyme,[24] most of these mutations occur at sites remote from the active

[24] M. Nardini, D. A. Lang, K. Liebeton, K.-E. Jaeger, and B. W. Dijkstra, *J. Biol. Chem.* **275,** 31219 (2000).

FIG. 6. Schematic representation of the directed evolution of enantioselective enzymes (lipase variants), which catalyze the hydrolytic resolution of ester **1**.

center.[20,23] Other enzyme variants showing enhanced enantioselectivity are characterized by mutations occurring solely at remote positions.[23] Although remote effects with respect to other properties have been described previously in other enzymatic systems,[25–28] the present findings are the first involving enantioselectivity. Because this property has traditionally been associated with the active site,[2,29] these findings are surprising. Indeed, all previous attempts to enhance enantioselectivity of an enzyme by rational design and site-specific mutagenesis rational have concentrated on positions directly at the active site, in line with Emil Fischer's lock and key principle or more refined models such as induced fit.[30,31] Of course, remote mutations are to some extent statistically "favored": there are numerically more amino acids far away from the active site than close to it. In the present case, theoretical studies based on molecular modeling and quantum mechanics point to a relay mechanism, which leads to the creation of a new chiral-binding pocket at the active site.[32] Moreover, the transition state of the (S) substrate is stabilized by an additional H bond, which cannot form with the (R) ester. The model thus explains the experimentally observed increased activity and selectivity for the (S)-enantiomer of substrate **1**.[23]

The chiral selectivity of an enzyme can also be reversed through directed evolution.[10,11] For this purpose, we used the strategies described earlier but screened for (R)-selective mutants.[33] An enzyme variant with 11 amino acid exchanges and showing $E = 30$ in favor of the (R) acid **2** was evolved. Again many of the amino acid substitutions occur at remote sites[33,34] (Fig. 7). Interestingly, these mutated sites differ from those evolved and identified in the case of the (S)-selective enzyme variants. (R) selectivity has yet to be explained on a molecular basis.

[25] H. Zhao and F. H. Arnold, *Protein Eng.* **12,** 47 (1999).

[26] A. Iffland, P. Tafelmeyer, C. Saudan, and K. Johnsson, *Biochem.* **39,** 10790 (2000).

[27] M. Kumar, K. K. Kannan, M. V. Hosur, N. S. Bhavesh, A. Chatterjee, R. Mittal, and R. V. Hosur, *Biochem. Biophys. Res. Commun.* **294,** 395 (2002).

[28] P. K. Agarwal, S. R. Billeter, P. T. R. Rajagopalan, S. J. Benkovic, and S. Hammes-Schiffer, *Proc. Natl. Acad. Sci. USA* **99,** 2794 (2002).

[29] G. P. Horsman, A. M. F. Liu, E. Henke, U. T. Bornscheuer, and R. J. Kazlauskas, *Chem. Eur. J.* **9,** 1933 (2003).

[30] D. Rotticci, J. C. Rotticci-Mulder, S. Denman, T. Norin, and K. Hult, *ChemBioChem.* **2,** 766 (2001).

[31] Y. Hirose, K. Kariya, Y. Nakanishi, Y. Kurono, and K. Achiwa, *Tetrahedron Lett.* **36,** 1063 (1995).

[32] M. Bocola, N. Otte, K.-E. Jaeger, M. T. Reetz, and W. Thiel, *ChemBioChem.* **5,** 214 (2004).

[33] D. Zha, S. Wilensek, M. Hermes, K.-E. Jaeger, and M. T. Reetz, *Chem. Commun. (Cambridge, U.K.)* 2664 (2001).

[34] M. T. Reetz, *Pure Appl. Chem.* **72,** 1615 (2000).

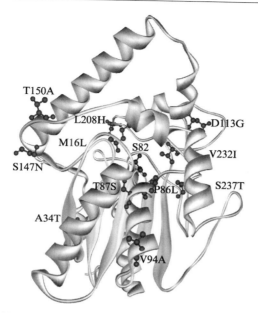

FIG. 7. Structure[24] of the wild-type lipase from *P. aeruginosa* showing the active site S82 and the 11 mutations of the most *R*-selective variant.[33]

Development of Efficient High-Throughput *ee* Assays

The *ee* assay used in the aforementioned studies is rather crude and has a capacity of only 500–800 samples per day.[12,14,17,35,36] Moreover, it is restricted to the particular case of the hydrolytic kinetic resolution of chiral esters. In order to perform directed evolution of enantioselective enzymes efficiently and generally, other *ee*-screening systems must be developed. In the author's laboratories half a dozen screening systems have been designed. Other groups have also contributed to the development of high-throughput *ee* assays, including those active in the area of combinatorial asymmetric transition metal catalysis.[14] Presumably the latter assays can be modified to include enzyme-catalyzed processes. Table I[12,17,37–60]

[35] M. T. Reetz, *in* "Methods in Molecular Biology" (F. H. Arnold and D. Georgiou, eds.), Vol. 230, p. 259. Humana Press, Totowa, NJ, 2003.

[36] M. T. Reetz, *in* "Methods in Molecular Biology" (F. H. Arnold and D. Georgiou, eds.), Vol. 230, p. 283. Humana Press, Totowa, NJ, 2003.

[37] L. E. Janes and R. J. Kazlauskas, *J. Org. Chem.* **62,** 4560 (1997).

[38] L. E. Janes, A. C. Löwendahl, and R. J. Kazlauskas, *Chem. Eur. J.* **4,** 2324 (1998).

TABLE I
HIGH-THROUGHPUT *ee*-ASSAYS

Detection system/description; application	Reference
UV/VIS; kinetic resolution of *p*-nitrophenol esters	12, 17
UV/VIS/Quick E test; kinetic resolution of esters	37
UV/VIS/pH indicator; kinetic resolution of esters	38
UV/VIS/enzyme coupled; kinetic resolution of acetates	39
UV/VIS/enzyme coupled; alcohols	40
UV/VIS/enzyme immunoassays; alcohols	41
MS/diastereomer formation; many compounds	42
MS/labeled compounds; kinetic resolution or desymmetrization of compounds bearing enantiotopic groups	43–49
NMR/flow-through cell; essentially any compound	50
FTIR/labeled compounds; kinetic resolution and desymmetrization	51
Circular dichroism; most compounds	52, 53
Fluorescence/capillary array electrophoresis; amines, alcohols, etc.	54
Fluorescence/enzyme coupled; kinetic resolution of esters	55
Fluorescence; kinetic resolution of alcohols	56
Fluorescence/DNA microarrays; amino acids, etc.	57
GC/special construction; volatile compounds	58
IR thermography; kinetic resolution in general	59
HPLC/many compounds	60

[39] M. Baumann, R. Stürmer, and U. T. Bornscheuer, *Angew. Chem.* **113,** 4329 (2001); *Angew. Chem. Int. Ed.* **40,** 4201 (2001).

[40] P. Abato and C. T. Seto, *J. Am. Chem. Soc.* **123,** 9206 (2001).

[41] F. Taran, C. Gauchet, B. Mohar, S. Meunier, A. Valleix, P. Y. Renard, C. Créminon, J. Grassi, A. Wagner, and C. Mioskowski, *Angew. Chem.* **114,** 132 (2002); *Angew. Chem. Int. Ed.* **41,** 124 (2002).

[42] J. Guo, J. Wu, G. Siuzdak, and M. G. Finn, *Angew. Chem.* **111,** 1868 (1999); *Angew. Chem. Int. Ed.* **38,** 1755 (1999).

[43] M. T. Reetz, M. H. Becker, H.-W. Klein, and D. Stöckigt, *Angew. Chem.* **111,** 1872 (1999); *Angew. Chem. Int. Ed.* **38,** 1758 (1999).

[44] G. DeSantis, K. Wong, B. Farwell, K. Chatman, Z. Zhu, G. Tomlinson, H. Huang, X. Tan, L. Bibbs, P. Chen, K. Kretz, and M. J. Burk, *J. Am. Chem. Soc.* **125,** 11476 (2003).

[45] S. A. Funke, A. Eipper, M. T. Reetz, N. Otte, W. Thiel, and G. van Pouderoyen, *Biocatal. Biotransform.* **21,** 67 (2003).

[46] D. Zha, A. Eipper, and M. T. Reetz, *ChemBioChem.* **4,** 34 (2003).

[47] F. Cedrone, S. Niel, S. Roca, T. Bhatnagar, N. Ait-Abdelkader, C. Torre, H. Krumm, A. Maichele, M. T. Reetz, and J. C. Baratti, *Biocatal. Biotransform.* **21,** 357 (2003).

[48] W. Schrader, A. Eipper, D. J. Pugh, and M. T. Reetz, *Can. J. Chem.* **80,** 626 (2002).

[49] M. T. Reetz, M. H. Becker, D. Stöckigt, and H.-W. Klein, German Patent DE-A 19913858.3 (1999).

[50] M. T. Reetz, A. Eipper, P. Tielmann, and R. Mynott, *Adv. Synth. Catal.* **344,** 1008 (2002).

[51] P. Tielmann, M. Boese, M. Luft, and M. T. Reetz, *Chem. Eur. J.* **9,** 3882 (2003).

[52] R. Angelaud, Y. Matsumoto, T. Korenaga, K. Kudo, M. Senda, and K. Mikami, *Chirality* **12,** 544 (2000).

summarizes current high-throughput *ee*-screening systems. The area has been reviewed critically[14–16,35,36] and only a few key points are emphasized here.

When choosing an *ee* assay for a particular study, several factors need to be considered, including cost, degree of throughput, and precision. Some of the assays have a precision not better than ±10%, which is acceptable in the early phases of a project directed toward enhancing enzyme enantioselectivity. However, once the *ee* has reached 90%, further improvement requires a more precise assay. The most practical of the current assays are based on mass spectrometry,[42–49] NMR spectroscopy,[50] infrared spectroscopy,[51] and circular dichroism.[52,53] These *ee*-screening systems allow 1000 to 10,000 samples to be evaluated per day with a precision ranging between ±3 and ±10%. In all cases the instruments, robotics, and software are available commercially and require little adaptation. No single *ee* assay is universal, and some are complementary. For certain reactions analysis is only possible with modified forms of gas chromatography[58] or HPLC[60]; although high-throughput is not possible, in favorable cases 300–700 samples can be evaluated per day.

Theoretically, selection rather than screening should be used for the directed evolution of enantioselective enzymes. Such systems would not only be closer to natural evolution, they would also allow for much larger libraries (10^6–10^8 mutants!). However, the design of an appropriate process is difficult and remains a challenge for the future. A biological system in which the optical density of bacterial cultures is used to express growth as a function of (*R*) and (*S*) substrates in the media has been reported,[61] but is still a special form of screening. Perhaps phage display can be adapted for the selection of improved enantioselective substrate binding,[62] although it can be argued that enantioselective binding alone is not sufficient for efficient asymmetric enzyme catalysis.

[53] M. T. Reetz, K. M. Kühling, H. Hinrichs, and A. Deege, *Chirality* **12**, 479 (2000).
[54] M. T. Reetz, K. M. Kühling, A. Deege, H. Hinrichs, and D. Belder, *Angew. Chem.* **112**, 4049 (2000); *Angew. Chem. Int. Ed.* **39**, 3891 (2000).
[55] J.-L. Reymond and D. Wahler, *ChemBioChem.* **3**, 701 (2002).
[56] E. R. Jarvo, C. A. Evans, G. T. Copeland, and S. J. Miller, *J. Org. Chem.* **66**, 5522 (2001).
[57] G. A. Korbel, G. Lalic, and M. D. Shair, *J. Am. Chem. Soc.* **123**, 361 (2001).
[58] M. T. Reetz, K. M. Kühling, S. Wilensek, H. Husmann, U. W. Häusig, and M. Hermes, *Catal. Today* **67**, 389 (2001).
[59] M. T. Reetz, M. H. Becker, K. M. Kühling, and A. Holzwarth, *Angew. Chem.* **110**, 2792 (1998); *Angew. Chem. Int. Ed.* **37**, 2647 (1998).
[60] M. T. Reetz, F. Daligault, B. Brunner, H. Hinndis, and A. Deege, submitted.
[61] M. T. Reetz and C. J. Rüggeberg, *Chem. Commun. (Cambridge, U.K.)*, 1428 (2002).
[62] M. J. Dröge, C. J. Rüggeberg, A. M. van der Sloot, J. Schimmel, R. S. Dijkstra, R. M. D. Verhaert, M. T. Reetz, and W. J. Quax, *J. Biotechnol.* **4**, 19 (2003).

Other Examples of Directed Evolution of Enantioselective Enzymes

The case study presented here is a representative investigation, which clearly shows that it should be possible to apply the appropriate molecular biological methods and *ee*-screening systems to other reactions and other enzymes. Indeed, several studies in this new area of research have appeared recently, although in some cases only a single round of epPCR was reported, which is not an evolutionary process.[29,44,45,63–70] Prominent examples include the directed evolution of an enantioselective aldolase by Fong et al.,[63] reversal of enantioselectivity of a hydantoinase in the preparation of unnatural methionine as reported by May et al.,[64] and the optimization of a nitrilase in the desymmetrization of a prochiral dinitrile as published by scientists at Diversa.[44] From a methodological point of view, the latter study is noteworthy because it involves a strategy of mutagenesis not yet mentioned in this article, namely systematic saturation mutagenesis of each amino acid position in an enzyme. Upon screening this collection of small libraries, a mutant nitrilase displaying enhanced enantioselectivity (*ee* = 98.1%) relative to the wild type (*ee* = 87.8%) was identified.[44] In independent work in the author's laboratory, systematic saturation mutagenesis was applied to the lipase from *Bacillus subtilis* for improving the desymmetrization of a *meso*-diacetate.[45] This strategy is practical because it provides a platform of enzyme variants, which can be used for screening on other substrates. If the selected mutants require further improvement, the methods of directed evolution can be applied subsequently. In our *B. subtilis* studies, remote effects on enantioselectivity were once again observed.[45] However, we do not suggest that only remote effects are essential in improving enantioselectivity.

Additional studies in the author's laboratories on the directed evolution of enantioselective enzymes include the application of a new recombinant method for gene synthesis (assembly of designed oligonucleotides) in the lipase-catalyzed (*B. subtilis*) desymmetrization of a *meso*-diacetate,[46] the kinetic resolution of chiral epoxides using the epoxide hydrolase from *Apergillus niger*,[47] and the desymmetrization of 4-hydroxycyclohexanone

[63] S. Fong, T. D. Machajewski, C. C. Mak, and C.-H. Wong, *Chem. Biol.* **7,** 873 (2000).

[64] O. May, P. T. Nguyen, and F. H. Arnold, *Nature Biotechnol.* **18,** 317 (2000).

[65] U. T. Bornscheuer, J. Altenbuchner, and H. H. Meyer, *Biotechnol. Bioeng.* **58,** 554 (1998).

[66] E. Henke and U. T. Bornscheuer, *Biol. Chem.* **380,** 1029 (1999).

[67] G. W. Matcham and A. R. S. Bowen, *Chim. Oggi.* **14**(6), 20 (1996).

[68] M. T. Reetz, C. Torre, A. Eipper, M. Lohmer, M. Hermes, B. Brunner, A. Maichele, M. Bocola, M. Arand, A. Cronin, Y. Genzel, A. Archelas, and R. Furstoss, *Org. Lett.* **6,** 177 (2004).

[69] R. Carr, M. Alexeeva, A. Enright, T. S. C. Eve, M. J. Dawson, and N. J. Turner, *Angew. Chem.* **115,** 4955 (2003); *Angew. Chem. Int. Ed.* **42,** 4807 (2003).

[70] M. W. Peters, P. Meinhold, A. Glieder, and F. H. Arnold, *J. Am. Chem. Soc.* **125,** 13442 (2003).

via the Baeyer–Villiger reaction employing a cyclohexanone monooxygenase as the asymmetric catalyst.[71,72] The latter investigation shows for the first time that it is possible to apply the methods of directed evolution to enantioselective oxidation processes, an area of considerable interest to synthetic organic chemists.[71,72] In this system, the use of whole cells rather than an isolated enzyme has two major practical advantages: the enzymatic problem of cofactor regeneration is circumvented and available atmospheric oxygen is used as the oxidant.

Conclusions

The directed evolution of enantioselective enzymes for use in synthetic organic chemistry is a new area of research that promises to become a viable alternative to traditional "rational" forms of protein engineering. Indeed, it is also an alternative to asymmetric transition metal catalysis and to catalytic antibodies. The main advantage of this fundamentally new approach to asymmetric catalysis is the fact that knowledge of the structure of the enzyme or of its mechanism is not required. Nevertheless, the process is rational in a different sense because it relies on an evolutionary process. Once enantioselectivity has been evolved, important structural and mechanistic lessons that are not accessible by traditional approaches can be learned. A prime example is the remote effects discovered during the directed evolution of enantioselective enzymes. By applying theoretical methods based on molecular modeling and quantum mechanics to these mutants, the source of enantioselectivity has been revealed. The models suggest further experiments using site-specific mutagenesis, a development that heralds the emergence of an appropriate fusion between directed evolution and rational design.

The most intensely studied case of directed evolution of enantioselective enzymes involves the lipase from *P. aeruginosa* applied to the hydrolytic kinetic resolution of a chiral ester. The combination of epPCR at a high mutation rate, saturation mutagenesis, and DNA shuffling constitutes a viable strategy in the quest to evolve enantioselectivity. However, the development of even more efficient strategies for exploring sequence space for enantioselectivity (and other properties such as activity and stability) are needed, possibly in combination with novel selection systems. The general concept delineated in this article can be applied to other enzymes, including additional lipases, esterases, aldolases, hydantoinases, epoxide hydrolases, nitrilases, and cyclohexanone monooxygenases. A number of

---

[71] M. T. Reetz, B. Brunner, T. Schneider, T. Schulz, C. M. Clouthier, and M. M. Kayser, submitted.
[72] M. Freemantle, *Chem. Eng. News* **81,** 31 (2003).

high-throughput *ee* assays are now available to suit the needs of almost any future project. Finally, the use of directed evolution of enantioselective hybrid catalysts bearing implanted synthetic transition metal centers may open new perspectives in catalysis.[23,73]

Acknowledgments

The author thanks Karl-Erich Jaeger, Bauke W. Dijkstra, Wim J. Quax, Roland M. Furstoss, Jacques C. Baratti, Margaret M. Kayser, and Walter Thiel for fruitful collaborative efforts. This work was supported by the Fonds der Chemischen Industrie and the European Community (projects QLK3-2000-00426 and QLRT-2001-00519).

[73] M. T. Reetz, M. Rentzsch, A. Pletsch, and M. Maywald, *Chimia* **56**, 721 (2002).

[22] Control of Stereoselectivity in Phosphotriesterase

By SUK-BONG HONG and FRANK M. RAUSHEL

Introduction

Phosphotriesterase (PTE), originally isolated from the soil bacterium *Pseudomonas diminuta*, catalyzes the hydrolysis of a wide range of organophosphate compounds, including agricultural insecticides and chemical warfare agents.[1] This enzyme has therefore received considerable attention as a reagent for the detoxification and remediation of organophosphorus nerve agents. However, nearly all of the most toxic chemical warfare agents, such as sarin, soman, and VX, have a stereogenic phosphorus center, and their inherent ability to inactivate the enzyme acetyl cholinesterase (AChE) varies substantially among the individual stereoisomers of these compounds. This observation indicates that the adverse biological activities of specific nerve agents are therefore highly dependent on the absolute configuration and the size of the substituents attached to the phosphorus center.

It has been shown that PTE catalyzes the hydrolysis of the insecticide paraoxon (I) at a rate that approaches the diffusion-controlled limit (Scheme 1).[2] A naturally occurring substrate for PTE has not, as yet, been identified. However, PTE has been shown to accept a substantial number of alterations in the makeup of the substituents attached to the central phosphorus core and to cleave these compounds with large catalytic rate enhancements. A fuller understanding of the overall reaction mechanism

[1] F. M. Raushel and H. M. Holden, *Adv. Enzymol. Relat. Areas Mol. Biol.* **74**, 51 (2000).
[2] D. P. Dumas, S. R. Caldwell, J. R. Wild, and F. M. Raushel, *J. Biol. Chem.* **264**, 19659 (1989).

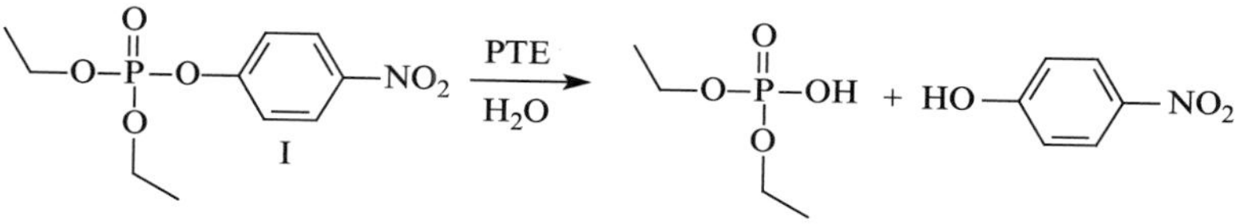

SCHEME 1.

of PTE and the elucidation of the asymmetric environment within the active site have helped to promote PTE as an effective tool for the remediation of hazardous chemical warfare agents and insecticides. Moreover, it has been demonstrated experimentally that the stereochemical constraints on substrate specificity can be manipulated systematically based on a rational reconstruction of the active site architecture.

Stereochemical Course for Substitution at Phosphorus Center

The mechanism of hydrolysis and overall stereochemistry of the reaction catalyzed by the bacterial phosphotriesterase was elucidated using ^{31}P nuclear magnetic resonance (NMR) spectroscopy with a chiral thiophosphonate substrate.[3] It was proposed initially that the hydrolysis reaction catalyzed by PTE could proceed via one of three possible reaction mechanisms: (a) a nucleophilic aromatic substitution by water on the C-1 carbon of the leaving-group phenol that would result in the direct cleavage of the C–O bond to yield diethyl phosphate and the substituted phenol; (b) two consecutive in-line displacement reactions at the phosphorus center that would occur through the formation of a phosphorylated-enzyme intermediate and subsequent hydrolysis by an activated water molecule; and (c) a single in-line displacement reaction at the phosphorus center by a direct nucleophilic attack of an enzyme-activated water molecule. ^{31}P NMR spectroscopy was used to distinguish mechanism a from mechanisms b and c. When paraoxon was hydrolyzed in the presence of a mixture of ^{16}O- and ^{18}O-labeled water, ^{18}O was found only in the product diethyl phosphate and not in the leaving-group phenol. This experiment demonstrated unambiguously that the hydrolysis reaction occurred with P–O bond cleavage rather than an attack on the aromatic ring of the phenol leaving group.

In order to distinguish mechanism b from mechanism c, chiral versions of the insecticide O-ethyl O-p-nitrophenyl phenylphosphonothioate (EPN) were synthesized. Preliminary experiments with racemic material demonstrated that only one of the two possible enantiomers was hydrolyzed at a significant rate. The identity of the more active substrate was

[3] V. E. Lewis, W. J. Donarski, J. R. Wild, and F. M. Raushel, *Biochemistry* **27**, 1591 (1988).

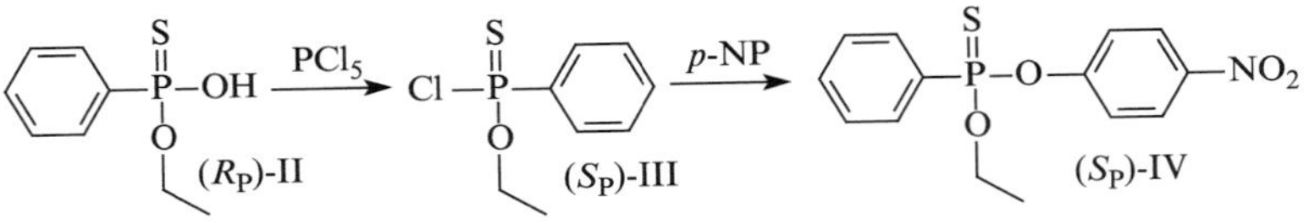

SCHEME 2.

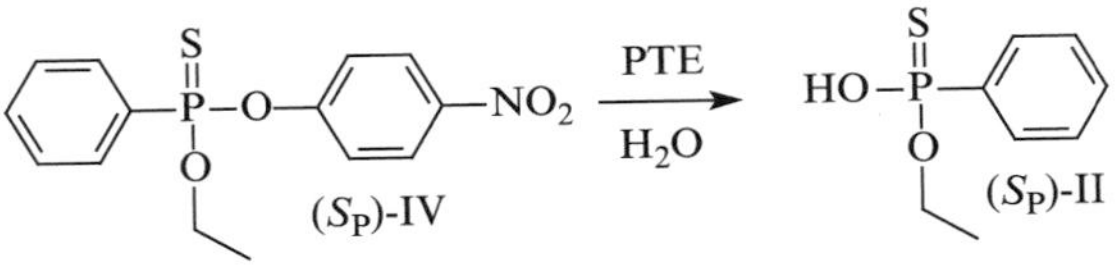

SCHEME 3.

established by the separate chiral synthesis of each enantiomer. In this effort a racemic mixture of ethyl phenylphosphonothioic acid (II) was synthesized and then separated into pure enantiomers by crystallization of the 1-phenylethylamine salts. The absolute configuration of the (R_P)- and (S_P)-ethyl phenylphosphonothioic acid was determined by X-ray diffraction analysis. The (R_P)-O-ethyl phenylphosphonothioic acid was activated by reaction with phosphorus pentachloride to produce the corresponding chlorophosphonothioate (III), which was then reacted with p-nitrophenolate, yielding the desired product (S_P)-EPN (IV). The chemical transformation from (R_P)-O-ethyl phosphonothioic acid (II) to (S_P)-EPN (IV) occurs by two consecutive inversions of configuration at phosphorus, and the reactions are summarized in Scheme 2.

The overall stereochemical course of the enzymatic hydrolysis reaction catalyzed by phosphotriesterase was determined using enantiomerically pure (S_P)-EPN as the initial substrate. Hydrolysis of this compound by PTE provided ethyl phenylphosphonothioic acid as the sole phosphorus-containing product. The addition of (S_C)-1-phenylethylamine as a chiral shift reagent resulted in a ^{31}P NMR spectrum of the reaction product that was clearly superimposable upon that of an authentic sample of (S_P)-O-ethyl phenylphosphonothioic acid, indicating that the S_P enantiomer of II was obtained by hydrolyzing (S_P)-EPN. These results were only consistent with a mechanism that proceeds with an overall inversion of configuration at the phosphorus center. Therefore, a phosphorylated enzyme intermediate is not formed and thus the reaction mechanism involved the direct attack of an activated water molecule on the phosphoryl center of the substrate. These reactions are summarized in Scheme 3.

Stereoselectivity of Wild-Type Phosphotriesterase

The three-dimensional X-ray structure of the Zn/Zn-substituted PTE with a bound substrate analog showed that a hydroxide from solvent and a modified lysine bridged the two metal ions within the active site. Moreover, the metal-bound hydroxide was oriented toward an in-line attack at the phosphorus center of the bound analog.[4,5] The X-ray structure of the inhibitor-bound enzyme also revealed that hydrophobic residues contributed predominantly to the formation of active site pockets. In order to delineate the structural determinants to the substrate specificity of PTE, one or both of the two ethoxy substituents of the original substrate paraoxon (I) were replaced with all possible combinations of methyl, ethyl, isopropyl, and phenyl groups (see Scheme 4).[6] The individual enantiomeric phosphotriesters were synthesized using L-proline methyl ester as a transitional chiral ligand according to published procedures.[7] Racemic mixtures and individual enantiomers were subsequently tested as substrates for the wild-type PTE by monitoring the appearance of p-nitrophenol at 400 nm. When racemic mixtures of phosphotriesters were hydrolyzed by PTE, an initial fast release of ~50% of the total expected p-nitrophenol was produced, followed by a slower release of the remaining p-nitrophenol as the reaction proceeded to completion.

Although paraoxon (I) was the best substrate for the wild-type enzyme under these reaction conditions, it was found that PTE was quite tolerant of specific alterations to this basic structure. The kinetic parameters differed substantially with the size and stereochemical orientation of the substituents surrounding the phosphorus core. Moreover, the results clearly demonstrated that the regions of the active site cavity that interact with the nonleaving group substituents of the substrate played a significant role in determining the rate of substrate turnover. For all of the racemic mixtures tested within this series of compounds, the specific optical rotations for the slower substrates were always found to be for the (+)-enantiomers. The substrate reactivity of the enzyme toward the faster (−)-enantiomers of these phosphotriesters was 1- to 100-fold greater than that toward the slower (+)-enantiomers. In order to determine absolute stereochemistry for the fast and slow enantiomers, the (+)-ethyl phenyl p-nitrophenyl phosphate was treated with sodium methoxide using a procedure known

[4] J. L. Vanhooke, M. M. Benning, F. M. Raushel, and H. M. Holden, *Biochemistry* **35,** 6020 (1996).

[5] M. M. Benning, S.-B. Hong, F. M. Raushel, and H. M. Holden, *J. Biol. Chem.* **275,** 30556 (2000).

[6] S.-B. Hong and F. M. Raushel, *Biochemistry* **38,** 1159 (1999).

[7] T. Koizumi, Y. Kobayashi, H. Amitani, and E. Yoshii, *J. Org. Chem.* **42,** 3459 (1977).

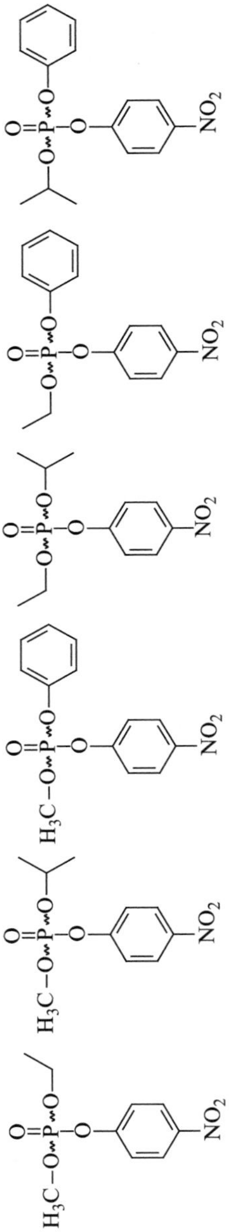

SCHEME 4.

TABLE I
RATIO OF ACTIVITIES FOR S_P/R_P ENANTIOMERS OF CHIRAL
ORGANOPHOSPHATE TRIESTERS

| | Enzyme | |
Compound	Wild type	G60A
V	1	12
VI	32	410
VII	90	13,000
VIII	10	241
IX	21	11,000
X	35	15,000

to occur with an inversion of configuration at the phosphorus center. The methyl ethyl phenyl phosphate product that was formed from this transformation was determined to be of the S_P configuration. This result demonstrates that the stereochemistry at the phosphorus center for the original slower organophosphate triester was of the R_P configuration. Therefore, S_P-($-$)-enantiomers are the preferred compounds for the series of substrates depicted in Scheme 4 for PTE. For the six pairs of chiral enantiomers used in this investigation, the ratios of k_{cat}/K_m for the S_P- and R_P-enantiomers are presented in Table I.

Systematic Alteration of Stereoselectivity by Recombination of
Active Site Residues

A comprehensive investigation of the stereochemical constraints on the substrate specificity for PTE in conjunction with the three-dimensional structure of PTE with a bound substrate analog identified three overlapping subsites that were responsible for the binding and orientation of the substrate within the active site. These subsites for substrate binding have been designated as *small, large,* and *leaving group* pockets. These subsites (presented in Fig. 1) reflect the observed stereoselectivity for the wild-type PTE by depicting the size of the substituents as spheres of different diameters.

The amino acids that come together within the active site to form the interior walls for these three subsites were identified from the crystal structure of PTE. The *small* subsite is formed from the side chains of Gly-60, Ile-106, Leu-303, and Ser-308. The *large* subsite contains His-254, His-257, Leu-271, and Met-317, whereas the residues that form the *leaving group* subsite are Trp-131, Phe-132, Phe-306, and Tyr-309. Systematic amino acid substitutions within the small pocket, the large pocket, and

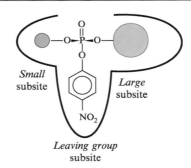

Fig. 1. The substrate-binding cavity of PTE.

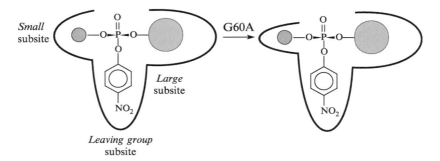

Fig. 2. Constriction of the small subsite to enhance stereoselectivity.

the leaving group pocket generated a small library of enzyme mutants that were characterized for alterations in the substrate profiles using the small library of modified organophosphate triesters.[8,9] The initial objective was to determine if the observed stereoselectivity of the wild-type enzyme could be *enhanced* without having to compromise the substantial turnover rates for the initially preferred enantiomer. It was anticipated that this design feature could be realized by constricting the size of the small pocket. Substantial success was achieved with the replacement of Gly-60 with a single alanine residue as presented in Fig. 2. The relative kinetic parameters for the G60A mutant with the six pairs of racemic substrates that appear in Scheme 4 are presented in Table I. For these substrates the observed stereoselectivity has been enhanced up to 500-fold. This mutant

[8] M. Chen-Goodspeed, M. A. Sogorb, F. Wu, S.-B. Hong, and F. M. Raushel, *Biochemistry* **40**, 1325 (2001).

[9] M. Chen-Goodspeed, M. A. Sogorb, F. Wu, and F. M. Raushel, *Biochemistry* **40**, 1332 (2001).

is able to distinguish between a methyl and an ethyl substituent by about an order of magnitude. This mutant has proven to be quite effective in the kinetic resolution of racemic organophosphates on a preparative scale.

Relaxation of the stereoselectivity observed with the wild-type PTE was achieved by enlargement of the small subsite in order to accommodate bulkier substituents within the active site of the protein. Replacement of Ile-106, Phe-132, and/or Ser-308 with alanine and/or glycine residues resulted in the loss of the chiral preference for the initially preferred S_P-enantiomers. These substitutions appreciably increased the rate of phosphotriesterase for the initially slower R_P-enantiomers. For example, with the I106G mutant the k_{cat}/K_m values for compounds IV and VII increased by factors of 74 and 35, respectively. For the I106G/S308G mutant, the value of k_{cat}/K_m for compound IX increased from 5.2×10^6 to $2.2 \times 10^8 \ M^{-1} \ s^{-1}$.

The most difficult transformation attempted was the reversal of the stereoselectivity exhibited by the wild-type enzyme. This required enhancement in the rate of turnover for the initially slower R_P-enantiomers while simultaneously reducing the rate of turnover for the initially faster S_P-enantiomers. The engineering strategy for this objective is presented in Fig. 3 and involved the enlargement of the small subsite and reduction in the size of the large subsite. In order for this plan to be effective, specific alterations to the active site had to be additive with one another. The expansion of the small subsite had already demonstrated that the initially slower R_P-enantiomers could be made to turn over at a faster rate. A mutagenic survey of those residues within the large subunit identified His-254 and His-257 as the most likely candidates for contraction of the large subsite via substitution with bulkier side chains. For example, with the mutant H257Y, the value of k_{cat}/K_m for S_P-VII decreased by a factor

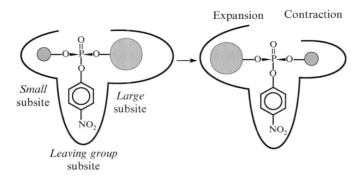

FIG. 3. Strategy for the reversal of stereoselectivity.

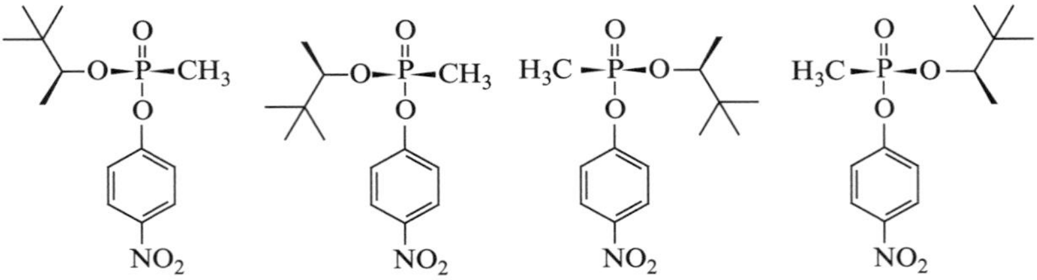

$S_P S_C$-XI $S_P R_C$-XI $R_P S_C$-XI $R_P R_C$-XI

SCHEME 5.

of 20, relative to the wild-type enzyme, but was essentially unchanged for the corresponding R_P-enantiomer. When this mutation (H257Y) was combined with alterations to the small subsite, a reversal in stereoselectivity was achieved for most of the racemic mixtures presented in Scheme 4. With the mutant I106G/F132G/S308G/H257Y the R_P-enantiomers are preferred over the S_P-enantiomers by factors of 190, 80, and 460 for compounds VII, IX, and X, respectively.

These results demonstrated quite clearly that significant alterations to the relative substrate profiles could be achieved with a small number of amino acid changes that are located exclusively within the active site. The most dramatic differences in stereoselectivity were achieved with compound X. The G60A mutant exhibited a preference for the S_P-enantiomer over the R_P-enantiomer by a factor of 15,000, whereas the I106G/F132G/S308G/H257Y mutant had a preference for the R_P-enantiomer over the S_P-enantiomer by a factor of 460. Between these two extremes there is nearly a seven orders of magnitude change in the observed substrate preference for one enantiomer over the other!

Another approach has proven to be effective in the identification of mutants of PTE that are improved for the hydrolysis of specific organophosphates nerve agents.[10] Scheme 5 shows the structures of four stereochemical analogs (XI) of the nerve agent soman. The toxicity of soman and the analogs shown in Scheme 5 are very much dependent on the stereochemistry at the chiral phosphorus and carbon centers. Unfortunately, the $S_P S_C$ isomer is the most toxic but the poorest diastereomeric substrate for the wild-type PTE. In an effort to improve the turnover rate for the most toxic isomer, two amino acid positions within the wild-type sequence (His-254 and His-257) were randomized to all possible combinations of the 20 amino acids (for a total of 400 possible mutants). This library

[10] C. R. Hill, W.-S. Li, J. B. Thoden, H. M. Holden, and F. M. Raushel, *J. Am. Chem. Soc.* **125,** 8990 (2003).

of mutants was screened in a whole cell format and a single mutant stood out from the rest, H254G/H257W. This mutant was two orders of magnitude better than the wild-type enzyme for the hydrolysis of S_PS_C-XI. This mutant was subsequently used as the parent enzyme, and L303 was replaced with all 20 natural amino acids. Of these mutants the substitution with threonine produced another order of magnitude increase in substrate turnover for the target substrate. The H254G/H257W/L303T mutant was improved in the value of k_{cat}/K_m by nearly three orders of magnitude. Again these studies demonstrated that a relative small number of mutations that were focused exclusively within the active site could elicit significant improvements in enzymatic turnover for specific target substrates.

Enzymatic Syntheses of Chiral Organophosphate
   and Organophosphothioates

Most of the published work on the chemical synthesis of chiral phosphotriesters has involved the use of stereospecific alcoholysis of an optically active intermediate or treatment with a chiral-resolving ligand, resulting in lengthy chemical procedures and the very time-consuming chiral separation of diastereomers.[7,11] The toxicity of many organophosphates depends on the chirality at the phosphorus center. Therefore, it is not surprising that much attention has been paid to the development of more efficient preparations of pure enantiomeric organophosphorus compounds for use as enzyme substrates and inhibitors. New approaches to the synthesis of chiral organophosphorus compounds have been developed using the engineered mutants of phosphotriesterase that facilitate the kinetic resolution of racemic mixtures of organophosphate triesters.[12] This approach is a direct consequence of the ability of the mutant G60A to leave the slow R_P-enantiomers intact when presented with a racemic mixture of phosphotriesters.[8] Likewise, hydrolysis of racemic mixtures of phosphotriesters catalyzed by the mutant I106G/F132G/H257Y enable the isolation of the corresponding S_P-enantiomers.[9] Treatment of racemic organophosphate triesters in 100 mM CHES buffer (pH 9.0) containing 5–20% acetonitrile with the G60A mutant enzyme yielded R_P triesters with >95% enantiomeric excess and isolate yields of >70%. When racemic mixtures of phosphotriesters were hydrolyzed by the I106A/F132A/H257W mutant, the S_P-enantiomers were obtained with >96% enantiomeric excess with >60% yield.

[11] C. R. Hall, T. D. Inch, G. J. Lewis, and R. A. Chittenden, *J. Chem. Soc. Chem. Commun.* 720 (1975).
[12] F. Wu, W.-S. Li, M. Chen-Goodspeed, M. A. Sogorb, and F. M. Raushel, *J. Am. Chem. Soc.* **122,** 10206 (2000).

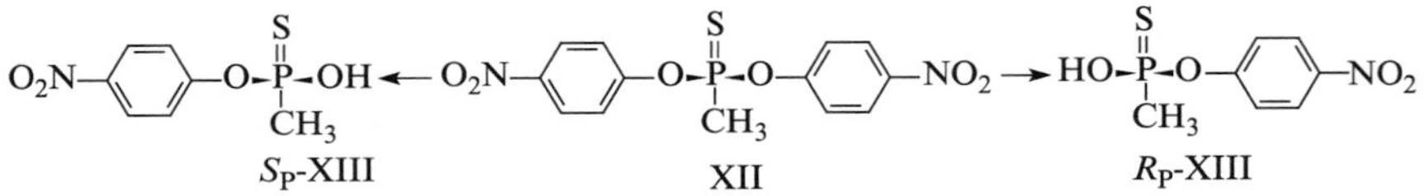

SCHEME 6.

The library of PTE mutants has also been exploited for the preparation of chiral organothiophosphates from prochiral starting materials.[13] For example, enzymatic hydrolysis of bis-p-nitrophenyl methyl phosphothioate (XII) would form the R_P isomer of XIII via cleavage of the proR substituent and the S_P isomer of XIII via cleavage of the proS substituent, as the stereochemical course of the reaction is known to proceed with a net inversion of configuration. These reaction products are illustrated in Scheme 6. Product stereochemistry was confirmed by ^{31}P NMR spectroscopy with a chiral shift reagent and chemical transformation to products of known stereochemistry. The wild-type PTE was found to cleave the proR substituent as predicted by analogy with prior stereochemical investigations.[6] The prochiral stereoselectivity catalyzed by the selected engineered mutants of PTE was reversed due to the alterations in the size of the substrate-binding cavity. The mutant I106A/F132A/H257W was found to cleave the proS substituent and thus reversed the stereoselectivity of the wild-type enzyme for the substrate bis-p-nitrophenyl methylphosphonate compared with the wild-type PTE.

Concluding Remarks

The active site of the bacterial phosphotriesterase has served as a predictable template for the modulation of substrate and stereoselectivity via site-directed mutagenesis. In this regard the stereoselectivity of PTE has been *enhanced, relaxed,* and *reversed* with remarkably few changes in the overall amino acid sequence. These modified proteins have proven to be of practical utility in the preparative isolation of chiral phosphorus products via the kinetic resolution of racemic mixtures.

Acknowledgments

This work was supported in part by the National Institutes of Health (GM 33894, GM 68550) and the Robert A. Welch Foundation (A-840).

[13] W.-S. Li, Y. Li, C. M. Hill, K. T. Lum, and F. M. Raushel, *J. Am. Chem. Soc.* **124,** 3498 (2000).

Section IV

Applications: Evolution of Biosynthetic Pathways

[23] Manipulation and Analysis of Polyketide Synthases

By Pawan Kumar, Chaitan Khosla, and Yi Tang

Introduction

Polyketides (PKSs) span a large range of medicinally important compounds, such as antibiotics, anticancer agents, immunosuppressants, and cholesterol-lowering agents.[1] Polyketides are produced in nature by microorganisms and may have suboptimal properties for their appropriate usage. Therefore, it is necessary to understand how their synthesis takes place in nature so that suitability and potency can be enhanced by the engineered derivatization of natural compounds, as well as the efficient production at economically relevant quantities. In type I PKSs individual domains catalyzing various steps of polyketide biosynthesis are organized in a modular fashion.[2] Since the 1990s, a large number of type I PKS gene clusters have been discovered.[3-7] The modular arrangement of individual gene clusters and the similarity between different clusters have motivated the combinatorial biosynthesis of novel polyketides by adopting various strategies, such as knockouts, domain swapping, module fusions, and linker swapping.[8,9] Numerous *in vivo* and *in vitro* tools have been developed to design, construct, and characterize recombinant PKSs and their biosynthetic products.[10] This article describes some of the techniques that have been used successfully over the years to analyze and engineer polyketide biosynthesis.

Discussion in the first section is organized based on the host organism used to manipulate the polyketide synthase of interest. Most of our work

[1] D. O'Hagan, "The Polyketide Metabolites." Ellis Howard, Chichester, UK, 1991.

[2] S. Donadio, M. J. Staver, J. B. McAlpine, S. J. Swanson, and L. Katz, *Science* **252,** 675 (1991).

[3] J. F. Aparicio, I. Molnar, T. Schwecke, A. Konig, S. F. Haydock, L. E. Khaw, J. Staunton, and P. F. Leadlay, *Gene* **169,** 9 (1996).

[4] H. Motamedi and A. Shafiee, *Eur. J. Biochem.* **256,** 528 (1998).

[5] Y. Xue, L. Zhao, H. W. Liu, and D. H. Sherman, *Proc. Natl. Acad. Sci. USA* **95,** 12111 (1998).

[6] A. R. Gandecha, S. L. Large, and E. Cundliffe, *Gene* **184,** 197 (1997).

[7] J. Cortes, S. F. Haydock, G. A. Roberts, D. J. Bevitt, and P. F. Leadlay, *Nature* **348,** 176 (1990).

[8] D. E. Cane, C. T. Walsh, and C. Khosla, *Science* **282,** 63 (1998).

[9] C. Khosla, *Chem. Rev.* **97,** 2577 (1997).

[10] B. J. Rawlings, *Nature Prod. Rep.* **14,** 523 (1997).

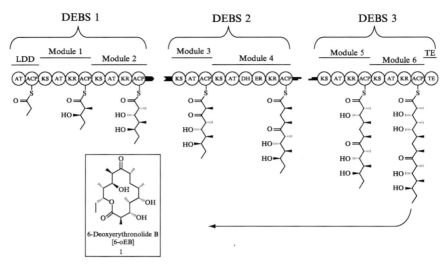

FIG. 1. Modular arrangements of 6-deoxyerythronolide B synthase. Reprinted with permission from Wu *et al.*, *Biochemistry*, **41**, (2002). Copyright (2002) American Chemical Society.

so far has involved the 6-deoxyerythronolide B synthase[11] (DEBS, Fig. 1); however, where relevant, this article highlights the use of other PKSs. The first section describes the genetic manipulation of PKS in actinomyces, either through chromosomal recombination in native hosts or through plasmid-based engineering in heterologous hosts (such as *Streptomyces coelicolor* or *Streptomyces lividans*). We will then discuss methods to express PKSs and produce polyketides in *Escherichia coli*. The second section discusses strategies used for hybrid PKS engineering. Subsequent sections review purification and enzymatic characterization tools used in our laboratory to analyze PKSs *in vitro*.

Biosynthesis of Polyketides

Manipulating PKS in Native Actinomycete Hosts

Prior to the use of heterologous hosts, characterization and engineering of type I PKS had, for the most part, been performed in natural producing organisms. Native hosts offer the advantages of a relatively high product

[11] C. Khosla, R. S. Gokhale, J. R. Jacobsen, and D. E. Cane, *Annu. Rev. Biochem.* **68**, 219 (1999).

yield and guaranteed functional protein expression. Furthermore, tailoring enzymes, such as oxygenases, methyltransferases, and glycosyltransferases, are all present, thus allowing full decoration of the PKS aglycone. As PKS gene sequences become available, homologous recombination is used frequently in constructing deletion, insertion, and hybrid PKS mutants at the chromosome level. We will not elaborate on the experimental details of genetic manipulations, as they have been well documented elsewhere.[12]

Several excellent examples of PKS engineering using homologous recombination in the native producer of erythromycin, *Saccharopolyspora erythraea*, have been reported.[13–16] Weber and coauthors[14] transformed libraries of integrative plasmids into *S. erythraea* and successfully knocked out *eryF*, which encodes a C6 oxygenase (not part of the modular PKS). The resulting strain was able to produce a fully accessorized erythromycin A without a hydroxyl group at C6. Leadlay and co-workers constructed a bimodular system using a double-crossover recombination. A 20-kb stretch of *S. erythraea* chromosomal DNA spanning from ACP2 to ACP6 was deleted.[13] The resulting truncated PKS was able to produce the expected triketide lactone with high efficiency. A different bimodular construct containing the entire *eryAI* produced the same lactone in a heterologous host (see later).

Both starter unit and extender unit specificities have been modified in polyketide producing actinomycetes using homologous recombination. For example, Kuhstoss *et al.* replaced the loading module of the spiramycin PKS in *Streptomyces ambofaciens* with that of the tylosin PKS.[17] Similarly, Leadlay and co-workers replaced the loading didomain at the N terminus of DEBS 1 in *S. erythraea* with the loading domain from the avermectin PKS[18] found in *Streptomyces avermitilis*.[19] In addition to propionate, the modified PKS inserted a large assortment of α-branched starter units into

[12] D. Hopwood *et al.*, "Genetic Manipulation of Streptomyces: A Laboratory Manual." The John Innes Foundation, Norwich, 1985.
[13] J. Cortes, K. E. Wiesmann, G. A. Roberts, M. J. Brown, J. Staunton, and P. F. Leadlay, *Science* **268**, 1487 (1995).
[14] J. M. Weber, J. O. Leung, S. J. Swanson, K. B. Idler, and J. B. McAlpine, *Science* **252**, 114 (1991).
[15] M. Oliynyk, M. J. Brown, J. Cortes, J. Staunton, and P. F. Leadlay, *Chem. Biol.* **3**, 833 (1996).
[16] D. L. Stassi *et al.*, *Proc. Natl. Acad. Sci. USA* **95**, 7305 (1998).
[17] S. Kuhstoss, M. Huber, J. R. Turner, J. W. Paschal, and R. N. Rao, *Gene* **183**, 231 (1996).
[18] H. Ikeda, T. Nonomiya, M. Usami, T. Ohta, and S. Omura, *Proc. Natl. Acad. Sci. USA* **96**, 9509 (1999).
[19] A. F. Marsden, B. Wilkinson, J. Cortes, N. J. Dunster, J. Staunton, and P. F. Leadlay, *Science* **279**, 199 (1998).

6-DEBS. Accessory proteins were able to transform the unnatural aglycons into the corresponding erythromycin A analogs.[20]

Katz and co-workers replaced an acyltransferase domain (AT4) found in the second module of DEBS2 with an ethylmalonyl-CoA-specific domain found in the niddamycin PKS cluster.[16] The resulting host was able to produce the expected 6-desmethyl-6-ethylerythromycin A in addition to erythromycin A when precursor molecules such as diethyl ethylmalonate were included in the growth media. The authors then performed a second chromosomal modification in which the *Streptomyces collinus* crotonyl-CoA reductase (*ccr*) gene was inserted under the control of a strong *ermE** promoter separate from the DEBS PKS genes. When overexpressed *in vivo*, *ccr* was able to synthesize butyryl-CoA, which can be converted to ethylmalonyl-CoA for utilization by the hybrid PKS. The *S. erythraea* strain equipped with the *ccr* gene produced the modified erythromycin A at levels comparable to that of wild type. Similar experiments have also been performed with malonyl-CoA-specific acyltransferase domains inserted into various DEBS modules in place of the cognate methylmalonyl-CoA-specific domains.[21] More recently, the methoxymalonyl-CoA-specific domains of the FK520 PKS have been replaced with both malonyl-CoA-specific and methylmalonyl-CoA-specific domains in *Streptomyces hygroscopicus.*[22]

Heterologous Expression of PKS

Streptomyces coelicolor *or* Streptomyces lividians *as Hosts for PKS Expression*

Despite the successful use of homologous recombination in engineering PKS hosts and polyketide products, genetic manipulation at the chromosomal level is a challenging task for someone unfamiliar with specialized microbiology techniques. Native hosts are often resistant to standard DNA transformation protocols and have slow doubling times. In addition, marker-independent gene replacement by homologous recombination is a low-frequency and consequently time-consuming process. To address these limitations, a host/vector system was developed in the model actinomycete, *S. coelicolor.*[23] Advantages of this system in studying PKSs include the following: (1) The *Streptomyces* host strain has been well studied, and a large repertoire of genetic tools are well established.[12] For example, DNA

[20] M. S. Pacey *et al.*, *J. Antibiot. (Tokyo)* **51,** 1029 (1998).

[21] X. Ruan *et al.*, *J. Bacteriol.* **179,** 6416 (1997).

[22] C. D. Reeves, L. M. Chung, Y. Liu, Q. Xue, J. R. Carney, W. P. Revill, and L. Katz, *J. Biol. Chem.* **277,** 9155 (2002).

[23] R. McDaniel, S. Ebert-Khosla, D. A. Hopwood, and C. Khosla, *Science* **262,** 1546 (1993).

transformation and *E. coli* conjugation can be performed with high efficiencies, and a range of selectable markers can be used. (2) PKS genes are encoded on a suitable shuttle vector that has a low copy number in *S. coelicolor* (where regulated gene expression is sought), but a high copy number in *E. coli* (where all DNA manipulation is performed). (3) The primary endogenous polyketide pathway (encoding actinorhodin biosynthesis) has been deleted from the host genome, thereby presenting a "clean" background. As a result, the phenotype associated with an introduced heterologous PKS gene cluster can be established readily and the potential of cross talk is minimized. (4) PKS genes can be expressed in a "natural"-like manner under control of the (deleted) *act* promoters.[24] The entire genome sequence of *S. coelicolor* has been reported,[25] thereby providing a blueprint of the metabolic regulatory pathways of the host and paving the way for the use of genomic and proteomic tools to enhance polyketide biosynthesis.

Our laboratory has made extensive use of this engineered strain of *S. coelicolor* (CH999) as a heterologous host for polyketide biosynthesis. This strain was initially developed for the biosynthesis and analysis of type II aromatic polyketides.[26] The PKS clusters involved in the biosynthesis of actinorhodin (*act*) and undecylprodigiosin pigment (*whiE*) were deleted from the chromosome using homologous recombinations. The resulting strain is effectively void of major polyketide production. CH999 has been instrumental in the elucidation of individual protein functions in the type II PKS systems. The production of pigmented aromatic polyketides can usually be observed visually in this host. Type I polyketides (including 6dEB) are generally colorless compounds and their biosynthesis cannot be visualized directly. Chromosomal deletion derivatives of other hosts have also been constructed, including *S. lividans*.[27] *S. erythraea*,[28,29] *Streptomyces venezuelae*[30] (which produces methymycin and picromycin), and *Amycolatopsis mediterranei*[31] (which produces rifamycin). Among them, *S. lividans* strain K4-114 has been used extensively for the engineered biosynthesis of polyketides.[22,32,33]

[24] M. A. Fernandez-Moreno, J. L. Caballero, D. A. Hopwood, and F. Malpartida, *Cell* **66,** 769 (1991).

[25] S. D. Bentley *et al.*, *Nature* **417,** 141 (2002).

[26] R. McDaniel, S. Ebert-Khosla, D. A. Hopwood, and C. Khosla, *Nature* **375,** 549 (1995).

[27] R. Ziermann and M. C. Betlach, *Biotechniques* **26,** 106 (1999).

[28] C. J. Rowe, J. Cortes, S. Gaisser, J. Staunton, and P. F. Leadlay, *Gene* **216,** 215 (1998).

[29] P. F. Long *et al.*, *Mol. Microbiol.* **43,** 1215 (2002).

[30] Y. J. Yoon, B. J. Beck, B. S. Kim, H. Y. Kang, K. A. Reynolds, and D. H. Sherman, *Chem. Biol.* **9,** 203 (2002).

[31] Z. Hu, D. Hunziker, C. R. Hutchinson, and C. Khosla, *Microbiology* **145,** 2335 (1999).

Shuttle Vectors

GENERAL CHARACTERISTICS OF A *STREPTOMYCES* SHUTTLE VECTOR. After choosing the appropriate host for heterologous expression, vectors containing the desired PKS genes can be transformed into the host using standard protocol, such as PEG-induced protoplast fusions.[12] The parent vector of numerous PKS constructs made in our laboratory, pRM5(23) (Fig. 2), contains the following essential features: (1) the plasmid contains a *ColEl* replicon for genetic replication in *E. coli*, which simplifies cloning. It also contains a SCP2* *Streptomyces* replicon for propagation in CH999 or K4-114. (2) The plasmid contains distinct selection markers for both *E. coli* (Amp/Kan) and *Streptomyces* (*tsr*, thiostrepton resistance). (3) The PKS genes of interest are cloned under the control of the *actI* promoter, whose activity is in turn regulated by *actII-ORF4*, also present on the same plasmid. Under this promoter/activator pair, high levels of PKS gene expression are observed at the onset of the *Streptomyces* stationary phase of mycelia growth (3–4 days on solid media and ~2 days in liquid cultures). The closely related host–vector systems consisting of CH999/pRM5 or K4-114/pRM5 have been used extensively in the study of both type I and type II polyketide biosynthesis. Derivatives of pRM5 containing other selectable markers in *Streptomyces*, such as apramycin, are also available for use in multi-plasmid experiments[34] (see later). Other host–vector pairs have also been reported by other workers[28,31]; their main features are similar to those described earlier.

CLONING CONSIDERATIONS. Actinomyce genomes are highly GC rich (~70%), thus a large number of AT-containing restriction sites are absent from most PKS genes, despite their large sizes (each chain extension module in a PKS is approximately 5 kb). Commonly used restriction sites in molecular biology, such as *Bam*HI (1, number of sites found in the DEBS gene cluster), *Xba*I (0), *Eco*RI (1), *Hind*III (0), *Spe*I (0), *Pac*I (0), *Nsi*I (0), *Nhe*I (2), *Pst*I (0), and *Nde*I (0), are rarely found in PKS genes of actinomyce origin, thus making these sites the most frequently used for the restriction cloning of individual domains, modules, and subunits. Insertion of some of these sites at the junctions between domains has paved the way for the combinatorial biosynthesis of 6dEB analogs.[35] In pRM5, unique

[32] Y. Kato, L. Bai, Q. Xue, W. P. Revill, T. W. Yu, and H. G. Floss, *J. Am. Chem. Soc.* **124**, 5268 (2002).

[33] R. Reid, M. Piagentini, E. Rodriguez, G. Ashley, N. Viswanathan, J. Carney, D. V. Santi, C. R. Hutchinson, and R. McDaniel, *Biochemistry* **42**, 72 (2003).

[34] Q. Xue, G. Ashley, C. R. Hutchinson, and D. V. Santi, *Proc. Natl. Acad. Sci. USA* **96**, 11740 (1999).

[35] R. McDaniel, A. Thamchaipenet, C. Gustafsson, H. Fu, M. Betlach, and G. Ashley, *Proc. Natl. Acad. Sci. USA* **96**, 1846 (1999).

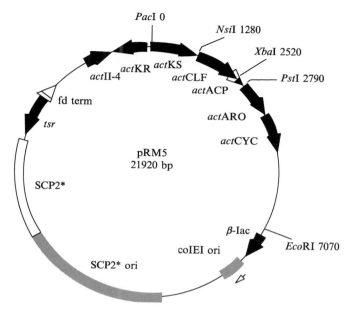

Fig. 2. The *E. coli/S. coelicolor* shuttle vector pRM5.

sites available for cloning are *Pac*I, *Xba*I, *Nsi*I, and *Eco*RI. In addition, *Pme*I, *Spe*I, *Mfe*I, *Nde*I, and *Nhe*I are absent and can be inserted to provide further cloning flexibility. The *Pac*I site is present downstream of the *act*I promoter and is the preferred site at the 5′ ends of foreign PKS genes. Multicistronic operons can be assembled under control of the *act*I promoter, provided that each gene contains its own ribosome-binding site (RBS). A commonly used RBS in our laboratory is GGAGGxxxxxcatATG. For some PKS genes, the *Nde*I site (underlined) is inserted at the start codon, which allows facile subcloning of the entire gene into a pET-derived *E. coli* vector (e.g., pET21, Novagen). Protein expression can be verified from the resulting *E. coli* construct, thereby providing a quick confirmation of ORF integrity (even if the protein is insoluble in *E. coli*).

The shuttle vector portion of pRM5 is ~14 kb in size. Due to the large sizes of PKS genes, expression plasmids derived from this vector are correspondingly large. For example, pCK12,[36] which only contains *eryAI*, is

[36] C. M. Kao, G. Luo, L. Katz, D. E. Cane, and C. Khosla, *J. Am. Chem. Soc.* **116**, 11612 (1994).

nearly 25 kb in length. Ligating two large DNA fragments is a challenging task. Smaller, self-replicating variants of the vector are often obtained, leading to a high number of background clones. The following protocols are used in our laboratory to ameliorate this problem.

1. We use a parent vector that has dual selection markers in *E. coli*. We have constructed a pRM5 derivative that carries resistance genes for both carbenicillin and kanamycin at either end of the insert. This vector greatly reduces background caused by erroneous plasmids.

2. Whenever possible, avoid gel purification of the digested vector. This is especially workable when the insert is smaller or equal in size to the insert liberated from the parent plasmid pRM5. If the insert is the same size as the original cassette, make sure a readily distinguishable restriction digest can be performed on the candidate clones. After *complete* digestion of the parent plasmid, ethanol precipitate the digestion mix directly and redissolve in a small amount of TE buffer (10 μl). Purify sufficient amounts of insert DNA and perform ligation at 16° overnight. An insert:vector ratio of 10:1 (molar ratio) is usually desired.

3. Avoid vortexing or harsh pipetting, as this can shear large plasmids. Mix DNA by tapping the tube gently with a finger.

Even with the most careful protocols, inserting an entire type I PKS cluster into a pRM5–derived shuttle vector is significantly more demanding and therefore requires unconventional cloning techniques. For example, the DEBS PKS spans over 30 kB and is one of the smaller type I PKS clusters. The first heterologous expression vector used for 6dEB biosynthesis, pCK7,[37] is constructed in *E. coli* using a homologous recombination step (Fig. 3). Briefly, the entire *eryAI* gene, a tetracycline resistance gene, and a 4-kb stretch of 3′ region of the *eryAIII* gene are inserted between the *Pac*I and *Eco*RI sites in pRM5 to yield the recipient vector pCK5. A donor vector (pCK6) is constructed by inserting a 4.1-kb stretch of the 3′ region of *EryAI* and the entire *EryAII* and *EryAIII* genes into the temperature-sensitive plasmid pSC101[38] (Cam resistance, viable at 30°, arrest at 44°). The *E. coli* strain harboring both pCK6 and pCK7 is subjected to four rounds of selections (Amp^R, Cam^R at 30°; Amp^R, Cam^R at 44°, Amp^R at 30°, and Amp^R, Cam^S, Tc^S at 44°) to yield the double recombinant pCK12, which contains the full set of DEBS genes. Transforming pCK7 into CH999 results in the biosynthesis of 6dEB. The aforementioned cloning procedures enables direct construction of large PKS genes and their mutants in

[37] C. M. Kao, L. Katz, and C. Khosla, *Science* **265**, 509 (1994).
[38] C. M. Hamilton, M. Aldea, B. K. Washburn, P. Babitzke, and S. R. Kushner, *J. Bacteriol.* **171**, 4617 (1989).

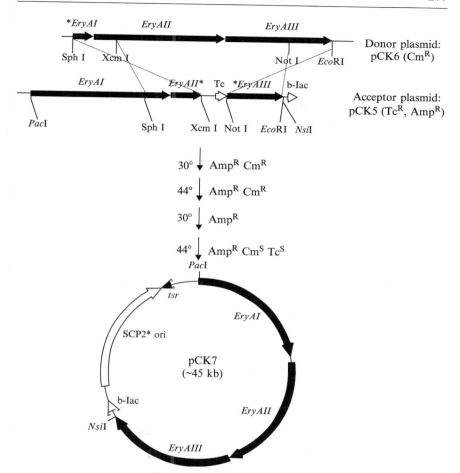

Fig. 3. Genetic strategy employed in the construction of pCK7. The recombinant was selected from *E. coli.* CH999 transformed with pCK7 produces 6-DEB.

E. coli. For example, pCK15,[39] which contains the full DEBS gene except module six, was also constructed using the *in vivo* recombination approach. *S. coelicolor* CH999 transformed with pCK15 produced the expected 12-member ring lactone (8*R*, 9*S*)-9,10-dihydro-8-methyl-9-hydroxy-10-deoxymethynolide.

[39] C. M. Kao, G. Luo, C. M. Kao, D. E. Cane, and C. Khosla, *J. Am. Chem. Soc.* **117,** 9105 (1995).

In-gel ligation techniques have been used extensively in constructing shuttle vectors. A number of such protocols are available in the literature and on the internet. This section describes briefly one version of the procedure that has been used in our laboratory.

1. Separate digested vector and insert DNA using SeaPlaque GTG agaraose (FMC). We use TAE buffer with a low EDTA concentration (0.1 mM), as EDTA chelates Mg^{2+} required for T4 ligase activity.

2. Excise DNA pieces from gel. Minimize gel slice volume and minimize UV exposure.

3. Set up ligation mixture [e.g., 15 μl H_2O, 8 μl 5× ligation buffer, 1 μl T4 DNA ligase (4 U/μl)]. The final ligation mixture volume is 40 μl.

4. Melt the gel slices at 70°, add desired volumes of vector, and insert into ligation mix (e.g., 5 μl vector and 15 μl insert) as soon as the gel slices melt. Mix gently by tapping the tube before the mixture solidifies. (Do not vortex or pipette mix.) Incubate overnight.

5. Melt the reaction mixture at 70°, dilute with 40 μl H_2O, and transform 20 μl into 100-μl aliquots of calcium-competent cells. Usually ~100 colonies can be observed for final plasmid sizes between 30 and 50 kb.

After construction of the target shuttle vector, DNA can be used to transform a chosen *Streptomyces* host. Methylated DNA (from XL-1 or most cloning strains) can be used to transform *S. lividans* directly, but must be passed through a methylase-deficient *E. coli* host before being transformed into *S. coelicolor*. We use *E. coli* strain GM2163 (New England Biolabs) for this purpose because of its relatively high transformation efficiencies. Other methylase-deficient hosts such as ET12567 are also suitable.

Escherichia coli *as a Host for Functional PKS Expression*

Escherichia coli is arguably the best characterized microorganism. It is extremely simple to handle with excellent growth characteristics. *E. coli* is intrinsically a clean host without any PKS production capabilities, thus providing an excellent opportunity for fundamental studies of PKSs. Furthermore, developing fermentation processes for *E. coli* is considerably easier and economical than that for most natural biological sources, such as the actinomycetes. *E. coli* also offers a faster turnaround for protein production as compared to actinomycete hosts (1 day vs 4 days). This removes the bottleneck in production and screening of a large number of proteins while engineering PKSs for novel properties.

The widely used *E. coli* strain BL21(DE3) was engineered to address the following limitations.

a. *Escherichia coli* does not produce polyketides naturally and therefore most polyketide precursors are not available in the cell. For example, biosynthesis of macrolides such as 6dEB requires methylmalonyl-CoA as an extender unit, which is not present in *E. coli.*[40]

b. A phosphopantetheine arm needs to be transferred from CoA to an ACP as a posttranslational modification in order for the ACP to be active.[41–43] Broad spectrum phosphopantetheinyl transferases (PPTases), which catalyze this reaction on ACP domains from modular PKSs, are absent from *E. coli.*[44]

It was therefore essential to develop a metabolically engineered *E. coli* strain that produced methylmalonyl-CoA from a simple carbon source, such as propionate, and that expressed a suitable phosphopantetheinyl transferase to convert apo ACPs into holo form. Pfeifer *et al.*[45] inserted the *sfp* phosphopantetheinyl transferase gene into the *prp* operon of BL21(DE3) using homologous recombination. The *sfp* gene, as well as the endogenous *prpE* gene (encoding propionyl-CoA synthetase), is under control of the T7 promoter. The resulting BAP1 strain was unable to catabolize propionate because of the deletion of *prpRBCD*, but could accumulate propionyl-CoA upon feeding of propionate due to intact prpE function. Propionyl CoA carboxylase (pcc) was expressed through a plasmid, which converted propionyl-CoA into methylmalonyl-CoA.

Cloning Strategies. In our laboratory, *E. coli* is used as a host to express individual PKS proteins for structural and mechanistic studies[46–59] and also to coexpress multiple PKS proteins for polyketide production.[45,60] We use

[40] T. Haller, T. Buckel, J. Retey, and J. A. Gerlt, *Biochemistry* **39,** 4622 (2000).
[41] R. D. Simoni, R. S. Criddle, and P. K. Stumpf, *J. Biol. Chem.* **242,** 573 (1967).
[42] T. C. Vanaman, S. J. Wakil, and R. L. Hill, *J. Biol. Chem.* **243,** 6420 (1968).
[43] J. Elovson and P. R. Vagelos, *J. Biol. Chem.* **243,** 3603 (1968).
[44] R. H. Lambalot, A. M. Gehring, R. S. Flugel, P. Zuber, M. LaCelle, M. A. Marahiel, R. Reid, C. Khosla, and C. T. Walsh, *Chem. Biol.* **3,** 923 (1996).
[45] B. A. Pfeifer, S. J. Admiraal, H. Gramajo, D. E. Cane, and C. Khosla, *Science* **291,** 1790 (2001).
[46] R. S. Gokhale, S. Y. Tsuji, D. E. Cane, and C. Khosla, *Science* **284,** 482 (1999).
[47] R. S. Gokhale, D. Hunziker, D. E. Cane, and C. Khosla, *Chem. Biol.* **6,** 117 (1999).
[48] J. Lau, D. E. Cane, and C. Khosla, *Biochemistry* **39,** 10514 (2000).
[49] S. Y. Tsuji, D. E. Cane, and C. Khosla, *Biochemistry* **40,** 2326 (2001).
[50] S. J. Admiraal, C. T. Walsh, and C. Khosla, *Biochemistry* **40,** 6116 (2001).
[51] N. Wu, S. Y. Tsuji, D. E. Cane, and C. Khosla, *J. Am. Chem. Soc.* **123,** 6465 (2001).
[52] S. C. Tsai, L. J. Miercke, J. Krucinski, R. Gokhale, J. C. Chen, P. G. Foster, D. E. Cane, C. Khosla, and R. M. Stroud, *Proc. Natl. Acad. Sci. USA* **98,** 14808 (2001).
[53] N. Wu, D. E. Cane, and C. Khosla, *Biochemistry* **41,** 5056 (2002).
[54] H. Lu, S. C. Tsai, C. Khosla, and D. E. Cane, *Biochemistry* **41,** 12590 (2002).
[55] G. F. Liou, J. Lau, D. E. Cane, and C. Khosla, *Biochemistry* **42,** 200 (2003).

pET28 and pET21 as vectors for protein expression. The PKS cassette is generally cloned between *Nde*I and *Eco*RI sites, yielding N term(pET28), C term(pET21), or both N and C term(pET28) 6xHis tags. For cloning two cassettes in the same plasmid, *Pst*I/*Pac*I (pET28) or *Nsi*I/*Pac*I(pET21) sites are inserted by replacing the fragment between *Bpu*1102I–*Dra*III of these vectors with a polylinker possessing the *Bpu*1102I, *Nsi*I, *Pst*I, *Pac*I, and *Dra*III sites. The first cassette is then introduced between *Nde*I and *Eco*RI. The second cassette is first cloned in a pUC plasmid between *Nde*I and *Eco*RI with *Nsi*I or *Pst*I and *Pac*I flanking above cloning sites. A ribosomal binding site and appropriate His tags are also cloned in frame. Now, the second cassette is cloned *Nsi*I/*Pac*I or *Pst*I/*Pac*I in the aforementioned modified expression plasmids from the aforementioned shuttle plasmids.

An alternative strategy used in our laboratory to assemble multicistronic operons involves the use of *Xba*I and *Avr*II sites, which have compatible overhangs. The first cassette is inserted between a unique *Xba*I and other unique site. Another cassette is cloned between *Xba*I and *Avr*II with the RBS downstream of *Xba*I and a stop codon before *Avr*II in a shuttle vector. This piece is digested with *Xba*I and *Avr*II and is inserted between *Xba*I in the plasmid carrying the first gene. *Xba*I/*Avr*II fusion results in a noncleavable site with either enzyme. This process is repeated and any number of cassettes are inserted in the same vector.

Culture Condition Optimization. Escherichia coli BAP1 cells are transformed by the electroporation technique. For the transformation of multiple plasmids containing different resistance markers, plasmids are first mixed together followed by the transformation of BAP1 cells with the mixture. Colonies containing the plasmid(s) are selected over LB plates with the appropriate resistance(s). Overnight starter cultures are grown in 5 ml volume at 30°. Cells are pelleted and resuspended in 1 mL LB media. Ten milliliters of LB is then inoculated by 100 μl of the aforementioned culture in a number of 50-ml falcon tubes. Tubes are shaken at 250 rpm at 37° until OD_{600} reaches a value of 0.6. The culture is then cooled in a water bath to room temperature. Varying amounts of isopropyl-β-D-thiogalactopyranoside (IPTG) are then added (typically 10, 100, and 1000 μM) to

[56] C. N. Boddy, T. L. Schneider, K. Hotta, C. T. Walsh, and C. Khosla, *J. Am. Chem. Soc.* **125**, 3428 (2003).

[57] P. Kumar, Q. Li, D. E. Cane, and C. Khosla, *J. Am. Chem. Soc.* **125**, 4097 (2003).

[58] M. Hans, A. Hornung, A. Dziarnowski, D. E. Cane, and C. Khosla, *J. Am. Chem. Soc.* **125**, 5366 (2003).

[59] K. Watanabe, C. C. Wang, C. N. Boddy, D. E. Cane, and C. Khosla, *J. Biol. Chem.* (2003).

[60] K. Watanabe, M. A. Rude, C. T. Walsh, and C. Khosla, *Proc. Natl. Acad. Sci. USA* **100**, 9774 (2003).

induce protein production, and the tubes are shaken further at 13, 18, 22, and 30° to find the optimal temperature of protein production. Protein production is induced for varying amounts of time (between 8 h for high temperatures to 48 h for low temperatures). Cells are then harvested and resuspended in 1 ml of 100 mM sodium phosphate, pH 7.2, 1 mM EDTA. Cells are disrupted by sonication. Cell debris and insoluble proteins are removed by spinning the lysates in 1.5-ml Eppendorf tubes at 13,000 rpm for 15 min. The whole cell lysate and the soluble fraction of proteins are compared by running a SDS–PAGE gel. The temperature and the IPTG concentration corresponding to high protein expression, as well as larger soluble fraction, are used as optimal conditions for protein and/or product production on a larger scale. For most PKS proteins, an IPTG concentration of 100 μM and an induction temperature of 22° have been found to be optimal in achieving highly active proteins with high yields.

PKS Engineering

DEBS PKS Domain Engineering

DEBS PKS contains six modules and 28 catalytic domains.[2,61] Among the individual catalytic units, the AT domains and the reductive (KR/DH/ER) domains are responsible for extender unit selection and polyketide chain tailoring, respectively.[2] These domains contribute extensively to the structural diversity among type I polyketides.[62] Domain swapping between different DEBS modules or different PKSs has been accomplished mostly in DEBS module 2,[63–67] module 5,[2,16] and module 6,[35,68] as well as in other DEBS modules.[15,69] Chimeric DEBS PKSs containing different AT and/or reductive domains have yielded a library of unnatural 6-DEB analogs, demonstrating the surprising tolerance of the overall DEBS PKS to

[61] S. Donadio and L. Katz, *Gene* **111**, 51 (1992).

[62] B. J. Rawlings, *Nature Prod. Rep.* **16**, 425 (1999).

[63] R. McDaniel, C. M. Kao, H. Fu, P. Hevezi, C. Gustafsson, M. Betlach, G. Ashley, D. E. Cane, and C. Khosla, *J. Am. Chem. Soc.* **119**, 4309 (1997).

[64] R. McDaniel, C. M. Kao, S. J. Hwang, and C. Khosla, *Chem. Biol.* **4**, 667 (1997).

[65] C. M. Kao, M. McPherson, R. McDaniel, H. Fu, D. E. Cane, and C. Khosla, *J. Am. Chem. Soc.* **120**, 2478 (1998).

[66] C. M. Kao, R. McDaniel, R. McDaniel, H. Fu, D. Cane, and C. Khosla, *J. Am. Chem. Soc.* **119**, 11339 (1997).

[67] D. Bedford, J. R. Jacobsen, G. Luo, D. E. Cane, and C. Khosla, *Chem. Biol.* **3**, 827 (1996).

[68] L. Liu, A. Thamchaipenet, H. Fu, M. Betlach, and G. Ashley, *J. Am. Chem. Soc.* **119**, 10553 (1997).

[69] S. Donadio, J. B. McAlpine, P. J. Sheldon, M. Jackson, and L. Katz, *Proc. Natl. Acad. Sci. USA* **90**, 7119 (1993).

TABLE I

AT DOMAINS OF DIFFERENT EXTENDER UNIT SPECIFICITY AND REDUCTION DOMAINS OF
DIFFERENT CATALYTIC FUNCTIONS FOUND IN TYPE I PKSs

AT specificities		Reductive domains	
Malonyl-CoA	RAPS AT2	No reduction	DEBS KR3[b]
Methylmalonyl-CoA	DEBS ATs	β-keto reduction	DEBS KR5
			RAPS KR4[c]
Ethylmalonyl-CoA	Nid AT5[a]	β-OH dehydration	RAPS DH/KR4
Methoxymalonyl-CoA	FK520 AT8	Enol reduction[d]	RAPS DH/ER/KRI

[a] Niddamycin.
[b] A nonfunctional ketoreductase.
[c] RAPS KR4 reduces a β-keto moiety with opposite stereochemistry to that of
   DEBS KR5.
[d] Complete reduction from a β-keto moiety to an alkylacyl moiety.

individual domain changes in these particular modules.[11] A partial compilation of AT domains with different CoA specificity and reductive domains with different tailoring properties are listed in Table I.

Leadlay and co-workers compared amino acid sequences of the methylmalonyl-specific DEBS AT1 and the malonyl-specific RAPS AT2 and noted a divergent region between two proteins that may determine extender unit specificities.[15] Replacing the 850-bp fragment from DEBS AT1 in a construct containing *eryAI* with the corresponding fragment from RAPS AT2 resulted in biosynthesis of the triketide lactone (2S, 3S, 5R)-2-methyl-3,5-dihydroxy-*n*-heptanoic acid δ-lactone, the predicted product of the hybrid *eryAI*. Similar substitutions using alternative domain junctions have been performed in each of the DEBS modules, as well as in other modular PKSs by several laboratories. Notwithstanding the simplicity and power of this approach, it suffers from the limitation that the heterologous AT domain perturbs domain–domain interactions in the module, thereby impairing their catalytic efficiency. More recently, Reeves et al.[70] mutated several residues in DEBS AT4 (methylmalonyl specific) to consensus residues found in malonyl-specific ATs and observed biosynthesis of the 6-dEB analog with the expected modification; however, some amount of 6-dEB was also produced, suggesting that the engineered AT domain had relaxed specificity. An alternative approach to altering the extender unit specificity of individual modules involves inactivation of the endogenous

[70] C. D. Reeves, S. Murli, G. W. Ashley, M. Piagentini, C. R. Hutchinson, and R. McDaniel, *Biochemistry* **40,** 15464 (2001).

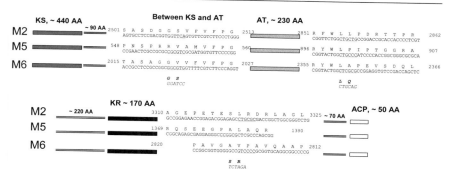

FIG. 4. Sequence alignments between DEBS modules 2, 5, and 6. The individual catalytic domains are shown. The junctions between domains where restriction sites are introduced are shown. For each of the modules, the indicated restrictions sites are introduced (underlined) to facilitate domain shuffling.

AT domain via site-specific mutagenesis, followed by coexpression of a kinetically fast type II acyl transferase in *trans*.[71] This strategy has the advantages of minimal perturbations on the tertiary and quaternary structure of modules and a clean alteration of extender unit specificity, but is restricted in scope by the limited repertoire of stand-alone acyl transferases, most of which have malonyl-CoA specificity.

In order to make chimeric PKSs in an efficient and combinatorial fashion, one must be able to clone various AT or reduction domains into DEBS PKS as cassettes.[35] Toward this end, unique restriction sites have been introduced at junctions between (1) KS and AT (e.g., *BamHI*), (2) AT and KR/DH/ER (e.g., *PstI*), and (3) KR/DH/ER (e.g., *XbaI*). The restriction sites are introduced using standard polymerase chain reaction (PCR) mutagenesis techniques. Activity of the entire DEBS PKS carrying amino acid substitutions encoded by the new restriction sites has been verified to be identical to the wild-type protein. Substitution domains can be amplified from the original cluster with primers containing the flanking restriction sites. Defining junction points for introducing artificial sites or designing PCR primers can be aided by performing multiple sequence alignments between the homologous modules.[61] An abbreviated example of such an alignment between DEBS modules 2, 5, and 6 is shown in Fig. 4. The locations of the engineered sites have been indicated in the alignment. *BamHI* and *PstI* sites are located immediately before and after the AT domains, whereas the *XbaI* site is introduced immediately before the

[71] P. Kumar, A. T. Koppisch, D. E. Cane, and C. Khosla, *J. Am. Chem. Soc.* **125,** 14307 (2003).

putative ACP domains. The engineered sites are introduced to coincide with the correct reading frames of the original genes so that changing a gene cassette would not affect the translation of the target protein. Hybrid PKS modules constructed using these methods are subcloned first, inserted into the shuttle vector (e.g., pCK7), and followed by transformation into *S. coelicolor* or *S. livians* for product analysis.

Using the aforementioned cloning strategy, McDaniel et al.[35] were able to generate single, double, and triple mutants of DEBS PKS containing AT and reductive domains from different sources. Most of the engineered DEBS PKSs were able to produce the predicted 6-DEB analogs at isolatable quantities. Over 50 6-DEB analogs were biosynthesized in this effort, a feat considered to be a milestone in the combinatorial biosynthesis of type *I* polyketides.

Xue and co-workers[34] developed a more flexible, multiple plasmid approach for PKS combinatorial biosynthesis. Each of the three DEBS genes was cloned between either *PacI* and *EcoRI* or *PacI* and *XbaI* sites in a pRM5-derived shuttle vector. The vectors have the same *Streptomyces* SPC2* origin of replication, but with different selection markers (for thiostrepton, hygromycin, and apramycin). Modifications to each of the DEBS genes, either through site-directed mutagenesis or through cassette-mediated domain shuffling, can be made separately on relatively smaller plasmids. A major advantage of this method is that mutant genes for each of the DEBS proteins can be constructed and combined rapidly. A potential rate-limiting step is the efficiency of transforming three plasmids into a *Streptomyces* host. Sequential transformation of plasmids is a time-consuming process, which can undermine the advantages of this approach.

DEBS PKS Module Engineering

An alternative strategy for combinatorial biosynthesis is to rearrange intact modules within a PKS or combine modules from different PKSs.[46] Early attempts at module swapping experiments failed mainly because communication between modules facilitated by the natural linkers between them was ignored. Both intrapeptide linkers (such as that between DEBS M1 and M2) and interpeptide linkers (such as that between the C terminus region of DEBS M2 and the N terminus region of DEBS M3) have been shown to be important in establishing polyketide chain transfer between two modules in DEBS PKS[72] and other type I PKSs.[60] Flexible cloning strategies have been developed to preserve the natural linkers when repositioning PKS modules (see later).

---

[72] R. S. Gokhale and C. Khosla, *Curr. Opin. Chem. Biol.* **4,** 22 (2000).

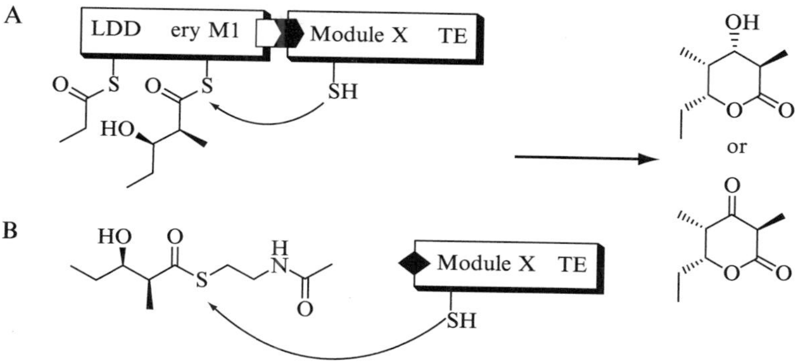

FIG. 5. Assaying individual module activity with either (A) bimodular fusion with DEBS module 1 or (B) a free-standing module. In the bimodular construct, the linker normally found between DEBS modules 1 and 2 is preserved. A N-terminal linker as those found before DEBS M3 and DEBS M5 is usually inserted at the N terminus of the free standing module.

A first step in module engineering is to determine the substrate specificity and activity of a module outside of its natural context, either as a free-standing protein or as a fusion to DEBS module 1 (Fig. 5).[59] When testing a free-standing module, N-acetylcysteamine (SNAC)[73] esters are typically used as electrophile donors and turnover is monitored by quantifying the production of a triketide product.[51] Alternatively, acyl-ACPs, generated via chemoenzymatic procedures, can also be used as electrophile donors. Relative to equivalent SNAC esters, they have significantly higher specificity (due to a considerably lower K_m, but sometimes also due to a higher k_{cat}).[53] Individual modules are expressed with N-terminal interpeptide linkers (e.g., the linker before DEBS module 3). Adding the N-terminal linker sometimes restores activities to modules that were otherwise inactive.[46] When expressed as a fusion to DEBS module 1, the targeted module is assayed for its ability to elongate the enzyme-bound diketide into a triketide lactone. The natural linker between DEBS modules 1 and 2, which is crucial for intermodule chain transfer, is preserved. In both assays, a thioesterase domain is often fused to the C termini of the target module to facilitate triketide lactone turnover, although some multimodular PKSs (e.g., RifA) can turn over even in the absence of an active release mechanism.[60] Invariably, a module that fails to produce a triketide product in

[73] J. A. Chuck, M. McPherson, H. Huang, J. R. Jacobsen, C. Khosla, and D. E. Cane, *Chem. Biol.* **4,** 757 (1997).

these assays is inactive when positioned in more complex environments, such as a part of a trimodular or larger PKS scaffold.[59] Assaying individual modules from DEBS, pikromycin (Pik), and rifamycin (Rif) PKSs has shown that while some modules are perfectly competent when transferred into a foreign context, other modules are incapable of processing unnatural substrates.[59]

Similar to the cassette shuffling strategy used in domain shuffling, modules are also cloned as interchangeable cassettes.[59] Cassettes encoding modules were designed so that (1) the N terminus of each module can be positioned after either an intrapeptide or an interpeptide linker; (2) a C-terminal linker can be inserted immediately following the module; (3) a thioesterase domain can be cloned immediately following a module instead of a linker, if needed; (4) all the hybrid constructs have the correct reading frames; and (5) the entire construct (linker–module–linker/TE) can be transferred readily to an *E. coli* expression vector. To satisfy these prerequisites, each of the DEBS module genes were outfitted with a *BsaBI* site before its KS domain and a *SpeI* site following its ACP domain. The *BsaBI* site was chosen because of its infrequent occurrence and the presence of a highly conserved tetrapeptide PIAI near the consensus start of most of the KS domains (4 out 6 in DEBS KSs, 8 out of 10 Rif KSs), which is partially encoded by the *BsaBI* site (see Fig. 6).[46] PCR mutagenesis was used to introduce the *SpeI* sites at the 3′ ends of the modules. *NdeI* sites and *EcoRI* sites were introduced at the 5′ and 3′ ends of each module, respectively, so that they can be conveniently cloned into the corresponding sites of pET-derived *E. coli* vectors. Thioesterases were cloned as *SpeI–EcoRI* cassettes and can be introduced downstream of each of the modules, if needed.

M1	1510	GGC	GAA	CCG	GTC	GCG	GTC	GTC	GCG	1533
		G	E	P	V	A	V	V	A	
M2	5925	GAC	GAG	CCG	ATC	GCG	ATC	GTC	GGC	5948
		D	E	P	I	A	I	V	G	
M3	88	TCC	GAC	CCG	ATC	GCC	ATC	GTC	AGC	111
		S	D	P	I	A	I	V	S	
M4	4462	TCC	GAG	CCC	ATC	GCC	ATC	GTC	GGC	4485
		S	E	P	I	A	I	V	G	
M5	112	GGT	GAG	CCG	ATC	GCG	ATC	GTC	GGC	135
		G	E	P	I	A	I	V	G	
M6	4465	GAC	GAG	CCG	ATC	GCG	ATC	GTC	GGC	4488
		D	D	P	I	A	I	V	G	

FIG. 6. Alignment of the beginnings of each module showing the semiconserved *BsaBI* site (GAT*NNNN*ATC) used in module-swapping experiments.

In Vitro Characterization of PKS Proteins

Protein Purification

Early Stage Purification. In a typical bacterial cell lysate, PKS proteins can be precipitated readily at relatively low ammonium sulfate concentrations. Harnessing this property, PKS proteins were originally prepared for *in vitro* analysis by ammonium sulfate precipitation.[74] The cell pellet is dissolved in the disruption buffer, which contains 250 mM sodium phosphate (pH 7.1), 0.3 M NaCl, 2 mM dithiothreitol (DTT), and protease inhibitors (1 mM benzamidine, 2 mM EDTA, 3 mg/liter leupeptin, 3 mg/liter pepstatin), and 30% glycerol. The presence of high phosphate and glycerol concentrations is crucial for retaining the high activity of PKS proteins, especially when purified from dilute lysates. Cells are disrupted by sonication (five bursts of 30 s, Branson sonifier) on ice. The lysate is centrifuged for 45 min–1.5 h at 19,200g to remove cell debris and insoluble proteins. Nucleic acids are precipitated with poly(ethylenimine) (0.2%) and removed by centrifugation (20 min at 23,100g). The protein solution is saturated with ammonium sulfate to 50% and precipitated overnight. After centrifugation (30 min, 31,100g), the pellet containing the DEBS proteins is redissolved in the appropriate buffer [typically 100 mM sodium phosphate (pH 7.1), 2 mM DTT, 2 mM EDTA, and 10–20% glycerol] and desalted on a Sephadex G25 M column using the appropriate buffer. This procedure yields close to two-fold purification and 90% PKS protein recovery.

Further Purification. To purify PKS proteins further, approximately 3 ml of the ammonium sulfate fraction saturated to 45% obtained from the aforementioned procedure can be loaded on a size-exclusion chromatography column (Biogel A, 150 ml, 3 × 75 cm).[75] PKS proteins are eluted using buffer A (100 mM sodium phosphate, pH 7.1, 2 mM DTT, 2 mM EDTA, 1 mM benzamidine, and 10% glycerol) with a flow rate of 0.4 ml/min in a volume range of typically 30–50 ml. Fivefold purification can be achieved by size-exclusion chromatography. Most PKS proteins have low calculated isoelectric points (pI); hence, it was expected that they would bind to an anion-exchange chromatography column. Pooled fractions from the aforementioned step can be applied to an anion-exchange chromatography column (resource Q; 6-ml column). A gradient from 0 to 0.22 M NaCl in buffer A is run at 1 ml/min for 20 min, followed by a shallow gradient from 0.22 to 0.28 M NaCl at 1 ml/min for 45 min. PKS proteins

[74] R. Pieper, S. Ebert-Khosla, D. Cane, and C. Khosla, *Biochemistry* **35,** 2054 (1996).
[75] R. Pieper, R. S. Gokhale, G. Luo, D. E. Cane, and C. Khosla, *Biochemistry* **36,** 1846 (1997).

elute in the range of 0.25–0.27 M NaCl. As high as 70-fold purification can be obtained after this step. Proteins are then concentrated to a final concentration of 1–2 mg/ml in buffer A using Centriprep 30 membranes (Amicon). It is also possible to purify PKS proteins further by sucrose gradient centrifugation. Proteins are equilibrated in buffer A (1 mg/ml) and centrifuged in a gradient of 10–40% sucrose in buffer A (Beckman Sw4 rotor, 30,000 rpm, 22 h). Fractions containing pure proteins are concentrated as described earlier. This additional step can yield 30-fold purification.

Current Purification. Although excellent purity can be achieved by the aforementioned multistep purification scheme, it is laborious and results in significant losses. Once expression protocols were develope in *E. coli,* shorter purification protocols followed. In particular, for individual modules, the Biogel A step was replaced with hydrophobic chromatography[46,49,51,53] or a Ni–NTA column,[47,50,52,54–59,76] and the sucrose gradient centrifugation was eliminated. For hydrophobic exchange chromatography, proteins are precipitated with ammonium sulfate, as discussed earlier. The ammonium sulfate pellet is then dissolved in buffer A (100 mM phosphate, pH 7.1, 2 mM DTT, 1 mM EDTA, and 20% glycerol). The resulting suspension is applied in 2.5-ml aliquots to a 9.1-ml gel filtration column (PD-10, Pharmacia) equilibrated with buffer B (buffer A + 1 M ammonium sulfate) and eluted with 3.5 ml of buffer B. This eluate is applied to a 30-ml hydrophobic interaction column (e.g., butyl-Sepharose 4-Fast Flow, Pharmacia). Elution is performed at 1 ml/min with a stepwise gradient starting from 100% buffer B to 40, 20, and 0%. Changes in the gradient are made when the absorbance at 280 nm approaches baseline. Ten-milliliter fractions are collected, and those containing DEBS proteins (typically 0% B) are pooled and applied to the anion-exchange column as described previously.

Currently, most PKS proteins are expressed with C-terminal, N-terminal, or dual 6xHis tags in our laboratory. Smaller proteins (e.g., monodomain and didomain proteins) are singly tagged, whereas most modules are doubly His tagged to increase their affinity with Ni ions. For Ni-NTA purification, cells are resuspended in Ni-NTA loading buffer (50 mM Tris, 300 mM NaCl, 1 mM EDTA, and 20% glycerol, pH 8.0), lysed at 1000 psi in a French press, and centrifuged for 45 min at 33,300g. The supernatant is batch loaded onto 5 ml of a Ni-NTA Superflow resin (Qiagen) for 45 min. The Ni-NTA matrix-bound protein is packed into a Flex column (Kontes), and the Ni-NTA resin is washed with 30 ml of loading buffer, followed by 15 ml of wash buffer (loading buffer + 10 mM imidazole). The protein is

[76] J. Lau, H. Fu, D. E. Cane, and C. Khosla, *Biochemistry* **38,** 1643 (1999).

eluted with 15 ml of elution buffer (loading buffer + 100 mM imidazole). DTT is added to a final concentration of 2.5 mM immediately after elution of the protein. The eluted protein is about 50–90% pure (depending on the expression of individual proteins) and is concentrated to 1 ml using Centriprep membranes with the appropriate molecular mass cutoff. Following concentration, 19 ml of buffer A [100 mM sodium phosphate (pH 7.2), 2.5 mM DTT, 1 mM EDTA, and 20% (v/v) glycerol] is added to the protein, and the resulting solution was applied to anion-exchange chromatography as described earlier.

Proteins obtained by Ni-NTA followed by anion exchange are generally 90–95% pure. Addition of an anion-exchange step also ensures the homogeneity of the protein, the Ni-NTA column alone is nonspecific to the proper folding state of the bound protein. A protease recognition site (usually thrombin or factor X A) is typically present to allow removal of the His tag, if needed. (It does not interfere with protein activity or module–module association.) When necessary, protease treatment is optimized in our laboratory using varying units of protease, varying temperature and time. After treatment, the protease and the cleaved His tag are removed by Ni column and/or anion-exchange chromatography.

Protein Purification for Physicochemical Assays. To study adduct formation between two different proteins, we have also prepared and purified FLAG-tagged proteins. For example, we have purified the ACP2 domain of the DEBS system.[57] For purification of FLAG-tagged ACP proteins, cells are harvested and processed up to the PEI precipitation step as described earlier. The protein solution is applied onto an anion-exchange column as described previously. The ACP elutes between 0.2 and 0.4 M NaCl. These fractions are pooled and applied to a Flex column packed with FLAG agarose (Sigma). After loading, the protein column is washed (buffer A + 300 mM NaCl) until no protein is detected by the Bradford assay (Bio-Rad). The ACP is then eluted with 20 ml of 100 μg/ml FLAG peptide (Sigma) in the wash buffer. The eluted protein is >95% pure and is concentrated and buffer exchanged into storage buffer (100 mM sodium phosphate, 1 mM EDTA, and 20% glycerol).

Activity Assays

Turnover Assays for Individual Modules. As mentioned earlier, electrophilic substrates are typically synthesized as SNAC esters,[77] which mimic the phosphopantetheine arm of ACP domains. These assays are widely used to interrogate the tolerance or specificity of a module toward a given

---

[77] J. R. Jacobsen, C. R. Hutchinson, D. E. Cane, and C. Khosla, *Science* **277**, 367 (1997).

substrate.[46,49,51] In such assays, various concentrations of the substrate are incubated with the enzyme (1 μM), saturating amounts of extender acyl-CoA ($\sim$500 μM), and NADPH (4 mM, added only if the module has a ketoreductase domain) in a reaction buffer (400 mM sodium phosphate, pH 7.2, 1 mM EDTA, 2.5 mM DTT, 20% glycerol). Even under these relatively optimized conditions, only a fraction of the limiting substrate is transformed into the polyketide product. Typical turnover numbers are in the 0.1- to 1-min^{-1} range, although lower turnover numbers can also be measured.

Kinetic assays can also be used to investigate intermodular chain transfer. These assays have been very useful in investigating the role of various structural features, such as interpolypeptide linkers[49,51,53,57] and ACP-KS[51,53] pairs in protein–protein interaction. Channeling of intermediates is considered efficient if the turnover number is in the 0.1- to 1-min^{-1} range. It has been observed that a natural and efficient bimodular system shows a K_D on the order of 1–10 μM. Therefore, in bimodular assays, minimally 1 μM of each module is used with otherwise similar conditions as single module turnover assays. Products are extracted and analyzed as discussed earlier.

Partial Activity Assay for Minimal PKS Module. To investigate the properties of hybrid modules, it is useful to assay the properties of individual domains in addition to those of the entire module. Examples of such assay formats are described next.

KS DOMAINS. The ketosynthase accepts a polyketide chain from the upstream ACP and catalyzes decarboxylative condensation between this substrate and an extender unit attached to the ACP in the same module. Thus far, it has not been possible to express an isolated KS domain as an intact functional protein or to refold insoluble KS proteins into an active state. As a result the properties of a KS domain must be explored within the context of an intact module.

The steady-state loading of a KS by the electrophilic substrate can be variable. Stoichiometric loading of the KS domain of DEBS module 2 has been demonstrated,[73] although other modules show variable loading.[59] If a ^{14}C-labeled substrate is available, the kinetics of this priming reaction can also be probed.[58]

The condensation reaction is probed most reliably by the multiple turnover assay of the intact module, although such an assay has the potential of masking the KS activity with other enzymatic activities in the module. Moreover, due to the high hydrolytic activity of the AT domain, chain elongation cannot be interrogated in a wild-type module that has been preloaded with extender units. To address this bottleneck, Hans et al.[58] inactivated the AT domain by mutating the active site serine to an alanine. The resulting module, expressed and purified from BL21(DE3) cells in an apo form, can be primed directly with any desired extender unit using the Sfp phosphopantetheinyl transferase. Extender units are added to the modules by incubating 0.1 μM Sfp PPTase, 30 μM module, 50 μM extender acyl-CoA, and 10 mM MgCl$_2$ in a Sfp reaction buffer (100 mM sodium phosphate, pH 6.6) at 30° for 30 min. The completion of acylation is analyzed by SDS–PAGE autoradiography. By adjusting the pH of the reaction to 7.2 in 400 mM sodium phosphate with 1 mM EDTA, 2.5 mM DTT, and 20% glycerol, the Sfp-catalyzed reaction is slowed down dramatically. The electrophile is then added at saturating concentrations, and chain elongation is monitored. Twenty microliters of the reaction mixture is quenched every 20 s in 250 μl ethyl acetate. Ethyl acetate from the extract is then evaporated. Dried products are resuspended in 20 μl ethyl acetate and analyzed by thin-layer chromatography. Typically an active module yields a turnover number of one per minute.

AT AND ACP DOMAINS. The acyl transferase domain in a module catalyzes the transfer of an acyl moiety from acyl-coenzyme A to the phosphopantetheine arm of the ACP domain. In the past, several unsuccessful attempts have been made to purify AT as an isolated protein in a soluble form. A didomain expression format has been developed for overexpressing intact AT domains,[55,76] in which the AT in the loading didomain of DEBS can be replaced successfully with extender AT domains of DEBS and RAPS systems. Because ATs are capable of transferring acyl groups from CoA thioesters to N-acetylcysteamine, this property can be used to quantify the substrate specificity of ATs. In such assays, AT-ACP didomains or modules are used in their apo form so that ATs are incapable of acylating the ACP. Typically, 0.1–1 μM of the apo didomain or module and 50–500 μM are incubated with saturating amounts of N-acetylcysteamine (typically 10 mM) in 100 mM sodium phosphate, pH 7.0, and 20% glycerol at 20°. Time points are collected by quenching 10 μM of sample volume in 25 μM of acetone. The mixtures are applied to thin-layer chromatography in 89% ethyl acetate, 10% 2-propanol, and 1% acetic acid, and acyl-SNAC products are quantified on a phosphoimager.

To probe AT-ACP interactions in an intact module, holo PKS modules (10 μM, in 100 mM sodium phosphate, pH 7.2, 20% glycerol) are incubated with 100 μM of acyl-CoA.[58] Time points are taken by quenching the reaction mixture with 1 volume of SDS–PAGE loading buffer without any thiol reagent, such as DTT or 2-mercaptoethanol. To distinguish between labeling of AT and ACP, the sample is treated with 0.3 M hydroxylamine after denaturation, which specifically cleaves the acyl-S-ACP thioester from ACP. Samples are then run on SDS–PAGE gel. The gel is subsequently dried and analyzed using a phosphoimager. Approximately 50%

loss of label would be observed if the transfer reaction is not impaired due to loss of the AT–ACP interaction.[58]

Protein Chemical Assays

CROSS-LINKING. In our laboratory, protein cross-linking experiments using bifunctional reagents are performed to probe the oligomeric state of proteins and the relative orientation of domains/monomers in homo/heterooligomeric complexes.[57,75] We typically use 1,3-dibromopropanone, ethylene glycol bis(succinimidyl succinate) (EGS) or bis(malaeimidohexane) (BMH) as cross-linkers, which react specifically with sufhydryl groups of two cysteine residues, amino groups of two lysine residues, or again sufhydryl groups of two cysteine with 5-, 16-, and 16-Å-long spacer arms, respectively. Usually, 1–10 μM of the enzyme(s) is incubated with 2–10 mM of the cross-linker. Reactions are quenched with 10–25 mM DTT in the case of BMH and 1,3 dibromopropanone and 50 mM Tris at pH 7.6 in the case of EGS, and protein samples are denatured and analyzed by SDS–PAGE gel for the formation of cross-linked species. For enhanced sensitivity and unambiguous identification of the desired adduct in the case of heteroligomeric complexes, the two proteins are tagged differently (typically 6xHis and FLAG) and analyzed by Western blots with antibodies specific to poly-His and FLAG tag.[57]

LIMITED PROTEOLYSIS. In addition to cross-linking experiments used to probe the orientation of domains in a module, limited proteolysis experiments are also useful in visualizing flexible portions in a protein.[58,78] We incubate 1 μM PKS module with 1:100 (w/w) trypsin in proteolysis buffer (100 mM phosphate, pH 8, 150 mM MgCl$_2$) at 0°. Time points are taken by quenching the reaction by adding SDS–PAGE buffer and heating the samples to 70° for 5 min. The proteins are separated by SDS–PAGE, and polypeptides of interest are extracted from the gel and subjected to mass spectrometric analysis. Enhanced sensitivity toward the proteolysis of hybrid modules compared to their wild-type counterparts is a good indicator of nonnative conformational changes in the engineered PKS proteins.[58]

Conclusions

Techniques similar to those described here have been employed diversely in various laboratories working with modular PKSs. *In vivo* systems developed in our laboratory and similar systems developed elsewhere have been used successfully in the biosynthesis of natural and hybrid

[78] J. F. Aparicio, P. Caffrey, A. F. Marsden, J. Staunton, and P. F. Leadlay, *J. Biol. Chem.* **269**, 8524 (1994).

polyketides. Although reconstitution of natural or hybrid PKS systems in heterologous hosts is the ultimate goal of the biosynthetic engineer, the *in vitro* analytical procedures described earlier are an important bridge for achieving that goal. Biochemical examinations of PKS proteins have provided and continue to provide useful insights into the scope and limitations of alternative engineering strategies. At the same time these protein chemical studies provide a technical and conceptual foundation for the structural biology of modular PKSs in the coming years.

[24] Reactions Catalyzed by Mature and Recombinant Nonribosomal Peptide Synthetases

By Uwe Linne and Mohamed A. Marahiel

Introduction

Polypeptide synthesis in nature occurs either by deciphering the genetic code on the ribosome or is driven nonribosomally on a protein template by megaenzymes called nonribosomal peptide synthetases (NRPSs).[1] Using the latter process, the vast majority of peptide antibiotics and many other substances of pharmaceutical importance produced by microorganisms are synthesized (such as penicillin, vancomycin, bacitracin, and cyclosporin). Irrespective of origin, these multienzymes are composed of modules, whereby each module is responsible for the recognition, activation, modification, and incorporation of one specific building block into the growing product chain (compare Fig. 1). Such modules can be further subdivided into domains, each harboring one special catalytic function. NRPSs are, from their modular organization and the mechanism of product assembly, related to polyketide synthases (PKSs).[2] However, this article focuses on NRPSs, although in nature NRPSs and PKSs are often combined in single biosynthetic pathways (e.g., epothilone synthetase).

In NRPSs, adenylation (A) domains[3] are responsible for the substrate recognition and activation as aminoacyl-*O*-AMP under ATP consumption. Subsequently, the activated amino acid is transferred to a 4′-phosphopantetheine (4′-Ppant) moiety and is bound as a thioester under the release of AMP. This 4′-Ppant cofactor itself is tethered covalently to an invariant

[1] D. Schwarzer, R. Finking, and M. A. Marahiel, *Nature Prod. Rep.* **20,** 275 (2003).

[2] D. Schwarzer and M. A. Marahiel, *Naturwissenschaften* **88,** 93 (2001).

[3] E. Conti, T. Stachelhaus, M. A. Marahiel, and P. Brick, *EMBO J.* **16,** 4174 (1997).

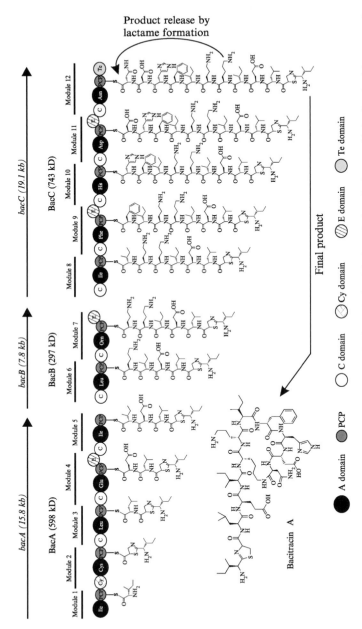

Fig. 1. Principle of an NRPS assembly line. The schematic biosynthesis of the branched cyclic decapeptide bacitracin A is presented. Three enzymes (BacA, BacB, and BacC) containing 12 modules and a total of 40 domains process the growing peptide chain along the protein template. During synthesis, the elongation intermediates remain tethered covalently as thioesters to the 4'-Ppant cofactors of the PCPs. The final product is released by macrocyclization.

serine residue of a peptidyl carrier protein [PCP or thiolation (T) domain].[4] The modification of inactive apo-PCPs to functional holo-PCPs is carried out posttranslationally by 4'-phosphopantetheine transferases such as Sfp[5,6] under the consumption of CoA. After activation and covalent binding of the substrates to the enzymes, condensation (C) domains catalyze the nucleophilic attack of the α amino group of the new building block onto the thioesterified carboxy group of the aminoacyl- or peptidyl-S-Ppant moiety bound to the PCP of the preceding module.[7] By the action of these essential domains (A domain, PCP, and C domain), in linear NRPSs,[8] the growing peptide chain is handed over from one module to the next until the termination module is reached (compare Figs. 1 and 2). There, a termination domain, most often a thioesterase (Te) domain in bacterial systems, catalyzes the release of the product from the enzyme template, mostly by intramolecular cyclization.[9] In addition to these essential domains, the modules can contain different optional domains. Among them are epimerization (E),[10,11] heterocyclization (Cy),[12,13] N-methylation (M),[14] and formylation (F) domains,[15] contributing significantly to the chemical diversity and therefore to the bioactivity of the corresponding secondary metabolites. The reactions catalyzed by these domains are presented in Fig. 3.

Due to the modular architecture of NRPSs and PKSs, as well as the pharmacological relevance of the assembled products, engineering approaches are a main focus of current academic and industrial research.[16–19]

[4] T. Stachelhaus, A. Hüser, and M. A. Marahiel, *Chem. Biol.* **3,** 913 (1996).

[5] R. H. Lambalot, A. M. Gehring, R. S. Flugel, P. Zuber, M. LaCelle, M. A. Marahiel, R. Reid, C. Khosla, and C. T. Walsh, *Chem. Biol.* **3,** 923 (1996).

[6] K. Reuter, M. R. Mofid, M. A. Marahiel, and R. Ficner, *EMBO J.* **18,** 6823 (1999).

[7] T. Stachelhaus, H. D. Mootz, V. Bergendahl, and M. A. Marahiel, *J. Biol. Chem.* **273,** 22773 (1998).

[8] H. D. Mootz, D. Schwarzer, and M. A. Marahiel, *Chembiochem.* **3,** 490 (2002).

[9] S. D. Bruner, T. Weber, R. M. Kohli, D. Schwarzer, M. A. Marahiel, C. T. Walsh, and M. T. Stubbs, *Structure (Camb.)* **10,** 301 (2002).

[10] T. Stein, B. Kluge, J. Vater, P. Franke, A. Otto, and B. Wittmann-Liebold, *Biochemistry* **34,** 4633 (1995).

[11] U. Linne, S. Doekel, and M. A. Marahiel, *Biochemistry* **40,** 15824 (2001).

[12] A. M. Gehring, I. Mori, R. D. Perry, and C. T. Walsh, *Biochemistry* **37,** 11637 (1998).

[13] T. A. Keating, D. A. Miller, and C. T. Walsh, *Biochemistry* **39,** 4729 (2000).

[14] R. Zocher, T. Nihira, E. Paul, N. Madry, H. Peeters, H. Kleinkauf, and U. Keller, *Biochemistry* **25,** 550 (1986).

[15] L. Rouhiainen, L. Paulin, S. Suomalainen, H. Hyytiainen, W. Buikema, R. Haselkorn, and K. Sivonen, *Mol. Microbiol.* **37,** 156 (2000).

[16] D. E. Cane, C. T. Walsh, and C. Khosla, *Science* **282,** 63 (1998).

[17] H. D. Mootz, N. Kessler, U. Linne, K. Eppelmann, D. Schwarzer, and M. A. Marahiel, *J. Am. Chem. Soc.* **124,** 10980 (2002).

Essential NRPS-Domains:

NRPS-Domains catalyzing product release:

Optional NRPS-Modification-Domains:

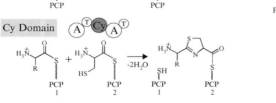

FIG. 2. Schematic presentation of the reactions catalyzed by NRPS domains.

[18] L. Tang, S. Shah, L. Chung, J. Carney, L. Katz, C. Khosla, and B. Julien, *Science* **287**, 640 (2000).

[19] C. J. Rowe, I. U. Bohm, I. P. Thomas, B. Wilkinson, B. A. Rudd, G. Foster, A. P. Blackaby, P. J. Sidebottom, Y. Roddis, A. D. Buss, J. Staunton, and P. F. Leadlay, *Chem. Biol.* **8**, 475 (2001).

NRPS chain elongation cycle with optional E domain:

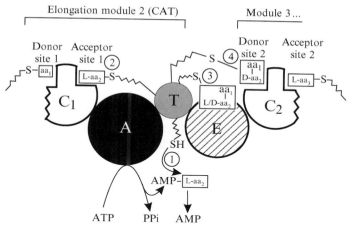

① Recognition, activation and thiolation of the amino acid

② Binding of aminoacyl-S-Ppant in the enantioselective acceptor site of N-terminal C domain

③ Epimerization of the substrate after formation of peptidyl-S-Ppant by the E domain

④ Handing over the substrate after formation and epimerization of peptidyl-S-Ppant to the donor site of the downstream C domain (C_2)

Fig. 3. Sequential presentation of a NRPS chain elongation cycle with an optional E domain. The occurring reactions and the different positions of the 4'-Ppant cofactor[1-4] are shown.

This article first describes the overexpression and purification of recombinant NRPSs and single domains as well as the common assays to determine their activities. Engineering strategies and suitable fusion sites of modules and domains are then outlined. Finally, some known limitations of engineering approaches and possibilities used to overcome them are discussed.

Experimental Procedures

Assays Used to Investigate Domain and Module Activities

Heterologous Expression and Enzyme Purification. Since the beginning of the genomic era and the publication of the first sequence of an A domain in 1988,[20] many entire gene clusters encoding NRPSs, PKSs, and mixed systems were discovered. However, the encoded large enzymes are produced by their natural hosts in low amounts and only at specific culture conditions. Therefore, the inducible overexpression in surrogate hosts, mainly *Escherichia coli*, represents one major route used to obtain entire NRPSs, modules, or domains for biochemical and structural studies in higher quantities. For a rapid purification of the enzymes, a C-terminal tagging works fine (e.g., His_6 tag, and Strep tag), whereby many different commercially available vector systems (e.g., pQE, Qiagen, Germany; pIBA, IBA, Germany; pET, Stratagene, Germany) can be utilized according to the manufacturers protocols. Subsequent overexpression of the corresponding genes can be achieved with common *E. coli* strains (e.g., M15, SG13009, or BL21-DE3α).

For overexpression, overnight cultures in LB medium containing 10 mM MgCl$_2$ are prepared at room temperatures (20–30°; rule of thumb: the lower the temperature, the higher the enzymes solubility and activity). After 12–16 h of incubation under usual growth conditions, these cultures normally have an OD$_{600}$ ≫ 1. Protein expression is then induced by the addition of a suitable inductor (the inductor depends on the expression system used) and the cultures are incubated for another 3 h. Subsequently, the cells are incubated on ice and harvested by centrifugation at 4° and 4500g for 10 min. If they are shock frozen with liquid nitrogen, the resuspended cell pellets (buffer A: 50 mM HEPES, 300 mM NaCl, pH 8.0) can be stored at −20° up to several months without loss of enzyme activity.

Cell lysis is performed efficiently with a French press according to the manufacturers' protocols, whereby it is important not to exceed a pressure of 1300 psig because large enzymes such as NRPSs may degrade. Single-step affinity chromatography is performed according to standard protocols (HEPES buffers work well, but other buffer systems have been used successfully). The protein concentrations are assigned using the method of Bradford or calculated UV-absorption coefficients of the enzymes. After dialysis against assay buffer [50 mM HEPES (pH 8.0), 200 mM NaCl, 1 mM EDTA, 2 mM DTE, 10 mM MgCl$_2$] the proteins are shock frozen in liquid nitrogen and can be stored at −80° for several months without significant (approximately 10%) loss of activity.

To express NRPSs that are inactive or insoluble when expressed in *E. coli*, different bacterial strains that are naturally producing NRPSs and are genetically accessible such as *Bacillus subtilis* or fungi-like yeast species or *Aspergillus nidulans* can be utilized, although the only successful expression of NRPSs described so far has been in the heterologous host *B. subtilis*.

[20] R. Weckermann, R. Furbass, and M. A. Marahiel, *Nucleic Acids Res.* **16,** 11841 (1988).

For this, Doekel et al.[21] developed an E. coli/B. subtilis shuttle vector system allowing the expression in both organisms without recloning. The four vectors pSD193, pSD270, pKE151, and pKE170 feature (1) gene sequences for stable chromosomal integration in B. subtilis KE30, (2) self-replication in E. coli and the possibility for a selection for ampicillin and kanamycin resistance, (3) a T5 P_{N25} promoter, which is accepted by both vegetative E. coli and B. subtilis RNA polymerases, (4) controllable expression by the combination of the promotor with two modified lac operator sequences (a chromosomal copy of lacI is available in the amyE locus of B. subtilis KE30), (5) a MCS with ATG start codons at optimal distance to the synthetic RBSII, and (6) plasmids pSD270 and pKE170 allow the C-terminal fusion of a His$_6$ tag to the translated proteins. However, it is still a big challenge to overexpress NRPS genes successfully, especially from fungi or Streptomycetes in heterologous hosts.

Another problem faced by researchers when manipulating NRPS bio-synthetic operons is the genetic inaccessibility of many natural producers. Therefore, strain B. subtilis KE30 was developed by Eppelmann et al.[22] It allows the stable chromosomal integration of entire biosysnthetic NRPS operons and their manipulation in vivo, as was demonstrated in the case of bacitracin production.

Priming of NRPSs: Apo- to Holoconversion. Heterologously expressed NRPSs are not modified efficiently posttranslationally by the E. coli 4'-Ppant transferases. The obtained amount of holoenzymes is normally less than 1–2%. Ensuring the formation of a high amount of posttranslational modification in vivo, a 4'-Ppant transferase has to be coexpressed in E. coli. Therefore, Sfp of B. subtilis or Gsp of B. brevis were chosen because Sfp was described to have a broad substrate specificity among all the known 4'-Ppant transferases.[6] E. coli expression strains were generated with chro-mosomal copies of the sfp or gsp gene, respectively.[23,24] Alternatively, modified pREP4 plasmids carrying a copy of sfp or gsp, respectively, under the control of a T7 promoter can be used efficiently for coexpression with peptide synthetases in BL21-λDE3, which has a T7 polymerase.[25] We however observed that holoenzymes, when compared to corresponding

[21] S. Doekel, K. Eppelmann, and M. A. Marahiel, FEMS Microbiol. Lett. 216, 185 (2002).
[22] K. Eppelmann, S. Doekel, and M. A. Marahiel, J. Biol. Chem. 11, 11 (2001).
[23] B. A. Pfeifer, S. J. Admiraal, H. Gramajo, D. E. Cane, and C. Khosla, Science 291, 1790 (2001).
[24] S. Gruenewald, H. D. Mootz, P. Stehmeier, and T. Stachelhaus, In vivo production of artificial nonribosomal peptide products in the heterologous host Escherichia coli. Applied and Environmental Microbiology in press (2004).
[25] S. Doekel and M. A. Marahiel, Chem. Biol. 7, 373 (2000).

enzymes that are stored in their apo forms, lose their aminoacylation activity more rapidly.

In vitro, the priming of heterologously expressed proteins is efficiently achieved by incubation with 200 μM CoASH and 25 nM recombinant *B. subtilis* 4'-Ppant transferase Sfp.[6] The reaction mixtures are allowed to preincubate for 15 min at 37° before any A domain substrate is added. Alternatively, chemically synthesized acetyl-, aminoacyl-, or peptidyl-CoA substrates can be transferred to the PCPs using Sfp.[26–28] It is therefore possible to bypass A domain selectivity.

A Domain Activity and Selectivity: ATP–PP$_i$ Exchange. A domains selectively recognize and activate amino and carboxy acids as (amino-) acyladenylates under the consumption of ATP and release of PP$_i$. This reversible reaction is a fast qualitative assay for A domain activity and selectivity. Adding ^{32}P-labeled inorganic PP$_i$ to the reaction mixture therefore leads to the formation of ^{32}P-ATP if the amino acid(s), which is present in the assay, is activated by the A domain. The ^{32}P-ATP can be separated easily by adding activated charcoal (Riedel de Haen, Germany) to the reaction mixture. Reaction mixtures in assay buffer (volume 50 μl) contain 600 nM enzyme and 1 mM (L- or D-)amino acid. The reaction is initiated by the addition of 50 μl of a second mixture containing 4 mM ATP, 0.4 mM tetrasodium pyrophosphate, and 0.15 μCi of tetrasodium [^{32}P]pyrophosphate in the assay buffer. After a 10-min incubation at 37°, the reaction is quenched by the addition of a stop mix containing 1.2% (w/v) activated charcoal, 0.1 M tetrasodium pyrophosphate, and 0.35 M perchloric acid. The charcoal is collected by centrifugation, washed once with 1 ml water, and resuspended in 0.5 ml water. Subsequently, the charcoal-bound radioactivity is quantified by liquid scintillation counting (LSC) and gives relative values for A domain activity and selectivity.

A quantitative assay for A domain activity (PP$_i$ release assay) was developed by Ehmann and co-workers.[29] PP$_i$ levels are measured using the continuous spectrophotometric assay furnished by the EnzChek Pyrophosphate assay kit (Molecular Probes). In this assay, PP$_i$ is hydrolyzed to inorganic phosphate by inorganic pyrophosphatase and phosphate production coupled to phosphorolysis of the guanosine analogue, 2-amino-6-mercapto-7-methylpurine ribonucleoside (MesG), catalyzed by the enzyme

[26] D. Schwarzer, H. D. Mootz, U. Linne, and M. A. Marahiel, *Proc. Natl. Acad. Sci. USA* **99**, 14083 (2002).

[27] S. A. Sieber, C. T. Walsh, and M. A. Marahiel, *J. Am. Chem. Soc.* **125**, 10862 (2003).

[28] S. L. Clugston, S. A. Sieber, M. A. Marahiel, and C. T. Walsh, *Biochemistry* **42**, 12095 (2003).

[29] D. E. Ehmann, C. A. Shaw-Reid, H. C. Losey, and C. T. Walsh, *Proc. Natl. Acad. Sci. USA* **97**, 2509 (2000).

purine nucleoside phosphorylase (PNP). The chromophoric product, 2-amino-6-mercapto-7-methylpurine, is monitored by absorbance at 360 nm.

Reactions (200 μl) contain 75 mM Tris–HCl (pH 7.5), 10 mM MgCl$_2$, 1 mM Tris-(2-carboxyethyl)phosphine, 5 mM ATP, 400 μM MesG, 0.2 units purine nucleoside phosphorylase, and 0.2 units inorganic pyrophosphatase. For the rate of release of acyl/aminoacyl-AMP, 1–2 μM enzyme is used with 2–10 mM substrate (amino or carboxy acids). Readings of absorbance at 360 nm are measured over a 6-min interval in a Perkin–Elmer Lambda 6 UV-VIS spectrophotometer, and slopes are correlated with a standard curve created with PP$_i$.

A Domain and PCP Interaction: Aminoacylation Assay. A reaction mixture (final volume 100 μl) containing 500 nM of holoenzyme in assay buffer and 1 mM ATP is equilibrated ($\sim$10 min) to 37°. Subsequently, ^{3}H/^{14}C-labeled substrate amino acids are added rapidly to a final concentration of 2.5 μM. At defined time points, samples are taken and quenched immediately by the addition of 1 ml ice-cold trichloroacetic acid (TCA) (10%). Alternatively, before addition of the TCA solution, 10 μl of a bovine serum albumin (BSA) solution (25 mg/ml) can be added to the reaction mixture. After 15 min of incubation on ice, samples are centrifuged (4°, 11,600g) for 20 min, washed twice with 1 ml ice-cold TCA (centrifugation of at least 2 min at 4°/11,600g is recommended each time), redissolved in 150 μl formic acid, and quantified by liquid scintillation counting (LSC).

The Calculation of Active Enzyme Concentration. A common margin of error is made when determining a protein concentration by the calculated extinction coefficient or by the method of Bradford. Furthermore, a distinct portion of the total enzyme amount may be inactive for some reason. According to our experience, aminoacylation activities may vary significantly (up to threefold) depending on the batch of protein utilized in the assays. As a consequence, for quantitative product formation assays it seems to be more reliable to calculate the enzyme concentration based on aminoacylation activity (see previous section) as an internal standard for the amount of active protein than to use the concentrations assigned previously by photometric protein concentration measurement.[30,31]

C Domain Activity: Amino Acid Transfer Assay(s). Condensation domains catalyze peptide bond formation between two amino acids or the growing peptide chain and an amino acid. The direct detection of peptide products is described later. The assay described here is used to detect the

[30] V. Bergendahl, U. Linne, and M. A. Marahiel, *Eur. J. Biochem.* **269**, 620 (2002).
[31] U. Linne, D. B. Stein, H. D. Mootz, and M. A. Marahiel, *Biochemistry* **42**, 5114 (2003).

transfer of an amino acid from one module to another, whereby the substrate of the upstream module has to be labeled radioactively. If the amino acid transfer should be detected *in cis*, the upstream module is allowed to covalently load its radioactively labeled substrate amino acid before addition of the substrate amino acid of the downstream module. Alternatively, if the transfer from one module to the other is studied *in trans*, both enzymes are preincubated separately with their substrates.

At defined time points, aliquots are taken and subsequently quenched and purified and enzyme-bound radioactivity is quantified by LSC. After a distinct time, depending on the module activity, a stable amount of enzyme-bound radioactivity is reached. Then, upon addition of the second substrate amino acid or aminoacylated enzyme, the labeled amino acid is transferred from the upstream to the downstream module if there is C domain activity. Now the 4'-Ppant cofactor of the upstream module is free and can be aminoacylated again with labeled substrate, resulting in an increase of total enzyme-bound radioactivity. If there is product release, normally the enzyme-bound radioactivity increases marginally and then decreases rapidly because the labeled substrate amino acid is exhausted after a while.

In a practical approach, 500 nM of each enzyme is preincubated separately with their substrate amino acids (2 μM ^{3}H/^{14}C-labeled amino acid as a substrate for the upstream module, 100 μM amino acid as a substrate for the downstream module). After 10 min of incubation, product formation is initiated by mixing equal volumes of reaction mixtures containing the upstream and the downstream modules, respectively. At various time points 200-μl aliquots are taken and transferred immediately to tubes containing 1 ml ice-cold TCA (10%). After 15 min of incubation on ice, samples are centrifuged (4°, 11,600g) for 20 min, washed twice with 1 ml ice-cold TCA (centrifugation of at least 2 min at 4°/11,600g is recommended each time), redissolved in 150 μl formic acid, and quantified by LSC.

Product Formation Assays. For the detection, identification, and quantification of peptide products, holoenzymes (500 nM each) in standard buffer are assayed at 37° with 100 μM of each substrate for up to 12 h (normally less than 3 h) in a final volume of 100 μl. If products are expected to remain enzyme bound, a thioester cleavage procedure has to be applied (see later). Alternatively, the reaction can be stopped by the addition of a 10-fold excess of methanol. After centrifugation (11,600g, 15 min), transfer the supernatant to fresh tubes. The solvent is removed under vacuum and the residue is dissolved in 100 μl 10% methanol (v/v). Subsequently, this solution is applied in common amounts to thin-layer chromatography (TLC), HPLC, or HPLC-MS, whereby HPLC-MS features the highest performance. TLC is especially applicative when radioactively labeled substrates are used, whereby products can be quantified using a two-dimensional radioactivity scanner such as RITA (Raytest, Germany). The selection of the development solution depends on the composition of the samples, Although a mixture of butanol/water/acetic acid/ethylacetate [1:1:1:1 (v/v)] works well for many peptide products or product mixtures.

HPLC separation of the reaction products can be achieved efficiently by reversed-phase chromatography, whereby the columns utilized and the gradients applied depend on the mixture that has to be separated. The stereochemistry of the peptide products can be confirmed by comparison with synthetic standards. This is especially recommended when E domains are present in the enzymes investigated or when D-amino acids are used as substrates. The quantification of products can be achieved easily by coinjection with defined amounts of synthetic standards or by comparison with a calibration curve.

E Domain Activity: Epimerization Assay. In nonribosomally synthesized peptides, D-amino acids are often present, significantly contributing to their biological activities. These D-amino acids can be incorporated by two mechanisms. First, like in cyclosporin synthetase, D-amino acids, which were generated previously by the action of external racemases, are activated selectively by the corresponding A domains, although when D-amino acids are found in nonribosomally synthesized peptides, E domains are present in the corresponding synthetases (compare Fig. 1). They are localized downstream of the PCPs within modules responsible for the incorporation of D-amino acids into the product. They epimerize 4'-Ppant-bound aminoacyl- or peptidyl intermediates.[32]

To monitor the epimerization activity of NRPSs containing E domains, holoenzymes are allowed to activate and covalently load [^{3}H/^{14}C]L- or [^{3}H/^{14}C]D-amino acids. Five hundred nanomolar holoenzyme in assay buffer is incubated with 2 μM labeled amino acid. Samples are taken at defined time points and precipitated by the addition of ice-cold 10% TCA. After a 15-min incubation on ice and centrifugation (4°, 11600g, 20 min), the pellet is washed twice with 1 ml ice-cold 10% TCA, 1 ml ether/ethanol [3:1 (v/v)], and 1 ml ether. Before removing the supernatant, 5 min centrifugation at 11,600g/4° is recommended each time. The pellet is dried at 37°. The ^{3}H/^{14}C-labeled amino acid bound to the precipitated enzyme as an acid-stable thioester is then hydrolyzed by the addition of 100 μl of a 100 mM potassium hydroxide solution and a 10-min incubation at 75°

[32] U. Linne and M. A. Marahiel, *Biochemistry* **39**, 10439 (2000).

("thioester cleavage"). Extraction of the cleaved amino acid is carried out by the addition of 1 ml methanol and centrifugation for 30 min at 4°/ 11,600g. After transferring the supernatant to fresh tubes, the solvent is removed under vacuum and the pellet is dissolved in 20 μl 50% ethanol (v/v) and applied to chiral TLC plates (ChiraDex Gamma, Merck, Germany) in the case of aminoacyl-S-Ppant epimerization and to silica gel 60 TLC plates (Merck, Germany) in the case of peptidyl-S-Ppant epimerization. The chiral TLCs are developed in acetonitrile/water/methanol [4:1:1 (v/v)] and the silica gel TLCs in butanol/water/acetic-acid/ethylacetate [1:1:1:1 (v/v)] as solvent. Radioactivity can be counted with a two-dimensional radioactivity scanner (Raytest, Germany) and quantified using the supplied RITA software. Alternatively, an autoradiography can be prepared or a phosphoimager can be utilized.

To test the specificity of E domains, Luo et al.[33] developed an E domain assay utilizing soluble aminoacyl-S-pantethein (aminoacyl-S-Pant) substrates, which are synthesized chemically. All aminoacyl-S-Pants were initially dissolved in 30 μM trifluoroacetic acid (TFA, pH 3) and stored at −80° to minimize hydrolysis. Reactions (100 μl each) contain 50 mM HEPES (pH 7.5), 20 μM apo-GrsA-ATE, and 0.5–24 mM various aminoacyl-S-Ppants, which are incubated at 30° and quenched at 30 min by adding 50 μl of 10% TCA (w/v). The precipitated protein is pelleted by centrifugation at 11,600g for 10 min. The supernatant is separated from the pellet and hydrolyzed by incubation with 100 μl of 100 mM potassium hydroxide for 30 min at 70°. The supernatant is then injected onto a chiral HPLC column (ChiralPAK WH, 4.6 × 250 mm, Chiral Technologies Inc.) for analysis. A 2.5 mM CuSO₄ buffer is employed to elute the column at a flow rate of 1 ml/min. The column eluant is monitored at UV 254 nm for aromatic amino acids and at 230 nm for all other amino acids.

Te Domains: Chemoenzymatic Cyclic Peptide Synthesis. Te domains are localized, most often at termination modules of NRPSs, which are approximately 300 amino acids in length and are responsible for the release of the final peptides from the enzyme templates. They often catalyze macrocyclization by an intramolecular attack of the N-terminal amino group or a side chain nucleophile onto the carboxy group of the 4'-S-Ppant-thioester, resulting in the formation of linear, cyclic, or branched cyclic lactones or lactames.[34]

In recombinant NRPSs, the fusion of Te domains to modules resulted most often in the release of linear peptides. There, the Te domain acts

[33] L. Luo, M. D. Burkart, T. Stachelhaus, and C. T. Walsh, *J. Am. Chem. Soc.* **123**, 11208 (2001).

[34] S. A. Sieber and M. A. Marahiel, *J. Bacteriol*, **185**, 7036 (2003).

unspecifically and catalyzes the nucleophilic attack of water to the peptidyl-4'-S-Ppant thioester connected covalently to the last modules PCP. In such systems, Te domain activity can be tested by a product formation assay (see earlier discussion).

Some assays have been developed to investigate Te domains activities and specificities as excised domains. Trauger and co-workers[35] established an assay that is based on chemically synthesized peptidyl-S-N-acetylcysteamine (peptidyl-SNAC) compounds. SNACs had been described previously as a soluble substrate analogue for ketosynthase (KS) domains of PKSs. Meanwhile, excised Te domains of tyrocidine (Te$_{Tyc}$), gramicidine S (Te$_{Grs}$), and surfactin (Te$_{Srf}$) biosynthetic systems have been characterized biochemically with such substrates.[35–37] The crystal structure of a prototype Te domain (Te$_{Srf}$) has been determined.[9] Among the excised Te domains found to be active with SNACs, Te$_{Tyc}$ showed a broad substrate specificity.[38–41] A library of different peptidyl-SNACs based on the tyrocidine chemical lead structure has been cyclized and was screened successfully for derivatives with improved or altered activities,[40] revealing the significant biotechnological potential of excised Te domains.

Experimentally, reactions containing 25 mM HEPES (pH 7.0), 50 mM NaCl, and 250 μM SNAC in a total volume of 50–400 μl are initiated by the addition of 1–5 μM enzyme at 25° (at 37° mainly an uncatalyzed SNAC hydrolysis occurs). At various time points, reactions are quenched by the addition of 25 μl 1.7% TFA/water (v/v), flash frozen in liquid nitrogen, and stored at −80°. The reactions are then thawed, 85 μl acetonitrile is added, and analysis is performed by reversed-phase HPLC(-MS) according to standard procedures. Identification of the products can be confirmed by ESI-MS or MALDI-TOF-MS. An additional MS/MS analysis of the cyclic products is recommended to verify their integrity.

Biochemical analysis of other Te domains, like these of bacitracin (Te$_{Bac}$) and fengycin (Te$_{Fen}$) showed no activity with SNAC substrates. Sieber et al.[27] demonstrated that peptidyl-CoAs can be utilized successfully as substrates for PCP Te bidomains. Thereby, apo-PCP-Te were modified with peptidyl-S-Ppant by the action of Sfp. This method has the advantage

[35] J. Trauger, R. Kohli, H. Mootz, M. Marahiel, and C. Walsh, *Nature* **407**, 215 (2000).
[36] R. M. Kohli, J. W. Trauger, D. Schwarzer, M. A. Marahiel, and C. T. Walsh, *Biochemistry* **40**, 7099 (2001).
[37] C. C. Tseng, S. D. Bruner, R. M. Kohli, M. A. Marahiel, C. T. Walsh, and S. A. Sieber, *Biochemistry* **41**, 13350 (2002).
[38] R. M. Kohli, J. Takagi, and C. T. Walsh, *Proc. Natl. Acad. Sci. USA* **99**, 1247 (2002).
[39] J. W. Trauger, R. M. Kohli, and C. T. Walsh, *Biochemistry* **40**, 7092 (2001).
[40] R. M. Kohli, C. T. Walsh, and M. D. Burkart, *Nature* **418**, 658 (2002).
[41] R. M. Kohli, M. D. Burke, J. Tao, and C. T. Walsh, *J. Am. Chem. Soc.* **125**, 7160 (2003).

of imitating the natural way of substrate presentation of the PCP-bound intermediate to the Te domain. However, the crucial disadvantage of the PCP-Te/peptidyl-CoA assay system is the fact that only one reaction cycle per enzyme is possible (no multiple turnover!). Nevertheless, the latter method allows the biochemical characterization of a large variety of different PCP-Te bidomains, including Te domains, which are inactive with peptidyl-SNAC substrates.

In the experimental approach, PCP-Te cyclization assays are carried out in 100 μl standard assay buffer at room temperature. The peptidyl-CoA concentration is varied from 20 to 60 μM and the PCP-Te concentration is 60 μM. Reactions are initiated by the addition of Sfp to give a final concentration of 5 μM, ensuring a fast posttranslational modification reaction. Reactions are quenched at various time points by the addition of 40 μl 4% TFA/H$_2$O (v/v) and can be analyzed by reversed-phase HPLC or, even better, HPLC-MS according to standard protocols. Identification of the products can be confirmed by ESI-MS or MALDI-TOF-MS. An additional MS/MS analysis of the cyclic products is recommended to verify their integrity.

Engineering Strategies

The Specificity-Conferring Code: Manipulating A Domain Specificity by Point Mutations. In engineering approaches of NRPSs, one main goal is to alter the amino acid sequence of the products because A domains are responsible for substrate selection, recognition, and activation. They function as gatekeepers of nonribosomal peptide synthesis. Therefore, alteration of their selectivity should lead to new products.

Starting with the first sequence of an A domain published by Weckermann et al.[20] in 1988, more than several hundred of such sequences are now available in databases. Enabled by the solution of the three-dimensional structure of the A domain of GrsA by Conti et al.[42] an amino acid-binding pocket was defined (residues 235, 236, 239, 278, 299, 301, 322, 330, 331, and 517, numbering corresponding to the A domain of GrsA).

Combining both structural and bioinformatic data of many A domains, the selectivity-conferring "nonribosomal" code was discovered by Stachelhaus et al.[43] in 1999. They first aligned the sequences of approximately 160 A domains with that of GrsA, identified the residues in each A domain forming the binding pocket, and then, in a second step, aligned only these stretches of 10 residues. Based on this, a clustering of A domains depending

[42] E. Conti, N. P. Franks, and P. Brick, *Structure* **4**, 287 (1996).
[43] T. Stachelhaus, H. D. Mootz, and M. A. Marahiel, *Chem. Biol.* **6**, 493 (1999).

| Domain | \|--- Position ---\| | | | | | | | | | | Biosynthetic template | Similarity |
	235	236	239	278	299	301	322	330	331	517		
Aad	E	P	R	N	I	V	E	F	V	K	AcvA	94%
Ala	D	L	L	F	G	I	A	V	L	K	CssA, Hts1	55%
Asn	D	L	T	K	L	G	E	V	G	K	BacA, CepA, Dae, Glg1, TycC	90%
Asp	D	L	T	K	V	G	H	I	G	K	BacC, SrfAB, LicB, LchAB	100%
Cys(1)	D	H	E	S	D	V	G	I	T	K	AcvA	96%
Cys(2)	D	L	Y	N	L	S	L	I	W	K	BacA, HMWP2	88%
Gln	D	A	Q	D	L	G	V	V	D	K	LicA, LchAA	100%
Glu(1)	D	A	W	H	F	G	G	V	D	K	FenA, FenC, FenE, PPS1, PPS3, PPS4	95%
Glu(2)	D	A	K	D	L	G	V	V	D	K	BacC, SrfAA	95%
Ile (1)	D	G	F	F	L	G	V	V	Y	K	BacA, BacC, LicC, LchAC	92%
Ile (2)	D	A	F	F	Y	G	I	T	F	K	FenB, PPS5	100%
Leu(1)	D	A	W	F	L	G	N	V	V	K	BacA, LicA, LchAA, LicB, LchAB, SrfAA, SrfAB	99%
Leu(2)	D	A	W	L	Y	G	A	V	M	K	CssA	100%
Leu(3)	D	G	A	Y	T	G	E	V	V	K	GrsB, TycC	100%
Leu(4)	D	A	F	M	L	G	M	V	F	K	LicA, LchAA, SrtAA	97%
Orn(1)	D	M	E	N	L	G	L	I	N	K	FxbC	100%
Orn(2)	D	V	G	E	I	G	S	I	D	K	BacB, FenC, GrsB, PPS1, TycC	98%
Phe	D	A	W	T	I	A	A	V	C	K	GrsA, SnbDE, TycA, TycB	88%
Pro	D	V	Q	L	I	A	H	V	V	K	GrsB, FenA, PPS4, SnbDE, TycB	87%
Ser	D	V	W	H	L	S	L	I	D	K	EntF, SyrE	90%
Thr/Dht	D	F	W	N	I	G	M	V	H	K	AcmB, Fxb, PPS2, PyoD, SnbC, SyrB, SyrE	91%
Tyr(1)	D	G	T	I	T	A	E	V	A	K	FenA, PPS2, PPS4	100%
Tyr(2)	D	A	L	V	T	G	A	V	V	K	TycB, TycC	80%
Tyr(3)	D	A	S	T	V	A	A	V	C	K	BacC, CepA, CepB	78%
Val(1)	D	A	F	W	I	G	G	T	C	K	GrsB, FenE, LicB, LchAB, PPS3, SrfAB, TycC	96%
Val(2)	D	F	E	S	T	A	A	V	Y	K	AcvA	94%
Val(3)	D	A	W	M	F	A	A	V	L	K	CssA	95%
Variability	3%	16%	16%	39%	52%	13%	26%	23%	26%	0%		

Wobble-like positions

FIG. 4. The specificity-conferring nonribosomal code of A-domains as elucidated by Stachelhaus et al.[43]

on their substrate specificities was discovered. The specificity-conferring code of A domains published by Stachelhaus et al.[43] is summarized in Fig. 4.

Interestingly, with the recent solution of the three-dimensional structure of the carboxylic acid (dihydroxybenzoate) activating A domain DhbE, May et al.[44] adapted the specificity-conferring code for A domains activating carboxylic acids.

The selectivity-conferring code can be used to predict substrate specificities of new A domains that have not been characterized biochemically before, although it is most valuable with respect to rational engineering approaches of NRPSs (compare Fig. 5). Eppelmann and colleagues[45] changed the selectivities of two A domains associated with surfactin biosynthesis in *B. subtilis in vitro* and *in vivo* by introducing point mutations into their binding pockets as predicted by the selectivity-conferring code. The specificity of the L-Glu activating initiation module was changed to L-Gln with sustainment of the catalytic efficiency simply by introducing a Gln at position 239 instead of a Lys. For the change of the substrate specificity of the second module of surfactin synthetase B from L-Asp to L-Asn,

[44] J. J. May, N. Kessler, M. A. Marahiel, and M. T. Stubbs, *Proc. Natl. Acad. Sci. USA* **99,** 12120 (2002).

[45] K. Eppelmann, T. Stachelhaus, and M. A. Marahiel, *Biochemistry* **41,** 9718 (2002).

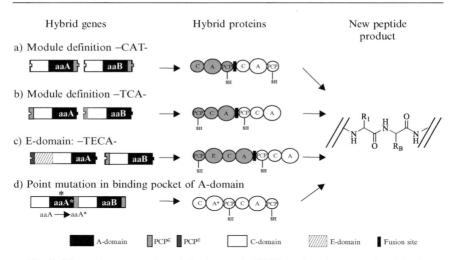

Fig. 5. Schematic presentation of the four main NRPS engineering strategies: (a) –CAT-module fusions; (b) –TCA-module fusions; (c) –TECA-fusion strategy for modules containing E-domains. Using this strategy, both the cognate PCP^E and the enantioselective downstream C-domain are the cognate ones of the E-domain. (d) Alteration of A-domain selectivity by introducing point mutations into the substrate-binding pocket.

three-point mutations were predicted to be necessary (Val^{299} → Ile, His^{322} → Glu, and Ile^{330} → Val). However, biochemical characterization of single, double, and triple mutants revealed that His^{322} represents a key position, whose mutation only was sufficient to give rise to the intended selectivity switch. Additionally, the gene fragment containing this single mutation (His^{322} → Glu) was introduced back into the surfactin biosynthetic gene cluster. The resulting *B. subtilis* strain was found to produce the expected so far unknown lipoheptapeptide [Asn^5] surfactin.

These initial results indicate that site-directed mutagenesis, guided by the selectivity-conferring code of NRPSs A domains, represents a powerful alternative for the genetic manipulation of NRPS biosynthetic templates and the rational design of novel peptide antibiotics. This seems to be at least true for conservative exchanges between similar amino acids of comparable sizes, such as Asp/Asn, Glu/Gln, or Trp/Phe/Tyr. However, it has to be proven in the future if more complex changes in substrate specificity than those described by Eppelmann *et al.*[45] can be achieved simply by introducing point mutations into the binding pockets of A domains or if it is better to swap whole A domains, C–A bidomains, or complete modules in such cases (compare also limitations caused by C domain selectivity, which is described later).

Identifying Suitable Fusion Sites for Domain and Module Swapping, Deletion, or Insertion. By the discovery of the module and domain structure of NRPSs at the beginning of the 1990s, researchers started to think about manipulations of the biosynthetic templates by domain and module swapping, deletion, or insertion. The proof of principle was done by Stachelhaus et al.[46] in 1995. They exchanged "domain-coding regions" in the surfaction biosynthesis operon against others of bacterial and fungal origin, resulting in hybrid genes encoding peptide synthetases with altered amino acid specificities and the production of peptides with modified amino acid sequences, although the product yields in these early studies were very low, indicating that the domain interactions were not as optimal as in the wild-type system. The main problem at that time was the limited number of sequences available in public databases that were used to identify the so-called "linker regions" between modules and domains. Therefore, the fusion sites used were within conserved domain regions.

With an increased number of sequences available, it was possible to identify regions between the domains of almost no conservation. These low-conserved interdomain regions were defined as the "linker regions." Interestingly, linkers can be defined clearly between all domains except between C and A domains, indicating that they belong together as a single unit. This makes sense, as the nucleophilic acceptor site of the C domain was described to be highly specific for the acceptor substrate, which is activated by the downstream A domain.[47] A role for the C domain in controlling the directionality of nonribosomal peptide synthesis, as well as the timing of epimerization, was revealed.[32] If the selectivity of the C domain nucleophilic acceptor site does not match that of the directly connected A domain or if it is deleted, an internal elongation module can be converted into an initiation module. The only difference between initiation and elongation modules seems to be that the latter harbors N-terminal cognate C domains trapping the aminoacyl-S-Ppant intermediates until its condensation with the upstream incoming thioester-bound intermediate is completed. For the described reasons, only two fusion sites remain to construct hybrid enzymes: The fusion site between A/T (module definition T-C-A) and T/C (module definition C-A-T). The suitability of both sites for efficient engineering approaches of NRPS biosynthetic templates has been demonstrated.[25,48]

[46] T. Stachelhaus, A. Schneider, and M. A. Marahiel, *Science* **269,** 69 (1995).
[47] P. J. Belshaw, C. T. Walsh, and T. Stachelhaus, *Science* **284,** 486 (1999).
[48] H. D. Mootz, D. Schwarzer, and M. A. Marahiel, *Proc. Natl. Acad. Sci. USA* **97,** 5848 (2000).

How can such linkers and domains be defined properly? For both A/T and T/C fusions, the linkers can be best defined by their location to the invariant serine residue of the PCPs in the "core motif T" LGG(H/D)S(L/I). The linker region between PCP and C domain starts ~35 amino acid residues downstream of this invariant serine and ends ~45 residues after it. By using a fusion site at residues 38/39 downstream of the invariant serine of the PCPs,[25] active hybrid enzymes were generated. The linker region between A and PCP domains consists of a poorly conserved approximately 9 amino acid stretch. The defined NMR structure of the PCP ends with a highly conserved Tyr (position 37 upstream of the Ser),[49] suggesting that the linker region starts with residue 38 upstream of the PCPs cofactor-binding site serine. The fusion site used was localized at residues 39/40 upstream of the serine.[48] The C-terminal end of the A domain, and therefore the starting point of the A–T interdomain linker, can additionally be defined by the location of core A10 (NGK) of the A domains. It starts approximately 11 amino acids downstream of the Lys (K) of this core motif.

Successful Examples for Module and Domain Fusions. In the literature, several rational engineering approaches of NRPS biosynthetic templates have been described. This article presents important examples describing general strategies and possibilities for an efficient engineering of NRPS assembly lines. As mentioned earlier, the proof of principle for a successful module swapping was done by Stachelhaus et al.[46] in 1995. In 1997, de Ferra et al.[50] described the *in vivo* production (*B. subtilis*) of truncated surfactin derivatives with the expected peptide sequence, where the "carboxy-terminal intrinsic thioesterase region" had been moved to internal modules.

A switch from the complex *in vivo* approach to more easy understandable and better controllable *in vitro* approaches was undertaken to investigate some basic engineering rules. *In vitro*, several successful domain and module fusions were reported based on the new linker definition described earlier. Thereby, the fusion strategies and the module fusions between PCP and the A domain,[25] as well as the fusion between PCP and the C domain,[48] resulted in short artificial hybrid enzymes revealing turnover numbers of ~0.1–2 min^{-1}, which is within the lower range of the catalytic efficiency of many natural assembly lines.

In all these artificial NRPSs, for a catalyzed product release, Te domains were fused directly to the last modules C-terminal end. For the Te

[49] T. Weber, R. Baumgartner, C. Renner, M. A. Marahiel, and T. A. Holak, *Struct. Fold Des.* **8**, 407 (2000).

[50] F. de Ferra, F. Rodriguez, O. Tortora, C. Tosi, and G. Grandi, *J. Biol. Chem.* **272**, 25304 (1997).

domain or Te-PCP-bidomain fusion, fusion sites based on the location of the invariant serine of the PCPs were again utilized successfully. In such short artificial systems forming di- and tripeptides, the Te domains catalyze only the hydrolysis instead of the intramolecular cyclization as they do most often in their natural context.

Things become more complex if modification domains are present. Fusion enzymes containing E domains were studied intensively.[11,31] Therefore, it is important to know that two types of E domains exist, as well as two types of PCPs. E domains localized at initiation modules like those of GrsA and TycA act on aminoacyl-S-Ppant substrates, wherease those localized at elongation or termination modules are peptidyl-S-Ppant epimerases. The latter ones were reported to be also capable of epimerizing aminoacyl-S-Ppant substrates, but with reduced efficiencies.[32,51] The timing of the epimerization reaction in elongation or termination modules is thereby controlled through the corresponding N-terminal C domains by the mechanism described earlier and takes place after peptide bond formation. Aminoacyl-E domains of initiation modules, however, epimerize their substrates directly after aminoacylation of the 4'-S-Ppant. Currently, it is still unclear if aminoacyl-E domains are also capable of epimerizing peptidyl substrates. Moreover, E domains are only active in the combination with a PCP^E, which is normally localized in front of an E domain. In contrast, they are inactive when connected to a PCP^C normally found attached to the N-terminal end of C domains. Obvious differences between PCP^E and PCP^C are the core T (a PCP^E reveals the core sequence LGGDSI, whereas a PCP^C shows the core sequence LGGHSL) and some other residues within α helices 2 and 3 of PCP.[11] In conclusion, if E domains are swapped, they should be swapped as whole modules or at least as PCP^E-E bidomains.

Very little is known about engineering approaches with other modification domains such as Cy, M and Ox domains. For example, Schauwecker et al.[52] reported domain swapping in actinomycin synthetase. The valine-activating A domain of module 2 was exchanged against an A–M bidomain [methylation (M) domains are inserted within cores A9 (LpxYM(I/V)P and A10 (NGK) of A domains] forming N-methylvaline. However, the natural E domain activity of module 2 was lost by this fusion, which can be due to two reasons: inability of the E domain to epimerize N-methylated substrates or the loss of a PCP^E in front of the E domain.

[51] L. Luo, R. M. Kohli, M. Onishi, U. Linne, M. A. Marahiel, and C. T. Walsh, Biochemistry 41, 9184 (2002).
[52] F. Schauwecker, F. Pfennig, N. Grammel, and U. Keller, Chem. Biol. 7, 287 (2000).

A Vector System for the Rapid and Directed Construction of Recombinant Fusion Enzymes. The construction of expression vectors for hybrid enzymes containing fused genes is laborious and time-consuming. With conventional strategies, several complex cloning procedures are necessary for the fusion of more than two domains or modules on the genetic level, which can result in problems and mistakes. The plasmid pMS, which is a derivative of the pQE60 vector (Qiagen, Germany), can be utilized as a genetic tool for the rapid and directed construction of recombinant fusion enzymes.[31] The modified MCS of pQE60 contains in pMS the restriction sites *Nco*I, *Bam*HI, *Kpn*I, and *Spe*I. The first gene fragment is cloned into pMS by insertion within the *Nco*I and *Bam*HI sites. The gene fragment that should be fused to it is subcloned into pMS itself by using the same restriction sites, with the exception that an additional *Bgl*II site is added by polymerase chain reaction directly behind the *Nco*I site. In a second step, the gene fragment is removed out of pMS by digestion with *Bgl*II and *Xba*I (or *Nde*I; *Xba*I and *Nde*I are localized 3′ to the MCS within the noncoding region of the vector) and ligated into the *Bam*HI and *Xba*I (or *Nde*I) digested vector containing the first gene fragment. Using this strategy, several gene fragments can be combined with each other. All gene fragments cloned once into pMS function as a library for an easy and fast domain and module recombination.

In Vivo Approaches. As described earlier, engineering approaches carried out *in vitro* are suitable for finding general engineering rules, although the final aim is to produce new potent products by fermentation processes. Therefore, it is necessary to transfer the results from *in vitro* to living systems. Early pioneer work resulted in very low yields, which were due to nonoptimized fusion sites, as we know today.[46] Inspired by the progress made *in vitro, in vivo* studies describe the production of new products in acceptable amounts.[17,22,45,53] However, the engineering attempts described in the literature so far are conservative. Two studies describe independently the decreasing of the ring size of surfactin, a natural cyclic lipoheptapeptide, by the deletion of one module, while another study deals with the exchange of Asp^5 to Asn^5 by introducing point mutations into the binding pocket of the corresponding A domain. Therefore, the new surfactin analogues produced are expected to show similar properties like the natural occurring surfactins. If whole biosynthetic templates are transferred from one organism to another, it was shown that the genes mediating resistance have to be cotransferred.

[53] H. Symmank, P. Franke, W. Saenger, and F. Bernhard, *Protein Eng.* **15,** 913 (2002).

Discussion: Limitations and Possible Solutions

Although immense progress has been made in the rational design of NRPS biosynthetic templates for the production of new bioactive compounds, there are still natural limitations that have to be overcome. C domain selectivities have been reported to cause problems. Its highly specific nucleophilic acceptor site is needed for the control of directionality of nonribosomal peptide synthesis, as well as for the timing of the epimerization reaction.[32] Therefore, as mentioned previously, C–A bidomains should always be swapped as a unit. The electrophilic donor site of the C domain is less substrate specific, although an enantioselectivity for the D-isomer has been reported if an E domain is proceeding. Furthermore, it was demonstrated that a C domain normally processing peptidyl-S-Ppant substrates with high stereospecificity is less enantioselective for aminoacyl-S-Ppant substrates.[28] This has to be kept in mind when swapping modules localized C-terminally to modules harboring an E domain or when swapping modules containing an E domain.

Concerning the modification domains, another limitation is E domain specificity. Although noncognate substrates are epimerized by E domains,[11] the observed efficiencies are low.[33] Also, the specialization of E domains for aminoacyl- or peptidyl-S-Ppant substrates, as well as the requirement for their combination with a PCP^E, has to be considered when performing engineering approaches with E domains. The proof of principle that swapping PCP^E-E bidomains represents an additional possibility to obtain new peptide products by directed engineering approaches was realized,[31] although product yields were low, probably because of the reduced epimerization activity toward the altered substrates.

While fungal NRPS biosynthesis systems normally consist of only one single megasynthetase (cyclosporin synthetase, which consists of 11 modules, is the largest known single peptide chain with a mass of ~1.4 MDa), bacterial ones are often dissected in several NRPSs. Very often, E domains are located at the C-terminal ends of these bacterial synthetases.

In the case of related PKSs, at the N- and C-terminal ends of modules, intermodular linker regions of approximately 60–70 amino acid stretches have been discovered.[54] However, in NRPSs, up until now nothing was known about such intermodular linkers. Though, recently it has been demonstrated on the example of the tyrocidine synthetases TycA, TycB, and TycC that there is a selective protein–protein interaction *in trans* between different NRPSs, which is mediated by the E domains. This can

[54] S. Y. Tsuji, N. Wu, and C. Khosla, *Biochemistry* **40**, 2317 (2001).

be used to build up *in trans* systems consisting of several small synthetases, whereas it implicates a limitation, because only modules recognizing each other properly can be combined *in trans*. Our latest results however revealed that it is possible to fuse modules *in cis*, which normally interact with each other *in trans*. The communication of two modules naturally not interacting with each other *in trans* can be forced simply by fusing the C-terminal end of the one synthetase with the N-terminal of the other.[31]

Because of the large variety of natural NRPS modules sequenced so far, it should be possible to find modules with all desired specificities for building up nearly any biosynthetic template. A limitation is that not all modules can be combined with each other, mainly because of different GC content or codon usage, which depend on the natural hosts. Additionally, many natural hosts are genetically not accessible and therefore not suitable for *in vivo* approaches. *E. coli* and *B. subtilis* are, with some limitations, still the most used heterologous hosts for NRPS gene expression. For future studies on the engineering and heterologous expression of entire gene clusters, it will be highly important to have organisms with efficient multiple drug transporters or other resistance mechanisms against the new products synthesized.

Due to space limitation, in this article, manipulations of the precursor biosysnthesis pathways and possible postsynthetic modification, reactions such as glycosylation or final modifications were not addressed. However, this should also be taken into consideration when expressing NRPSs and PKSs in heterologous hosts because unnatural substrates are often utilized by adenylation or acetyltransferase domains. For PKSs, such a metabolic engineering of the heterologous host *E. coli* has been reported.[23,55] In the field of postnonribosomal synthesis modification reactions, much of the interesting work is currently under investigation by the group of Walsh (Harvard Medical School, Boston).[56–59]

In summary, directed complex engineering approaches of NRPSs are still a scientific challenge. However, the broad variety of engineering tools described in this article are opening a large field of different possibilities and have resulted in several successful attempts *in vitro* and *in vivo*.

[55] K. Watanabe, M. A. Rude, C. T. Walsh, and C. Khosla, *Proc. Natl. Acad. Sci. USA* **100**, 9774 (2003).

[56] C. T. Walsh, H. Chen, T. A. Keating, B. K. Hubbard, H. C. Losey, L. Luo, C. G. Marshall, D. A. Miller, and H. M. Patel, *Curr. Opin. Chem. Biol.* **5**, 525 (2001).

[57] C. Walsh, C. L. Freel Meyers, and H. C. Losey, *J. Med. Chem.* **46**, 3425 (2003).

[58] C. T. Walsh, H. C. Losey, and C. L. Freel Meyers, *Biochem. Soc. Trans.* **31**, 487 (2003).

[59] H. C. Losey, J. Jiang, J. B. Biggins, M. Oberthur, X. Y. Ye, S. D. Dong, D. Kahne, J. S. Thorson, and C. T. Walsh, *Chem. Biol.* **9**, 1305 (2002).

Learning more about the secrets of this interesting class of enzymes should open more possibilities for more effective targeted engineering approaches.

[25] Engineering Carotenoid Biosynthetic Pathways

By Benjamin N. Mijts, Pyung C. Lee, and Claudia Schmidt-Dannert

Introduction

Carotenoids constitute a structurally diverse class of natural pigments, which are produced as food colorants, feed supplements, and, more recently, as nutraceuticals and for cosmetic and pharmaceutical purposes. Although microorganisms and plants synthesize more than 600 different carotenoid structures, only a few carotenoid structures can be produced in useful quantities. The discovery that carotenoids exhibit significant anti-carcinogenic activities and play an important role in the prevention of chronic diseases has triggered studies on the synthesis of new carotenoid structures and the economic production of these compounds in engineered cells.[1]

These pigments are derived from the general isoprenoid pathway by head-to-tail condensation of two C_{15} or C_{20} isoprenoid diphosphate precursors to form carotenoids with a C_{30} or C_{40} backbone, respectively, which then are modified further to yield distinct carotenoid structures. Genes encoding the enzymes involved in central carotenoid biosynthetic pathways have been cloned, and enzymes from different species have been shown to function cooperatively when combined in heterologous systems. The greatest diversity of carotenoid structures is produced by microorganisms, and Fig. 1 summarizes our knowledge of the genes involved in microbial carotenoid biosynthesis.

The availability of biosynthetic genes from different organisms that can function cooperatively to produce visually detectable products makes carotenoid biosynthesis an ideal model system to explore engineering strategies for the generation of organic compound diversity from a pathway composed of individual enzyme functions. Known examples for engineering natural product diversity are so far mostly restricted to the inherently modular polyketide pathways.[2,3]

[1] P. C. Lee and C. Schmidt-Dannert, *Appl. Microbiol. Biotechnol.* **60,** 1 (2002).
[2] A. deBoer and C. Schmidt-Dannert, *Curr. Opin. Chem. Biol.* **7,** 273 (2003).
[3] B. N. Mijts and C. Schmidt-Dannert, *Curr. Opin. Biotechnol.,* **14,** 597 (2003).

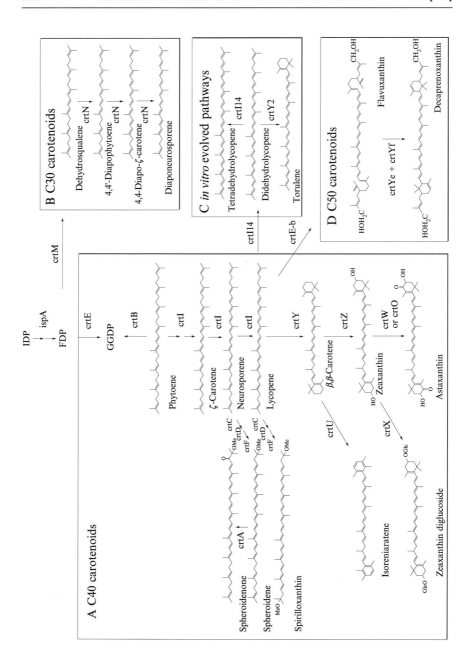

This article describes how combinatorial, *in vitro* evolution, and metabolic engineering strategies can be applied to large multienzyme pathways such as carotenoid biosynthesis to exploit its potential for the synthesis of new compounds. Special emphasis is placed on the methods and strategies employed to assemble pathways, analyze and identify compounds, and generate new biosynthetic routes.

Assembly of Large Multienzyme Pathways: A Modular Expression Vector System

Early examples of recombinant carotenoid production in noncarotenoid producing hosts, typically *Escherichia coli*, involved cloning and expression of entire carotenoid biosynthetic operons in single plasmids.[4–6] In order to fully exploit the potential of directed evolution, metabolic, and combinatorial engineering for the recombinant production of carotenoids, an expression system is required in which individual carotenoid pathway genes from different organisms can be combined in a modular fashion. In order to achieve this goal, we have constructed a series of compatible plasmids for modular carotenoid enzyme expression. These vectors include the pUC19-derived pUCMod, the pACYC184-derived pACMod, and the pBBR1MCS-2 plasmid.[7–9]

[4] N. Misawa, M. Nakagawa, K. Kobayashi, S. Yamano, Y. Izawa, K. Nakamura, and K. Harashima, *J. Bacteriol.* **172**, 6704 (1990).

[5] G. Sandmann, W. S. Woods, and R. W. Tuveson, *FEMS Microbiol. Lett.* **59**, 77 (1990).

[6] G. Schnurr, A. Schmidt, and G. Sandmann, *FEMS Microbiol. Lett.* **62**, 157 (1991).

[7] C. Schmidt-Dannert, D. Umeno, and F. H. Arnold, *Nature Biotechnol.* **18**, 750 (2000).

[8] S. J. Kwon, A. L. de Boer, R. Petri, and C. Schmidt-Dannert, *Appl. Environ. Microbiol.* **69**, 4875 (2003).

[9] P. C. Lee, A. Z. Momen, B. N. Mijts, and C. Schmidt-Dannert, *Chem. Biol.* **10**, 453 (2003).

FIG. 1. Microbial biosynthetic pathways known on a molecular level for C40 (A), C30 (B), and C50 (D) carotenoids and *in vitro*-evolved carotenoid pathways (C).[7] Enzymes are crtM, dehydrosqualene synthase; crtN, dehydrosqualene desaturase; crtE, GGPP synthase; crtB, phytoene synthase; crtI, phytoene desaturase; crtI14, *in vitro*-evolved phytoene desaturase; crtY, lycopene β-cyclase; crtY2, *in vitro*-evolved lycopene β-cyclase; CrtL-e, lycopene ε-cyclase; crtZ, β-carotene hydroxylase; crtW or crtO, β-carotene C(4) oxygenase; crtX, zeaxanthin glucosylase; crtU, β-carotene desaturase; crtC, hydroxyneurosporene synthase; crtD, methoxyneurosporene desaturase; crtF, hydroxyneurosporene-*O*-methyltransferase; crtA, spheroidene monooxygenase; crtEb, lycopene elongase; crtYe/Yf, heterodimeric decaprenoxanthin synthase. IDP, isopentenyl diphospate; FDP, farnesyl diphosphate; GGPP, geranylgeranyl diphosphate.

pUCMod was generated from pUC19 by deleting the *lacZ* fragment and introducing a new multiple cloning site (suitable for directional cloning of all selected carotenogenic genes), thereby changing the operator sequence to facilitate constitutive expression. Carotenogenic genes were amplified with polymerase chain reaction (PCR) primers containing appropriate restriction sites and providing an optimal Shine–Dalgarno sequence and a start codon. The PCR products were initially cloned individually into pUCmod. For coexpression of several carotenogenic genes, the entire expression cassettes (including coding region and *lac* promoter) were amplified from pUCmod and subcloned into different restriction sites of pACMod (pACYC194 devoid of the *Xba*I site) or pBBR1MCS-2. Thus, the combination of two or three compatible plasmids, each containing one or more modular carotenoid expression cassettes, allows the rapid assembly of different carotenoid pathways in *E. coli*.[7,9,10]

Combining Genes into Functional Pathways and Analysis of Carotenoid Products

Genes forming complete carotenoid pathways, when present on one or more compatible plasmids, can be transformed into *E. coli* host cells, resulting in the recombinant production of carotenoids. Recombinant *E. coli* cells producing carotenoids will produce distinctive color phenotypes, which are observed most readily when cells are pelleted from culture medium by centrifugation. A large number of factors influence the production of carotenoids in *E. coli*. Carotenoid yields are important not only in the potential scale-up of production processes, but also in producing sufficient quantities of products for laboratory-scale characterization and effective screening of recombinant libraries. The stability of carotenoid production is an important issue, as loss of color phenotype is often observed in recombinant, carotenoid-producing strains. This loss of phenotype suggests that carotenoid synthesis places considerable stresses on host cells, and plasmid or genomic mutations that eliminate carotenoid production are under strong positive selection. Although there is no way to completely eliminate this problem, it can be restricted by storing carotenoid-producing plasmids as purified DNA preparations and not as cell stocks, which may lose color phenotypes with repeated subculturing. In practice, whenever a plasmid(s) is transformed into *E. coli* to confer the desired phenotype, the authors chose the smallest colonies on agar plates for genetically stable seed cultures, as high carotenoid production is associated with slow growth of recombinants.

---

[10] P. C. Lee, B. N. Mijts, and C. Schmidt-Dannert, Submitted for publication.

Culture temperature has a significant influence on the levels of carotenoid production in *E. coli*. Generally, lower culture temperatures result in higher final carotenoid yields.[11,12] For example, torulene production from *E. coli* cells grown at 28° was two times higher than that at 37°.[10] Although it is unclear why this is the case, there appears to be a strong correlation between slower *E. coli* growth rates and higher carotenoid yield.

Medium composition has also been shown to influence greatly both the overall yield of carotenoids and, in some cases, the relative proportions of products. Secondary metabolite pathways in natural organisms, including carotenoid pathways, are generally thought to be regulated by physiological or environmental changes.[13] Many multistep secondary metabolite pathways are regulated in a coordinated fashion as operons. Even though the authors have used modular plasmid systems as opposed to the natural biosynthetic operons for the production of carotenoids in *E. coli* as described earlier, we observed that glucose severely suppresses carotenoid production in recombinant *E. coli*, irrespective of defined or complex medium.[10] In contrast, glycerol is found to improve carotenoid production levels in both media. Even the relative abundance of mixed carotenoid products can be influenced using different carbon sources.[10]

Effective analysis of recombinant carotenoids requires an efficient procedure to isolate carotenoids from cellular contaminants. For laboratory-scale experiments, this is generally achieved by direct extraction from wet cells with organic solvents followed by further organic extraction steps and, in some cases, open-column chromatography or thin-layer chromatography (TLC). Isolated carotenoids can then be analyzed further by TLC, high performance liquid chromatography (HPLC), liquid chromatography mass spectrometry (LC-MS), or nuclear magnetic resonance spectroscopy (NMR).

Initially, we utilized 100% acetone or a mixture of acetone and methanol (7:3) to obtain crude organic extracts containing carotenoids from wet *E. coli* cells. Extractions are performed by adding the solvent to pelleted and washed *E. coli* cells, sonicating samples in a sonicating water bath, and subsequent centrifugation. Carotenoid-containing supernatants are recovered, and extractions of cell pellets are repeated until no visible pigment is present. If these solvents are not effective in the extraction of pigments present in cell pellets, alternative solvents such as 100% methanol or mechanical disruption such as freeze-drying of cells prior to extraction is recommended. Carotenoid extracts may then be pooled and concentrated

[11] A. Ruther, N. Misawa, P. Boger, and G. Sandmann, *Appl. Microbiol. Biotechnol.* **48,** 162 (1997).
[12] M. Albrecht, S. Takaichi, N. Misawa, G. Schnurr, P. Boger, and G. Sandmann, *J. Biotechnol.* **58,** 177 (1997).
[13] G. A. Armstrong, *Annu. Rev. Microbiol.* **51,** 629 (1997).

by partially drying under a stream of nitrogen gas. The resulting pigment extracts are generally reextracted with an equal volume of ethyl acetate or hexane after the addition of 1/2 volume of salt water (15% NaCl) and are resuspended in 0.5–1 ml of hexane or methanol.

Ultraviolet-visible absorption spectrophotometry (UV-VIS) analysis of the crude extract allows a quick survey of the possible carotenoid structures present. Depending on the number of conjugated double bonds, acyclic or cyclic end groups, and/or the presence of additional modification of the carotenoid backbone, such as oxygen-containing groups, carotenoids exhibit distinct absorption maxima and spectra fine structures. Figure 2 (top) shows absorption spectra of two crude extracts from *E. coli* transformants expressing different pathways for the production of oxocarotenoids. A notable shift toward longer wavelengths of the absorption maximum concurrent with a loss of fine structure is observed for pathway B producing a novel purple oxocarotenoid spirilloxanthin[9] (see Fig. 4 for structure). The bottom part of Fig. 2 shows spectra of common acyclic and cyclic carotenoids.

Following UV-VIS inspection, a small aliquot of the carotenoid extracts was generally subjected to analytical TLC for initial analysis of the carotenoid compositions (Whatman silica gel 60 plates, 4.5 μm particle size, 200 μm thickness) using different solvent systems optimized for individual carotenoid properties.[9] For further purification of carotenoids, preparative TLC can be used. Preparative TLC is generally performed under the same conditions as analytical TLC and the resolved carotenoids are isolated and eluted with acetone or methanol.

For the analysis of carotenoids, generally 10–20 μl of the carotenoid extracts is applied to a reversed-phase Zorbax SB-C18 column and is typically eluted under isocratic conditions with a solvent system containing 90% (acetonitrile:H_2O, 99:1) and 10% (methanol:tetrahydrofurane, 8:2) at a flow rate of 1 ml/min equipped with a photodiode array detector. For the separation of polar xanthophylls and nonpolar carotenes, we used gradient conditions with solvent A (acetonitrile:H_2O, 85:15) and solvent B (methanol:tetrahydrofurane, 8:2). Figure 3 shows an example of the separation of acyclic oxocarotenoids produced in recombinant *E. coli* cells.[9]

For structural elucidation, carotenoids can be identified by a combination of HPLC retention times, absorption spectra, and mass fragmentation spectra. Normally, positive mass fragmentation spectra are monitored on a mass spectrophotometer equipped with an electron spray ionization (ESI) or atmosphere pressure chemical ionization (APCI) interface. If more structural information on the carotenoids is necessary, parent molecular ions can be fragmented further by MS/MS using different collision-induced

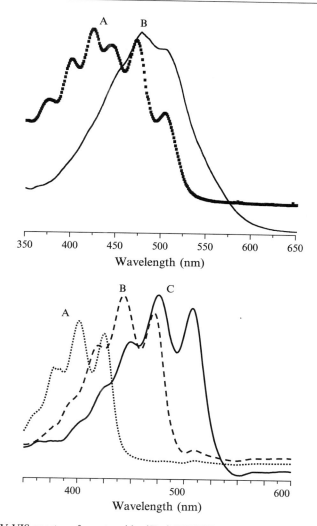

FIG. 2. UV-VIS spectra of carotenoids. (*Top*) UV-VIS spectrum of the total carotenoid extracts of (A) *E. coli* [pAC-crtE-crtB-crtI + pUC-crtA] or (B) *E. coli* [pAC-crtE-crtB-crtI14 + pUC-crtA]. (*Bottom*) Typical UV-VIS spectra of neurosporene (A), β-carotene (B), and lycopene (C).

dissociation energy (25–30%). MS and MS/MS spectra of a novel purple acyclic oxocarotenoid produced in recombinant *E. coli* are shown as examples in Fig. 4.[9] More detailed information on isolation, analysis, and

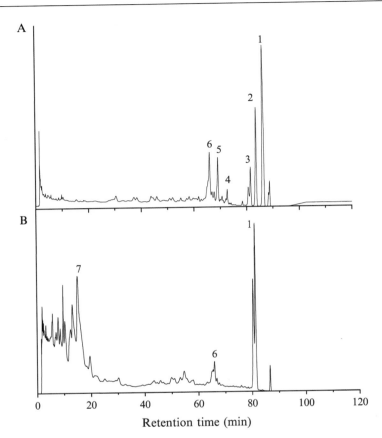

FIG. 3. Examples of typical HPLC chromatograms of total carotenoid extracts. Separations of oxocarotenoids produced by (A) *E. coli* [pAC-crtE-crtB-crtI + pUC-crtA] or (B) *E. coli* [pAC-crtE-crtB-crtI14 + pUC-crtA] are shown. Peaks were identified as follows: peak 1, ζ-carotene (λ_{max}: 377, 400, 424; M+ at m/e = 540.4); peak 2, neurosporene (λ_{max}: 419, 442, 470; M+ at m/e = 538.4); peak 3, lycopene (λ_{max}: 449, 475, 507; M+ at m/e = 536.4); peak 4, ζ-carotene-2-one (λ_{max}: 377, 400, 424; M+ at m/e = 556.4); peak 5, neurosporene-2-one (λ_{max}: 419, 442, 470; M+ at m/e = 554.4); peak 6, lycopene-2-one (λ_{max}: 449, 475, 507; M+ at m/e = 552.4); and peak 7, phillipsiaxanthin (λ_{max}: 516, 524; M+ at m/e = 596.3).[9]

spectroscopic identification of carotenoids can be found in the excellent volumes on carotenoids edited by Britton, Liaaen-Jensen, and Pfander.[14,15]

[14] G. Britton, S. Liaaen-Jensen, and H. Pfander, eds., "Carotenoids: Spectroscopy," Vol. 1B. Birkhauser, Basel, 1995.
[15] G. Britton, S. Liaaen-Jensen, and H. Pfander, eds., "Carotenoids: Isolation and Analysis," Vol. 1A. Birkhauser, Basel, 1995.

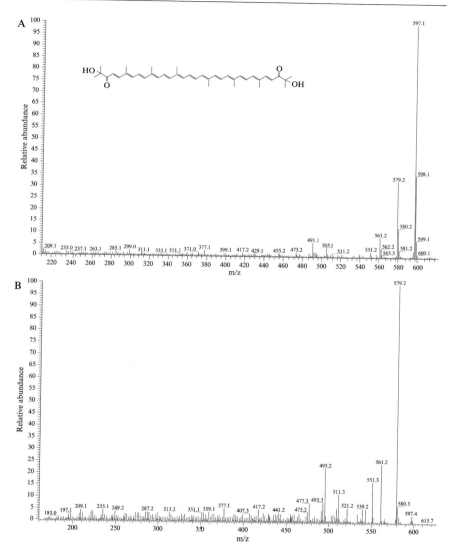

FIG. 4. Example of a typical mass fragmentation spectrum of a carotenoid. (A) APCI-MS and (B) APCI-MS/MS of a novel acyclic oxocarotenoid spirilloxanthin produced by recombinant *E. coli* transformants.[9]

Altering Activities of Carotenoid Enzymes within the Context of Assembled Carotenoid Pathways: Directed Evolution

Although protein crystal structures are available for some enzymes responsible for carotenoid precursor synthesis, such as prenyltransferases,[16] no structures of enzymes directly involved in carotenoid synthesis have been solved. Carotenoid biosynthesis in microorganisms takes place at the membrane, and the hydrophobic nature of many carotenoid enzymes renders purification and *in vitro* analysis of the enzymes difficult.[17–19] This lack of structural information greatly limits the possibility of rational engineering of protein structure to tailor carotenoid biosynthesis. Directed evolution, however, requires no structural information, and there are numerous examples of this approach being applied to the creation of improved protein functions.[20] Directed evolution is a technique by which mutations in a particular genetic sequence space are generated *in vitro*. Mutated sequences are then transformed into a recombinant host and variants with desired enzyme activities are selected. Multiple rounds of mutation and selection are utilized frequently to promote the desired enzyme properties. There are a number of approaches described to generate the library of variants required for directed evolution. The simplest of these is error-prone PCR in which the PCR reaction environment is altered to promote the semirandom misincorporation of bases.[21] This population of variant genes is then cloned into a vector of choice for subsequent screening. However, error-prone PCR generates only limited diversity and is biased toward particular mutation events. DNA shuffling is the *in vitro* recombination of homologous genes to generate a library of variants incorporating sequence fragments from a number of parent sequences.[22] Parent sequences in this approach can be generated by error-prone PCR or utilize naturally available homologous gene families. The genetic diversity obtained by this approach is considerable and this is the method of choice for many directed evolution experiments.

Effective directed evolution is dependent on a rapid, high-throughput and sensitive screening procedure to identify desired variants. The inherent chromatic property of carotenoids provides a simple and effective screen that satisfies these criteria. Changes in carotenoid structure are frequently

[16] L. C. Tarshis, M. Yan, C. D. Poulter, and J. C. Sacchettini, *Biochemistry* **33**, 10871 (1994).
[17] G. Schnurr, N. Misawa, and G. Sandmann, *Biochem. J.* **315**, 869 (1996).
[18] M. Albrecht, A. Ruther, and G. Sandmann, *J. Bacteriol.* **179**, 7462 (1997).
[19] C. Schneider, P. Boger, and G. Sandmann, *Protein Eng. Express. Purif.* **10**, 175 (1997).
[20] C. Schmidt-Dannert, *Biochemistry* **40**, 13125 (2001).
[21] D. W. Leung, E. Chen, and D. V. Goeddel, *Technique* **1**, 11 (1989).
[22] W. P. Stemmer, *Nature* **370**, 389 (1994).

reflected in changes in the visible color of recombinant colonies and the spectral properties of carotenoid extracts (Fig. 2). We have developed a simple screening procedure in which recombinant, carotenoid-expressing *E. coli* colonies are blotted on white nitrocellulose filters, transferred to fresh media plates with limited nutrients, and allowed to grow slowly at room temperature. This promotes high levels of carotenoid expression and allows colonies with altered color to be selected easily by visual screening. Although this approach allows the detection of many altered carotenoid structures, there is no indication of the amount and distribution of pathway intermediates accumulated and there is no way to infer the structures of carotenoids produced from color alone. Gene shuffling and the color screen described have been utilized to alter the enzymatic function of two carotenoid pathway enzymes—phytoene desaturase and lycopene cyclase.[7]

A library of phytoene desaturase (CrtI) variants was generated by DNA shuffling the corresponding genes from *Erwinia uredovora* and *Erwinia herbicola* on the pUCMod vector. This library was transformed into strains harboring *E. uredovora* GGDP synthase (CrtE) and phytoene synthase (CrtE) genes on pACMod, thus establishing a complete carotenoid biosynthetic pathway, and colonies were screened for color production. Both parent desaturase genes in this case produced only the orange–red carotenoid lycopene when transformed into *E. coli* harboring pAC-CrtE-CrtB. Approximately 30% of the 10,000 variant library transformants screened were colorless, presumably as a result of phytoene desaturase inactivation. A number of yellow colonies were observed, which suggest that carotenoids were produced with fewer conjugated double bonds than lycopene (11 conjugated double bonds). One of these colonies (I25) was found to produce β-carotene (7 conjugated double bonds) as a major product and neurosporene and lycopene as minor products (see Fig. 1 for pathway and structures). One pink colony was observed (I14), which when analyzed was found to produce the fully desaturated linear carotenoid 3,4,3′,4′-tetradehydrolycopene (15 conjugated double bonds). Sequence analysis of mutant I25 indicated that no recombination crossover events had taken place and that two amino acid substitutions of the *E. uredovora* desaturase had taken place—R332H and G470S. G470S occurs in a hydrophobic C-terminal domain that is well conserved among carotenoid desaturases and thought to be involved in substrate binding and dehydrogenation. Mutant I14 had four amino acid deviations from the *E. uredovora* CrtI sequence (P3K, T5V, V27T, L28V) resulting from the substitution of the N terminus (residues 1–39) with the equivalent region from *E. herbicola* CrtI. Two additional substitutions from nonrecombination events (F231L and A269V) were also present. Construction of chimeras indicated that only the N terminus crossover event was responsible for the altered enzyme activity. This region comprises a conserved FAD-binding domain. However, as with many carotenoid pathway enzymes, a lack of available protein structural data limits greatly the conclusions about changes in enzyme mechanism that can be drawn.

A similar DNA shuffling and screening approach was also utilized to modify the substrate specificity of the enzyme lycopene cyclase (CrtY) toward the more desaturated linear carotenoids produced by mutant I14. One bright-red recombinant (Y2) was identified that primarily produced the carotenoid torulene—the monocyclic derivative of 3,4-didehydrolycopene (Fig. 1). Sequence analysis indicated this variant had two amino acid substitutions from the *E. uredovora* parent sequence (R330H and F367S) that lie outside any conserved amino acid regions. These results demonstrate the utility of directed evolution approaches to modifying carotenoid biosynthesis enzymes and produce novel carotenoid structures in *E. coli*.

Probing the Catalytic Promiscuity of Carotenoid Modifying Genes as a Step Toward Novel Carotenoids Produced by Evolved Pathways

It has been demonstrated in many secondary metabolite biosynthetic pathways that enzymes that catalyze downstream modification reactions frequently have high flexibility in substrate specificity. This characteristic can be exploited to combine metabolic pathway enzymes from different source organisms in a recombinant host to create novel branches in metabolic pathways and synthesize previously inaccessible products. The process is often referred to as combinatorial engineering. Examples include the glycosyl transferases of polyketide antibiotics, which can frequently act on nonnatural substrates to produce novel glycosylation patterns.[2,3]

We have demonstrated previously, that *in vitro*-evolved torulene and tetradehydrolycopene pathways can be extended with downstream carotenoid-modifying enzymes in order to direct the synthesis of novel carotenoid structures. This combinatorial approach, based on the principle of promiscuity of carotenogenic enzymes from different source organisms, was also applied to produce novel unnatural C30 carotenoid structures in *E. coli*.[9] Extension of the known acyclic C30 diaponeurosporene pathway with enzymes known to be active on C40 substrates—monooxygenase CrtA and lycopene cyclase CrtY—yielded new oxygenated acyclic carotenoids and the unnatural cyclic C30 carotenoid diapotorulene, respectively. Extension of wild-type and evolved acyclic C40 carotenoid pathways with CrtA generated novel oxygenated carotenoids, including the violet

TABLE I

COMBINATORIAL CAROTENOID BIOSYNTHESIS AND THE RESULTING MAJOR
CAROTENOID ACCUMULATED IN *E. COLI* TRANSFORMANTS[a,7,9]

Gene combination	Major carotenoids
crtM + crtN	Diapolycopene, diaponeurosporene
crtM + crtN + crtY	Diapotorulene
crtM + crtN + crtA	Oxocarotenoids
crtE + crtB + crtI	Lycopene
crtE + crtB + crtI + crtA	Neurosporene-2-one,
	ζ-carotene-2-one, lycopene-2-one,
	neurosporene, ζ-carotene, lycopene
crtE + crtB + crtI14	Tetradehydrolycopene, lycopene
crtE + crtB + crtI14 + crtA	Phillipsiaxanthin, lycopene-2-one,
	uncharacterized oxocarotenoids
crtE + crtB + crtI14 + crtY	β-Carotene
crtE + crtB + crtI14 + crtY + crtO	Canthaxanthin, echinenone
crtE + crtB + crtI14 + crtY + crtU	Isorenariatene
crtE + crtB + crtI14 + crtY + crtZ	Zeaxanthin
crtE + crtB + crtI14 + crtY + crtZ + crtX	Zeaxanthin-β-diglucoside
crtE + crtB + crtI14 + crtY2	Torulene
crtE + crtB + crtI14 + crtY2 + crtO	Ketotorulene
crtE + crtB + crtI14 + crtY2 + crtU	Didehydro-β, β-carotene
crtE + crtB + crtI14 + crtY2 + crtZ	Hydroxytorulene
crtE + crtB + crtI14 + crtY2 + crtZ + crtX	Torulene-β-glucoside

[a] Carotenoid enzyme names are shown in Fig. 1. Novel carotenoids produced in *E. coli* are underlined.

carotenoid phillipsiaxanthin from tetradehydrolycopene. Extension of the β-carotene and torulene biosynthetic pathways with CrtU (β-ring desaturase) in *E. coli* generated aromatic carotenoids, including aromatic torulene. Extension of the torulene biosynthetic pathway with β-carotene hydroxylase (CrtZ), zeaxanthin glucosylase (CrtX), and β-carotene ketolase (CrtO) yielded new hydroxylated, glucosylated, and ketolated torulene derivatives, respectively. Table I summarizes carotenogenic gene combinations in *E. coli* and the resulting major carotenoids accumulated in the transformants.

However, this flexibility in substrate specificity can also be a disadvantage in directing the synthesis of a final product, as promiscuous modifying enzymes can take up intermediates of a multistep biosynthetic pathway and produce unwanted side products. The inability to predict which substrates will be accepted by carotenoid-modifying enzymes may also lead to unexpected enzymatic reactions and products.

Mining Genomes for New Carotenoid Activities

Considering the enormous structural diversity observed in natural carotenoids (more than 500 structures described in the literature[23]), there is still relatively little known about the genetic and enzymatic machinery responsible for the biosynthesis of many of these compounds. The ever-expanding array of microbial genome sequences provides an excellent resource to better understand the biosynthesis of natural carotenoids and possibly generate novel, useful carotenoid structures through the directed evolution and combinatorial engineering approaches described previously.

Putative carotenoid biosynthetic genes are distributed in a number of microbial genomes, including both pathogenic organisms and environmental isolates, and in most cases the biological activity of these genes remains uncharacterized. Carotenoid biosynthetic pathway genes are frequently clustered in microbial genomes, which can aid greatly in the identification of novel carotenoid pathway genes. Enzymes responsible for the synthesis of desaturated carotenoid backbone structures are members of the oxidoreductase family and show considerable homology in all known carotenoid-producing organisms. These genes are often closely associated with other carotenoid biosynthetic genes in genome sequences and can therefore be useful in identifying novel carotenoid operons.

In many cases, central carotenoid pathways genes may be located together on genome sequences but further modifying genes may be located at different genome regions. This may reflect the likelihood of the genetic

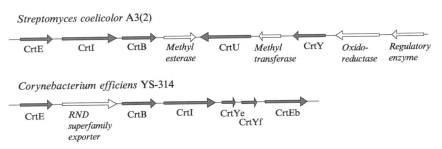

FIG. 5. Carotenoid biosynthetic operons from genome sequences of *Streptomyces coelicolor* A3(2) and *Corynebacterium efficiens* YS-314. Homologues to known carotenogenic enzymes are indicated with gray arrows: CrtE, geranyl geranyl pyrophosphate synthase; CrtB, phytoene synthase; CrtI, phytoene desaturase; CrtY, lycopene cyclase; CrtU, β-carotene desaturase; CrtYe/Yf, heterodimeric decaprenoxanthin synthase; CrtEb, lycopene elongase.

[23] H. Pfander, M. Gerspacher, M. Rychener, and R. Schwabe, eds., "Carotenoids: Key to Carotenoids." Birkhauser, Boston, 1987.

transfer of diverse modifying genes or suggest evolutionary adaptation of modifying genes from other cellular functions. Such nonclustered modifying genes, which are mostly responsible for the large diversity of carotenoid structures produced by microorganisms, are difficult to identify by sequence analysis alone. Functional screening, such as *in vivo* complementation of carotenoid pathways in heterologous hosts with cloned genes from genome sequences, will be necessary to broaden the current restricted repertoire of known carotenoid-modifying genes. Figure 5 shows an example of carotenoid gene clusters observed in a number of microbial genome sequences.

Conclusion

In addition to the inherent value of carotenoids as chemicals for medicine and industry, carotenoid biosynthesis is an excellent model system to develop and test tools for the creation of novel multienzyme pathways in recombinant organisms using combinatorial and directed evolution tools. Design of an efficient vector system for coordinated enzyme expression is critical in effectively reconstituting complex, branching biosynthetic pathways in heterologous hosts such as *E. coli*. With such a system in place, *in vitro* evolution and combinatorial gene combination can be powerful approaches to generate structurally diverse products and to explore the plasticity of downstream modifying enzymes. We have described a well-characterized system for the production of carotenoids in *E. coli*.

The lack of protein structural formation available for carotenoid biosynthetic pathway enzymes renders any rational approach to altering or investigating enzyme function impossible. This lack of structures may be the result of the proposed membrane localization of many carotenoid enzymes that can render crystallization difficult. Carotenoid biosynthesis is thought to occur in membrane-associated multienzyme complexes, which bridge the biophysical barrier between hydrophilic precursor molecules in the cytoplasm and hydrophobic carotenoids accumulated in cell membranes. The nature of these enzyme complexes remains largely unexplored and *in vitro* carotenoid enzyme reactions remain inefficient and often require lipid membrane extracts for function.

Section V

Devices, Antibodies, and Vaccines

[26] Directed Evolution of Bacteriorhodopsin for Device Applications

By Jason R. Hillebrecht, Kevin J. Wise, Jeremy F. Koscielecki, and Robert R. Birge

Introduction

Outside the laboratory, biological systems gain comparative advantage through an algorithm of mutation and natural selection, but evolution provides biological molecules that are optimized for device applications only if these phenotypes confer a selective advantage. This observation does not reflect an inherent shortcoming of natural selection but rather the significant differences in the requirements of biological organisms versus device environments. The environments of multicellular organisms are usually well buffered from fluctuations in solvent, solute, pH, ionic strength, and often temperature. Furthermore, there is a high turnover rate for cellular proteins; proteins are constantly being degraded and synthesized. For this reason, it is not necessary for a cell to produce proteins with long life spans. In contrast, devices require long-lived proteins with enhanced thermal stability and other altered properties either irrelevant or nonoptimal for the biological host. If scientists and engineers had access to fully rigorous theoretical models and sufficiently fast computers, new proteins could be designed *ab initio* and created using site-directed mutagenesis. In reality, neither software nor hardware are sufficiently sophisticated to predict which mutations will provide the desired optimizations, and experimental methods are required. One of the most efficient optimization protocols is directed evolution.[1–7] Directed evolution is defined as repeated iterations of screening and differential selection to achieve a particular goal. This technique has been explored in great detail for structure–function studies of enzymes and for optimizing enzymes for commercial applications.[8–14] However, directed evolution can also be used to optimize any protein

[1] M. Callahan and B. Jerpseth, *Methods Mol. Biol.* **57,** 375 (1996).

[2] H. Zhao, L. Giver, Z. Shao, J. A. Affholter, and F. H. Arnold, *Nature Biotechnol.* **16,** 258 (1998).

[3] K. Miyazaki and F. H. Arnold, *J. Mol. Evol.* **49,** 716 (1999).

[4] A. Sawano and A. Miyawaki, *Nucleic Acid Res.* **28,** e78 (2000).

[5] F. Arnold, P. L. Wintrode, K. Miyazaki, and A. Gershenson, *Trends Biochem. Sci.* **26,** 100 (2001).

[6] U. T. Bornscheuer, *Curr. Opin. Biotechnol.* **13,** 543 (2002).

[7] H. Tao and V. W. Cornish, *Curr. Opin. Chem. Biol.* **6,** 858 (2002).

[8] F. Arnold and J. C. Moore, *Adv. Biochem. Engineer.* **58,** 1 (1997).

provided a suitable expression system and optimization algorithm can be found.[15,16]

This article explores the use of directed evolution to optimize the light-transducing protein, bacteriorhodopsin (BR). This protein is synthesized by the salt marsh archaeon *Halobacterium salinarum* when oxygen availability drops below a level sufficient to sustain respiration. The organism then switches spontaneously to photosynthetic energy production by generating a purple membrane containing BR. When BR absorbs light it pumps a proton, and the pH gradient is utilized to convert ADP to ATP.[17,18] Because a salt-marsh organism is exposed to high thermal, solar, and aerobic stressors, natural selection has created a robust protein with excellent efficiency in converting light into molecular changes. For that reason, native BR has long been investigated for use in photonic devices, including holographic and three-dimensional memories, spatial light modulators, pattern recognition systems, artificial retinas, photovoltaic converters, and hybrid electrooptical devices.[16,19–25] In virtually all long-term studies, a modified form of the protein was eventually found that outperformed the native protein.[16,22,23,26,27] The purpose of this article is to review

[9] J. Hoseki, T. Yano, Y. Koyama, S. Kuramitsu, and H. Kagamiyama, *J. Biochem.* **126,** 951 (1999).
[10] M. Lehmann, L. Pasamontes, S. F. Lassen, and M. Wyss, *Biochim. Biophys. Acta* **1543,** 408 (2000).
[11] U. T. Bornscheuer and M. Pohl, *Curr. Opin. Chem. Biol.* **5,** 137 (2001).
[12] O. Kirk, T. V. Borchert, and C. C. Fuglsang, *Curr. Opin. Biotechnol.* **13,** 345 (2002).
[13] J. R. Cherry and A. L. Fidantsef, *Curr. Opin. Biotechnol.* **14,** 438 (2003).
[14] P. A. Dalby, *Curr. Opin. Struct. Biol.* **13,** 500 (2003).
[15] S. R. Whaley, D. S. English, E. L. Hu, P. F. Barbara, and A. M. Belcher, *Nature* **405,** 665 (2000).
[16] K. J. Wise, N. B. Gillespie, J. A. Stuart, M. P. Krebs, and R. R. Birge, *Trends Biotechnol.* **20,** 387 (2002).
[17] D. Oesterhelt and W. Stoeckenius, *Nature (Lond.) New Biol.* **233,** 149 (1971).
[18] J. A. Stuart and R. R. Birge, *in* "Biomembranes" (A. G. Lee, ed.), Vol. 2A, p. 33. JAI Press, London, 1996.
[19] T. Miyasaka, K. Koyama, and I. Itoh, *Science* **255,** 342 (1992).
[20] Z. Chen and R. R. Birge, *Trends Biotechnol.* **11,** 292 (1993).
[21] F. Hong, *in* "Molecular and Biomolecular Electronics" (R. R. Birge, ed.), Vol. 240, p. 527. American Chemical Society, Washington, DC, 1994.
[22] R. R. Birge, N. B. Gillespie, E. W. Izaguirre, A. Kusnetzow, A. F. Lawrence, D. Singh, Q. W. Song, E. Schmidt, J. A. Stuart, S. Seetharaman, and K. J. Wise, *J. Phys. Chem. B* **103,** 10746 (1999).
[23] N. Hampp, *Appl. Microbiol. Biotechnol.* **53,** 633 (2000).
[24] J. Xu, P. Bhattacharya, and G. Varo, *in* "Lasers and Electro-Optics Society 2001. 14th Annual Meeting, IEEE Annual Meeting Conference Proceedings," p. 833.
[25] P. Bhattacharya, J. Xu, G. Varo, D. L. Marcy, and R. R. Birge, *Opt. Lett.* **27,** 839 (2002).
[26] A. Druzhko, S. Chamorovsky, E. Lukashev, A. Kononenko, and N. Vsevolodov, *BioSystems* **35,** 129 (1995).
[27] J. E. Millerd, A. Rohrbacher, N. J. Brock, P. Smith, and R. Needleman, *Opt. Lett.* **24,** 1355 (1999).

current efforts to use directed evolution to improve photophysical characteristics and thermal stability of BR. Heterologous protein expression tools are now in place to increase the thermal stability of the protein using *Thermus thermophilus* as a host. Repeated iterations of mutagenesis, expression, and screening have already been used to alter photocycle characteristics in BR, and similar techniques are now being used to increase the thermal stability of the protein.

Optimizing Photophysical Properties in Bacteriorhodopsin

Directed evolution is best envisioned in terms of a hypothetical "mutational landscape."[16] Figure 1 shows the different types of mutational fields representative of both photophysical and thermal characteristics, with the surface height proportional to the quality of the mutation with reference to the desired property. Whereas the optimization of photophysical characteristics of BR is represented more accurately by fairly localized, irregular changes (Fig. 1a), thermal optimization of BR is best represented as involving a continuous and slowly varying landscape (Fig. 1b). The ragged "photochemical" landscape is a result of the nature of the characteristics being optimized; more fluid characters, such as thermal stability, involve much larger portions of the protein and can be accomplished via a variety of complementary mutations.

The photophysical optimization of BR for device applications involves both genetic and chemical modification of the protein. Chemical modification is not discussed here, but has been discussed elsewhere (see, e.g., Druzhko et al.[26] and Birge et al.[22]). This article focuses on progress made to methods used for the genetic modification of BR.

Bacteriorhodopsin has a number of intrinsic characteristics that make it a useful template for devices. The native protein is remarkably stable across a wide range of conditions.[20,28,29] The protein possesses unique photophysical properties that render it both suitable for device applications and amenable to optimization. The light-adapted protein undergoes a series of conformation and electronic changes upon illumination. A series of intermediates have been identified spectroscopically and are arbitrarily named bR (the light-adapted resting state), K, L, M, N, and O. Following absorption of a photon of light, bR converts photochemically to the K state. The remainder of the intermediates progress through a series of thermal states as shown in Fig. 2. Unless the O state is illuminated with

[28] Y. Shen, C. R. Safinya, K. S. Liang, A. F. Ruppert, and K. J. Rothschild, *Nature* **366,** 48 (1993).
[29] E. P. Lukashev and B. Robertson, *Bioelectrochem. Bioenerg.* **37,** 157 (1995).

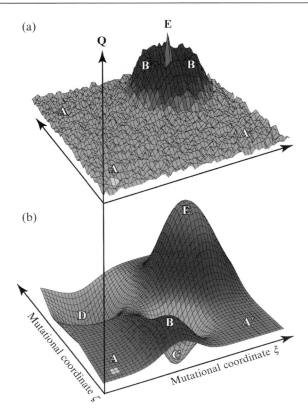

FIG. 1. (a) The mutational landscape expected for more variable characters such as photophysical optimizations. (b) The mutational landscape expected for optimization of a fluid trait, such as temperature stability. Starting at mutational coordinate A, troughs and local maxima must be examined (A', B, C, D) before reaching an optimization peak (E). A large number of variants must be screened to ensure that a global peak has been reached. (See color insert.)

red light, the O state will thermally decay back to bR. However, if the O state absorbs a photon of red light, it will enter the branched photocycle composed of the P1, P2, and Q intermediates.[30,31] The Q intermediate is the resting state of the branched photocycle and has a lifetime of over

[30] A. Popp, M. Wolperdinger, N. Hampp, C. Bräuchle, and D. Oesterhelt, *Biophys. J.* **65,** 1449 (1993).

[31] N. B. Gillespie, K. J. Wise, L. Ren, J. A. Stuart, D. L. Marcy, J. Hillebrecht, Q. Li, L. Ramos, K. Jordan, S. Fyvie, and R. R. Birge, *J. Phys. Chem. B* **106,** 13352 (2002).

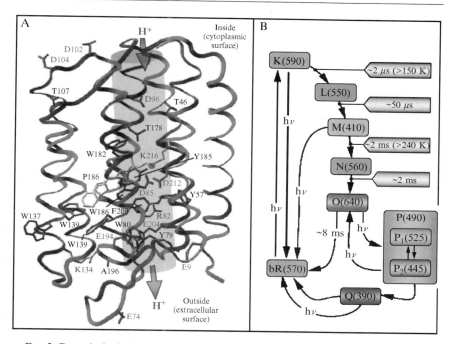

FIG. 2. Bacteriorhodopsin structure and photocycle. (A) BR is an ion pump that results in the net transfer of a proton from the intracellular to the extracellular surface. Key residues are highlighted, and the proton transfer channel is shown. (B) The main and branched photocycle in BR used for three-dimensional memory device applications. The branching reaction involves the O → P transition that is optimized using directed evolution. M and O state lifetimes, and yields can also be optimized using directed evolution. (See color insert.)

7 years at ambient temperature. The Q state can be converted back photo-chemically to bR via illumination with ultraviolet light. The protein therefore presents a unique bistable system, composed of the bR, K, L, M, N, and O states in the main photocycle and P1, P2, and Q states in the branched photocycle. The existence of this bistable system has led to the proposal that BR can be used as a medium for storing information in a three-dimensional paged memory.[22] Bacteriorhodopsin has also been a candidate for holographic device applications based on the refractivity difference and diffraction efficiency between the bR and the M states.[23] Some important residues involved in the photocycle of BR are shown in Fig. 2.

Due to their importance in different device applications, two states have been the target of genetic optimization: the M state and the O state. These states represent the most blue-shifted (M) and red-shifted (O) intermediates in the main photocycle and are detectable most readily by spectroscopic techniques. The M state has been tailored chemically and genetically for use in holographic devices.[32,33] The O state has been modified genetically for use in three-dimensional architecture bistable memory systems[16] because this state is the entry point to the branched photocycle composed of the P and Q intermediates.[30,31]

Photophysical Optimization Requirements

In order to optimize BR for device applications, a number of criteria must be satisfied. First, methods must exist to introduce genetic variation into the BR-coding sequence. Second, methods for expressing mutant proteins must be in place. Third, methods for screening and characterizing mutant proteins must exist. The ability to create protein variants is only useful if an efficient screening method is in place; high-throughput methods are required in detection of significant photophysical improvements. The methods described here can be applied to any protein system, given the following criteria are met: a method for constructing mutants, a method for expressing mutants, and a method for screening and characterizing mutants.

Despite significant progress in using site-directed mutagenesis to optimize BR for device applications (i.e., the D96N mutant is being used for holographic applications), more advanced mutagenesis methods were used to locate improvements in photophysical characteristics. Two photocycle intermediates were targeted for BR optimization. For holographic device applications, the M state was targeted for increased yield and lifetime of the M state. For three-dimensional memory usage, the O state was targeted for increased efficiency of the O $\rightarrow$ P transition and increased yield of the O state.

Constructing and Expressing Photophysical Variants

The generation of novel protein constructs starts with the generation of various DNA molecules. Mutagenesis technique specificity ranges from mutating an individual amino acid, a region of 15 amino acids, or mutation of the entire BR gene. These techniques are called site-directed mutagenesis (SDM), semirandom mutagenesis (SRM), and random mutagenesis (RM), respectively. An overview of the levels of mutagenesis used to modify BR can be found in Wise et al.[16]

[32] N. Hampp and T. Juchem, in "Bioelectronic Applications of Photochromic Pigments" (A. Dér and L. Keszthelyi, eds.), Vol. 335, p. 44. IOS Press, Szeged, Hungary, 2000.
[33] T. Juchem and N. Hampp, Opt. Lasers Engineer. **34**, 87 (2000).

Directed evolution is most successful when various mutagenesis techniques are used concomitantly. Umeno *et al.*[34] developed a technique that involves protection of a single amino acid using restriction site substitution at the desired site. For example, SDM can be used to identify single amino acids responsible for a given optimization. These residues can be protected from further mutation and the remainder of the protein can then be subjected to further rounds of random mutagenesis. Proteins can thus be modified in a directional "stepwise" fashion until a desired optimization is achieved.

Site-directed mutants are constructed using the Stratagene Quik-Change mutagenesis system. This kit is used to alter one amino acid per round and represents the most specific method for altering protein structure. Semirandom mutagenesis is used to modify a region of ~15 amino acids. All amino acids in a given 15 amino acid region are mutated randomly and multiple mutations are very common. For SRM on the BR gene, the sequence was split into 17 manageable segments. A total of 900 proteins were constructed, having an average of 2.6 amino acid changes per clone (these mutants also possessed an additional 1.1 silent amino acid changes per clone). Random mutagenesis can also be used to generate protein variants across the entire protein sequence.

Mutants are expressed using a plasmid called pBA1 (generously obtained from Dr. Mark P. Krebs). This plasmid has sequences necessary for the expression of BR in the native organism, *Halobacterium salinarum*. Mutants are transformed into *H. salinarum* using a procedure described previously,[16] and protein is isolated from colonies for screening. Mutants constructed via SDM or SRM are constructed and isolated using this system.

Screening of Photophysical Variants In Vitro

An efficient system must be in place in order to detect photocycle alterations. For BR, each protein was illuminated at 568 nm, and absorbance changes were recorded at 412, 568, and 650 nm. Each screening was completed on ~0.5 mg of isolated BR protein. Mutant proteins that show significant deviation from wild-type kinetics (i.e., lifetime and yield of the M and/or O state intermediates) were studied further using time-resolved spectroscopy spanning each nanometer per millisecond between 300 and 800 nm.

Figure 3 shows the kinetics summary of the SRM region spanning amino acids 149–163. A total of 77 mutants were screened, and a favorable

---

[34] D. Umeno, K. Hiraga, and F. Arnold, *Nucleic Acid Res.* **31**, e91 (2003).

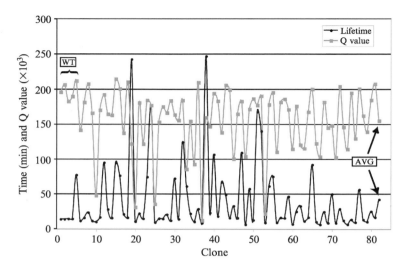

Fig. 3. Semirandom mutagenesis of the amino acid region 149–163. A total of 77 unique mutants were screened at 412 nm for photocycle deviations from wild-type protein in terms of M state lifetime and yield. The Q value is measured as the integral under each trace divided by the lifetime and standardized for variation in sample optical density. The Q value has been multiplied by 1000 for viewing purposes. A unique mutant, L149P/F154Y/T157S/S158L/E161V, is being tested in holographic memory devices. The first four points represent screens on the wild-type protein, and the last point is an average of all mutants.

M state mutant was located in this region (clone 38) and screened further. This mutant, L149P/F154Y/T157S/S158L/E161V, possessed a M state lifetime of approximately 300 ms, as compared to 5–10 ms in wild-type protein. This mutant represents an ideal candidate for use in holographic memories and is currently being tested in that regard.

Using a combination of SDM and SRM, mutants have been identified that increase the lifetime of the O state to greater than 4 s as compared to 10–15 ms in wild-type protein. A six-point mutation, A196S/I198L/P200T/E204A/T205Q/F208Y, was located in a photocycle screen of semirandom mutants in a region encompassing residues 194–208. Another mutant identified in a screen of the same region, I198F/T205A, was found that had a 100-fold improvement in O state yield, as measured by the maximal absorbance change at 650 nm. This mutant provides a large net yield of O state, a character essential for the optimization of BR for use in three-dimensional protein-based memory systems.

Directed evolution of BR for device application involves screening mutants constructed from SRM and RM techniques for a desired photophysical property (i.e., O state yield, M state lifetime), characterization and identification of these mutants, and using optimal mutants as a template for the next round of mutation. Mutants containing multiple amino acid changes are often dissected into single amino mutants using SDM to identify which amino acid(s) is most responsible for a given photophysical effect.

Identified M and O state mutants are used as starting platforms, or templates, for future rounds of directed evolution. Each generation results in the detection of proteins with increasingly optimized photophysical properties. In addition to usefulness in device work, these mutants also provide valuable basic knowledge in the photochemistry of protein systems.

Device Testing of Photophysical Mutants

Mutants of BR have been used in holographic applications for a number of years. The single mutant D96N has been used in development of the Fringemaker device, a holographic device based on a long M state lifetime.[32,33] Other long M state lifetime mutants have been identified in SRM screens of other regions (in addition to 149–163 mentioned earlier), and these mutants are currently being tested for their usefulness in holographic devices. For three-dimensional memory devices, O state mutants are currently being tested for O → P branching efficiency. A number of mutants have been identified that provide nearly a 100-fold improvement in the efficiency of the branched photocycle three-dimensional memory.[16]

Optimizing the Thermostability of Bacteriorhodopsin

Increasing the thermostability of BR requires expression and selection of protein variants in a high temperature environment. Directed evolution combines *in vivo* and *in vitro* techniques to increase the thermal versatility of BR variants. Advancements in thermostable vectors, antibiotic resistance genes, and the genetic characterization of extreme thermophiles have prompted the development of *in vivo* thermoselection systems to optimize mesophilic proteins (i.e., BR) for device applications.[35,36] Thermostability studies on BR are conducted heterologously using *T. thermophilus* HB27, a eubacterium strain that grows at temperatures up to 85°.

[35] M. Tamakoshi, M. Uchida, K. Tanabe, S. Fukuyama, A. Yamagishi, and T. Oshima, *J. Bacteriol.* **179,** 4811 (1997).
[36] T. Kotsuka, S. Akanuma, M. Tomuro, A. Yamagishi, and T. Oshima, *J. Bacteriol.* **178,** 723 (1996).

For *T. thermophilus* to serve as an *in vivo* screening platform for BR variants, a versatile, heat-stable expression vector is required. Moreno *et al.*[37] constructed a bifunctional vector system (pMKE1) capable of expression in both extremely thermophilic (*T. thermophilus*) and mesophilic (*Escherichia coli*) microorganisms. This shuttle vector is used to create, manipulate, and express genetically modified BR variants in *T. thermophilus*.

Construction of BR variants for expression in *T. thermophilus* is achieved using a combination of mutagenesis methods, as described earlier. To broaden the scope of this study, purified BR variants are also screened *in vitro* (purified protein) for increased thermostability relative to wild-type proteins. Mutants that retain structural stability at elevated temperatures are used as starting points for additional rounds of mutagenesis and thermoselection. Several iterations of thermoselection may be required before a BR variant with adequate thermostability can be used for device applications.

Construction of BR Variants in T. thermophilus

Thermus thermophilus HB27 is an aerobic, gram-negative eubacterium that grows optimally at temperatures ranging from 50° to 82°. These non-sporulating, rod-shaped bacteria grow in hot-spring ecosystems and industrial composts.[38] The presence of *Thermus* spp. in hot compost systems suggests that these heterotrophic bacteria may play a role in the degradation of organic waste. The commercial value of thermally stable macromolecules that break down waste products and demonstrate a resistance to organic solvents led to the extensive characterization of this extreme thermophile. The isolation and development of catalytic macromolecules with thermostable characteristics will enhance industrial fermentation systems, enzymes used for genetic engineering, and molecular electronic devices.[39]

Extreme thermophiles are not only unique in their ability to survive at high temperatures, but they also possess a high competence for the natural transformation of foreign DNA.[40] While most naturally competent bacteria do not transform plasmid DNA efficiently, *T. thermophilus* HB27 is relatively proficient at the process.[41] The natural transformation machinery

[37] R. Moreno, O. Zafra, F. Cava, and J. Berenguer, *Plasmid* **49**, 2 (2003).
[38] T. Beffa, M. Blanc, P. F. Lyon, G. Vogt, M. Marchiani, J. L. Fischer, and M. Aragno, *Appl. Environ. Microbiol.* **62**, 1723 (1996).
[39] K. Tabata, T. Kosuge, T. Nakahara, and T. Hoshino, *FEBS Lett.* **331**, 81 (1993).
[40] A. Friedrich, T. Hartsch, and B. Averhoff, *Appl. Environ. Microbiol.* **67**, 3140 (2001).
[41] Y. Koyama, T. Hoshino, N. Tomizuka, and K. Furukawa, *J. Bacteriol.* **166**, 338 (1986).

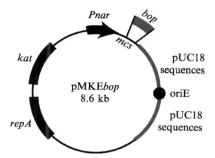

FIG. 4. The bifunctional shuttle plasmid pMKE*bop*. A 1.2-kb *bop* fragment is inserted into the multiple cloning site (*mcs*) located downstream of the inducible *Pnar* promoter region. Genes for the *Thermus* replication protein (*repA*), a kanamycin selection gene (*kat*), sequences derived from the *E. coli* overexpression vector pUC18, and an *E. coli* origin of expression comprise this 8.6-kb plasmid. Adapted from Moreno *et al.*[37]

used by extreme thermophiles is still not completely understood, but it is suggested that this mode of lateral gene transfer is essential to the survival and evolution of these extreme microorganisms.[42]

A fast growth cycle, a high competence for foreign DNA, and the development of thermostable selection markers (i.e., kanamycin resistance) make *T. thermophilus* HB27 an attractive candidate to host thermostability studies on mesophilic proteins such as BR.[35,36,43]

Increasing the thermal stability of BR requires a system that can express large numbers of mutant proteins efficiently at elevated temperatures. The pMKE1 plasmid (generously obtained from Dr. Jose Berenguer) contains the genetic elements required for expression in both *E. coli* and *T. thermophilus*.[37] Figure 4 shows the pMKE1 plasmid with the *bop* (bacterio-opsin coding gene) fragment inserted into the multiple cloning site (now called pMKE*bop*). This plasmid contains the coding elements for a *Thermus* replication protein (RepA), a thermostable kanamycin resistance cassette (*kat*), an *E. coli* origin of replication (oriE), and sequences derived from the *E. coli* overexpression vector pUC18.[37,44,45] These genetic elements enable pMKE1 to serve a dual role in the directed evolution of BR.

[42] A. Friedrich, C. Prust, T. Hartsch, A. Henne, and B. Averhoff, *Appl. Environ. Microbiol.* **68,** 745 (2002).

[43] M. Tamakoshi, A. Yamagishi, and T. Oshima, *Mol. Microbiol.* **16,** 1031 (1995).

[44] I. Lasa, P. Casta'n, L. A. Ferna'ndez-Herrero, M. A. de Pedro, and J. Berenguer, *Mol. Microbiol.* **6,** 1555 (1992).

[45] M. de Grado, M. Lasa, and J. Berenguer, *FEMS Microbiol. Lett.* **165,** 51 (1998).

Mutagenesis and the genetic manipulation of the *bop* gene are carried out in *E. coli*, whereas the expression and thermal selection of BR are hosted by *T. thermophilus* HB27::*nar*. The propensity of DNA uptake, quick growth time, and genetic characterization of *T. thermophilus* make this expression system an ideal tool for directed evolution studies on BR.

The HB27::*nar* strain is designed to control the transcriptional activity of pMKE1 through the induction of the *Pnar* promoter.[37] A genetic element known as the *nar* operon was cloned from the closely related HB8 stain and inserted into HB27, thereby creating HB27::*nar*.[46] The *nar* operon encodes a cluster of membrane-bound, protein components known as the respiratory nitrate reduction complex.[46a] This complex enables HB8 bacteria to grow anaerobically in the presence of nitrate. Upon exposing HB27::*nar* to low oxygen conditions and high concentrations of nitrate, the *nar* cluster releases a signal that activates the *Pnar* promoter. The induction of *Pnar* drives the expression of downstream genes (i.e., *bop*) located in the multiple cloning site of the plasmid. *Pnar* induction is an important mechanism for controlling the timing of *bop* expression in HB27::*nar*.

Inserting mutant *bop* fragments into the multiple cloning site of pMKE1 is the final step in the construction of a BR thermoselection vector (pMKE*bop*). A combination of mutagenesis techniques are used to create randomly mutated regions in the *bop* gene. Polymerase chain reaction-generated constructs are subcloned into pMKE1, thereby creating pMKE*bop*. Procedures for DNA manipulation, subcloning, and ligation are described in Sambrook *et al*.[47] Plasmids containing proteins other than BR can be constructed by replacing *bop* with the desired sequence of interest.

Expression and Screening of BR Variants

Thermoselection of BR variants *in vivo* requires that the pMKE*bop* expression vector be transformed into HB27::*nar* and induced to express the *bop* gene. Transforming pMKE*bop* into HB27::*nar* is consistent with the protocol described by de Grado *et al*.[48] To express the *bop* gene in HB27::*nar*, the *Pnar* promoter must be induced by the addition of nitrate

[46] S. Rami'rez-Arcos, L. A. Ferna'ndez-Herrero, and J. Berenguer, *Biochim. Biophys. Acta* **1396,** 215 (1998).

[46a] S. Rami'rez-Arcos, L. A. Ferna'ndez-Herrero, I. Mari'n, and J. Berenguer, *J. Bacteriol.* **180,** 3137 (1998).

[47] J. Sambrook, E. F. Fritsch, and T. Maniatis, "Molecular Cloning: A Laboratory Manual." Cold Spring Harbor Laboratory Press, Plainview, NY, 1989.

[48] M. de Grado, P. Castán, and J. Berenguer, *Plasmid* **42,** 241 (1999).

and anaerobic conditions. Transformant colonies are selected on kanamycin plates, picked, and grown to an optical density required for induction. Active incubation of the cultures is then ceased and nitrate in the form of KNO$_3$ (40 mM) is added.[37] Adding these elements to transformant cultures will trigger the induction of *Pnar* and the expression of downstream elements, specifically the *bop* gene.

Final formation of BR requires the addition of all-*trans* retinal to a final concentration of 2 μM.[49] Following induction and the addition of all-*trans* retinal, small aliquots from each transformant culture are transferred into a 96-well plate format and incubated at the desired selection temperature. This step is performed in duplicate so that control cultures (without all-*trans* retinal) can be monitored simultaneously (with all-*trans* retinal). Absorbance changes at 568 nm detect the presence of mature protein. Screening directly for BR in transformant cultures eliminates the need for protein isolation procedures, which can be both tedious and time-consuming. Samples with absorption peaks at 568 nm are sequenced and characterized for BR thermostability.

Over 900 BR variants are currently being tested for thermostability *in vitro* as well as *in vivo*. Purified, mutant proteins are subdivided into a 96-well plate format and incubated at temperatures starting at 50°. The temperature setting is then increased every 15 min by 1° and is monitored using a Biotek μQuant (96-well plate UV spectrophotometer). Mutants that retain their functional stability at temperatures greater than 65° (the denaturation temperature of monomeric BR) are used as starting points for additional rounds of directed evolution.[50] Our current goal is to find mutants that operate at or above 85° for extended periods of time.

Device Testing of Thermostable Mutants

Each round of thermal selection is divided into three phases, as shown in Fig. 5. Phase 1 includes the mutagenesis and genetic manipulation of *bop* gene fragments. Mutant *bop* fragments can be transformed into either *T. thermophilus* (*in vivo* screening) or *H. salinarum* (*in vitro* screening via protein isolation). Mutants that exhibit functional stability *in vitro* are sequenced and subjected to additional rounds of mutagenesis and selection

[49] V. Hildebrandt, M. Ramezani-Rad, U. Swida, P. Wrede, S. Grzesiek, M. Primke, and G. Bueldt, *FEBS Lett.* **243**, 137 (1989).
[50] C. Heyes and M. El-Sayed, *J. Biol. Chem.* **277**, 29427 (2002).

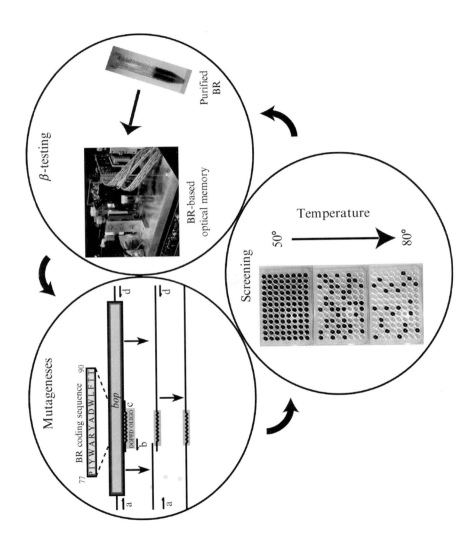

in vivo. Phase 2 includes the screening of BR variants in a 96-well plate format at the desired selection temperature. Several rounds of directed evolution may be necessary before a mutant with adequate thermostability for device applications is isolated. Upon purifying BR mutants with increased thermal diversity relative to wild type, these variants are used for β-testing in BR-based protein optical memories and holographic mediums. β-testing the protein in device applications constitutes the third and final phase of thermoselection. BR variants that fail to retain their functional stability in the β-testing phase are subjected to additional rounds of mutagenesis and thermoselection.

Concluding Remarks

Directed evolution is a powerful tool that is now being used to optimize photophysical properties in proteins. Photocycle optimization of BR has been achieved through the combined effect of site-directed mutagenesis, semirandom mutagenesis, high-throughput protein expression, and differential selection. Iterative cycles have increased the lifetime and yield of selected photocycle intermediates in BR while maintaining functional, stable proteins. The thermal stability of BR is achieved by expressing randomly mutated proteins in an extremely thermophilic background. By employing a heterologous expression system, the thermal stability of BR can be optimized far beyond levels possible in the native organism. These experiments provide a testament to the utility of directed evolution in biochemical systems and provide an optimization protocol for use in bioelectronic device optimization.

FIG. 5. Three phases of thermoselection. (1) Random and semirandom mutageneses are used to generate a random pool of *bop* gene fragments. Mutant constructs can be expressed in either *T. thermophilus* (*in vivo* screening) or *H. salinarum* (*in vitro* screening through protein isolation). (2) BR variants in a 96-well plate format are screened *in vivo* and *in vitro* for thermostability and functionality at the desired selection temperature. (3) Favorable BR variants are β tested in BR-based protein optical memories and holographic devices. Subsequent optimization of the protein may be attainable by subjecting favorable BR variants to additional rounds of mutagenesis and thermoselection.

[27] Engineering Antibody Affinity by Yeast Surface Display

By David W. Colby, Brenda A. Kellogg, Christilyn P. Graff, Yik A. Yeung, Jeffrey S. Swers, and K. Dane Wittrup

Introduction

Yeast surface display (YSD) is a powerful tool for engineering the affinity, specificity, and stability of antibodies, as well as other proteins. Since first described in 1997 by Boder and Wittrup,[1] YSD has been employed successfully in engineering a number of antibodies,[2,3] as well as T-cell receptors.[4–6] A recently reported large nonimmune single chain antibody library serves as a good starting point for engineering high-affinity antibodies.[7] Cloned variable genes from hybridomas or scFvs or Fabs from phage display libraries are also incorporated easily into a yeast display format. The original YSD protocols were described earlier,[8] but new and refined methods have been developed, in particular improved vectors, mutagenesis methods, and efficient ligation-free yeast transformation procedures. This article provides up-to-date protocols for engineering single chain antibodies by YSD.

Compared to other display formats, yeast surface display offers several advantages. One chief advantage to engineering protein affinity by YSD is that yeast cells can be sorted by fluorescence-activated cell sorting (FACS), allowing quantitative discrimination between mutants.[9] Further, FACS simultaneously gives analysis data, eliminating the need for separate steps of expression and analysis after each round of sorting. Without exception to date, equlibrium-binding constants and dissociation rate constants measured for yeast-displayed proteins are in quantitative agreement with those measured for the same proteins *in vitro* using BIAcore or ELISA.

[1] E. T. Boder and K. D. Wittrup, *Nature Biotechnol.* **15**, 553 (1997).
[2] M. C. Kieke, B. K. Cho, E. T. Boder, D. M. Kranz, and K. D. Wittrup, *Protein Eng.* **10**, 1303 (1997).
[3] E. T. Boder, K. S. Midelfort, and K. D. Wittrup, *Proc. Natl. Acad. Sci. USA* **97**, 10701 (2000).
[4] P. D. Holler *et al.*, *Proc. Natl. Acad. Sci. USA* **97**, 5387 (2000).
[5] M. C. Kieke *et al.*, *Proc. Natl. Acad. Sci. USA* **96**, 5651 (1999).
[6] M. C. Kieke *et al.*, *J. Mol. Biol.* **307**, 1305 (2001).
[7] M. J. Feldhaus *et al.*, *Nature Biotechnol.* **21**, 163 (2003).
[8] E. T. Boder and K. D. Wittrup, *Methods Enzymol.* **328**, 430 (2000).
[9] J. J. VanAntwerp and K. D. Wittrup, *Biotechnol. Prog.* **16**, 31 (2000).

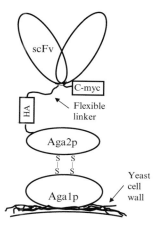

FIG. 1. Yeast surface display schematic. The single chain antibody is expressed as a fusion to the Aga2 mating protein. C-myc and HA epitope tags are present to quantify expression by immunofluorescence.

Traditional panning methods have also been successfully employed with YSD, including magnetic particle separation.[10] Other advantages arising from the yeast system include ease of use and the presence of the yeast endoplasmic reticulum, the later of which acts as a quality control mechanism, ensuring that only properly folded proteins reach the cell surface.

This article contains methods for displaying an antibody on yeast, creating mutant libraries, and sorting libraries for improved clones. The constructs and strains required for yeast surface display are described in the first section. The next section contains the method for creating large mutant libraries using homologous recombination, including the precise conditions used for error-prone polymerase chain reaction (PCR) using nucleotide analogues. Finally we include protocols for labeling yeast with fluorophores and sorting by FACS for improved affinity.

The Yeast Surface Display System

As the name implies, yeast surface display involves the expression of a protein of interest on the yeast cell wall, where it can interact with proteins and small molecules in solution. The protein is expressed as a fusion to the Aga2p mating agglutinin protein, which is in turn linked by two disulfide bonds to the Aga1p protein linked covalently to the cell wall (Fig. 1).

[10] Y. A. Yeung and K. D. Wittrup, *Biotechnol. Prog.* **18,** 212 (2002).

Expression of both the Aga2p–antibody fusion and Aga1p are under the control of the galactose-inducible GAL1 promoter, which allows inducible overexpression.

In order to use YSD, one must construct a yeast shuttle plasmid with the single chain antibody of interest fused to Aga2p. This can be derived from the pCTCON vector (Fig. 2) by inserting the open reading frame of the scFv of interest between the *Nhe*I and the *Bam*HI sites (both of which should be in frame with the antibody). The yeast strain used must have the Aga1 gene stably integrated under the control of a galactose-inducible promoter. EBY100 (Invitrogen) or one of its derivatives are suggested as hosts.

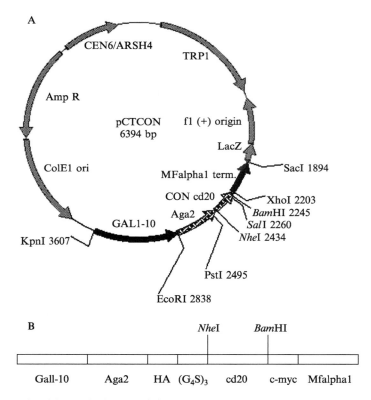

FIG. 2. Plasmid map of pCTCON. (A) CON cd20 is expressed from the plasmid as a fusion to the yeast mating protein Aga2. (B) The CON cd20 gene can be replaced with an scFv of interest using *Nhe*I and *Bam*HI sites.

Generating Large Mutant Antibody Libraries in Yeast

The most efficient way to make a mutant library in yeast is to use homologous recombination, thereby eliminating the need for ligation and *Escherichia coli* transformation.[11] In brief, cut plasmid and an insert containing the mutated gene are prepared separately, with significant homology (30–50 bp or more) shared by the insert and plasmid at each end. These DNA fragments are then taken up by yeast during electroporation and are reassembled *in vivo*. Libraries prepared by this method typically include at least 10^7 transformants and are often over 10^8 in diversity, which approximates the amount that can be sorted by state-of-the-art cell sorters in an hour.

The following section describes how to prepare scFv insert DNA with random point mutations by error-prone PCR using nucleotide analogues. However, this mutagenesis procedure may be replaced with DNA shuffling with slight modification using one of many published protocols.[12–14]

Preparation of Insert: Error-Prone PCR Using Nucleotide Analogues

Nucleotide analogue mutagenesis allows the frequency of mutation to be tuned based on the number of PCR cycles and the relative concentration of the mutagenic nucleotide analogues used during PCR.[15,16] The two analogues, 8-oxo-2'-deoxyguanosine-5'-triphosphate and 2'-deoxy-p-nucleoside-5'-triphosphate (8-oxo-dGTP and dPTP, respectively, TriLink Biotech), create both transition and transversion mutations. In order to ensure that some fraction of the library created is sufficiently mutated to generate improvements, but not so highly mutated so as to completely ablate binding, a range of several different mutagenesis levels are used in parallel. The conditions reported here are the ones typically used to create antibody libraries; these conditions give an error rate ranging from 0.2 to 5%.

If the gene to be mutated is already in pCTCON, then the following primers may be used to carry out the mutagenesis and subsequent amplification. These primers are designed to have >50 bp of homology to pCTCON for use during homologous recombination.

Forward primer: cgacgattgaaggtagatacccatacgacgttccagactacgctctgcag
Reverse primer: cagatctcgagctattacaagtcttcttcagaaataagcttttgttc

[11] C. K. Raymond, T. A. Pownder, and S. L. Sexson, *Biotechniques* **26**, 134 (1999).
[12] W. P. Stemmer, *Nature* **370**, 389 (1994).
[13] W. P. Stemmer, *Proc. Natl. Acad. Sci. USA* **91**, 10747 (1994).
[14] A. A. Volkov and F. H. Arnold, *Methods Enzymol.* **328**, 447 (2000).
[15] M. Zaccolo and E. Gherardi, *J. Mol. Biol.* **285**, 775 (1999).
[16] M. Zaccolo, D. M. Williams, D. M. Brown, and E. Gherardi, *J. Mol. Biol.* **255**, 589 (1996).

1. Set up six 50-μl PCR reactions

	Final concentration
10× PCR buffer (without MgCl$_2$)	
MgCl$_2$	2 mM
Forward primer	0.5 μM
Reverse primer	0.5 μM
dNTPs	200 μM
Template	0.1–1 ng
8-Oxo-dGTP	2–200 μM
dPTP	2–200 μM
dH$_2$0	to final volume
Taq polymerase	2.5 units

Of the six PCR reactions, two should contain 200 μM nucleotide analogues, two should contain 20 μM nucleotide analogues, and two should contain 2 μM nucleotide analogues.

2. Run the PCR for the number of cycles specified below. The cycles should have the following incubation temperatures and times: denature at 94° for 45 s, anneal at 55° for 30 s, and extend at 72° for 1 min. One should also include a 3-min denaturation step at 94° before the cycles begin and a 10-min extension step after the cycles are completed (the 10-min extension may be done on a heating block to run all reactions simultaneously).

Nucleotide analogue concentration (μM)	Number of PCR cycles
200	5
200	10
20	10
20	20
2	10
2	20

3. Run the entire mutagenic PCR products out on a 1% low-melt agarose gel. PCR products cycled 20 times are easily visible on a gel stained with SYBR Gold (Molecular Probes). Reactions cycled 10 times or less may not be visible on the gel; however, it is important to gel purify anyway to remove the nonmutated template before amplification (next step). Cut out and purify using the Qiagen gel purification kit following the manufacturer's protocol.

4. Amplify each reaction in the absence of nucleotide analogues to generate sufficient insert DNA for the transformation. Three 100-μl reactions should be set up for each mutagenic reaction, and 1 μl or more of

the gel purified product should be used as a template in the new reaction. Do not add nucleotide analogues. Cycle 25–30 times as you would for a normal PCR.

5. Optional. Gel purify the PCR products from step 4. Purification will eliminate many PCR artifacts from the library, but may also result in significant loss of PCR product.

6. Concentrate the PCR products using Pellet Paint (Novagen). After the pellet has dried, dissolve in water to a final concentration of 5 μg/μl. This protocol typically produces 40–100 μg of PCR product.

Preparation of Vector

1. Miniprep 10 μg or more of pCTCON.
2. Digest with *Nhe*I (New England Biolabs) for at least 2 h in NEB2 buffer.
3. Adjust salt concentration by adding one-tenth of the total volume of 1 *M* NaCl.
4. Double digest with *Bam*HI and *Sal*I for another 2 h to ensure complete digestion of pCTCON and reduce reclosure of the acceptor vector. (Note that the plasmid is cut in three places to ensure that the vector will not transform yeast cells in the absence of insert.)
5. Use the Qiagen nucleotide removal kit to purify DNA from enzymes, keeping in mind that a single column saturates with 10 μg DNA.
6. Concentrate DNA using the Paint Pellet reagent. After drying the pellet, dissolve in water to 2 μg/μl.

Preparation of Electrocompetent Yeast Cells

This protocol has been adapted from Meilhoc *et al.*[17] and generates enough cells for the transformation of ~60 μg of insert DNA and ~6 μg of vector, which typically produces ~5 × 10^7 yeast transformants.

1. Inoculate 100 ml of YPD to OD_{600} 0.1 from a fresh overnight culture of EBY100 (or appropriate yeast strain).
2. Grow cells with vigorous shaking at 30° to an OD_{600} of 1.3–1.5 (about 6 h).
3. Add 1 ml filter-sterilized 1,4-dithiothreitol (DTT, Mallinckrodt) solution (1 *M* Tris, pH 8.0, 2.5 *M* DTT). DTT is unstable and the solution must be made fresh just before use. Continue to grow with shaking at 30° for 20 min.

[17] E. Meilhoc, J. M. Masson, and J. Teissie, *Biotechnology* **8**, 223 (1990).

4. Harvest cells at 3500 rpm, 5 min, 4°. Discard supernatant. All centrifugation steps should be carried out in autoclaved centrifuge tubes or in sterile Falcon tubes.

5. Wash with 25 ml of E buffer (10 mM Tris, pH 7.5, 270 mM sucrose, 1 mM MgCl$_2$) at room temperature. Spin down again.

6. Transfer to two 1.5-ml microcentrifuge tubes and wash a second time with 1 ml of E buffer each. Spin down.

7. Resuspend both pellets in E buffer to a final combined volume of 300 μl. Any extra cells not used immediately may be frozen down in 50-μl aliquots for future use. Note that using frozen cells results in a 3- to 10-fold loss in transformation efficiency.

Electroporation

Electroporation is carried out using a Bio-Rad gene pulser device.

1. In a microcentrifuge tube, mix 0.5 μl vector (1 μg), 4.5 μl insert (9 μg), and 50 μl electrocompetent yeast cells. Add the mixture to a sterile 0.2-cm electroporation cuvette (Bio-Rad). Incubate on ice 5 min. Prepare additional cuvettes until all of the DNA is used.

2. Set gene pulser settings to 25 μF (capacitance) and 0.54 kV (voltage), which gives an electric field strength of 2.7 kV/cm with 0.2-cm cuvettes; the time constant should be about 18 ms with 55 μl volumes. The pulse controller accessory is not used.

3. Carry out pulsing at room temperature. Insert the cuvette into the slide chamber. Push both red buttons simultaneously until a pulsing tone is heard and then release.

4. After pulsing, immediately add 1 ml of room temperature YPD media[8] to the cuvette. Incubate at 30° for 1 h in 15-ml round-bottom Falcon tubes with shaking (250 rpm).

5. Spin down cells at 3500 rpm in a microcentrifuge. Resuspend in selective media (SD+CAA,[8] 50 ml/electroporation reaction). Plate out serial 10-fold dilutions to determine transformation efficiency. The library may be propagated directly in liquid culture without significant bias due to the repression of scFv expression in glucose-containing medium such as SD+CAA.[7]

Transformation efficiency should be at least $10^5/\mu$g, but is typically around $10^6/\mu$g. In addition to the electroporation mixture described here, one should perform a control where no insert is added and determine the transformation efficiency. This is the background efficiency and should be less than $\sim$1% of that obtained in the presence of insert DNA.

Equilibrium Labeling Protocol

Labeling yeast that are displaying an antibody or antibody library with a fluorescent or biotinylated antigen allows quantification of binding affinity and enables library sorting by FACS. Typically, a second fluorophore conjugated to an antibody is used to detect the epitope tag C-terminal to the scFv, which allows for the normalization of expression and eliminates nondisplaying yeast from quantification. A short protocol follows for labeling with a biotinylated antigen and the 9E10 monoclonal antibody against the C-terminal epitope tag c-myc. This protocol is for analytical labeling; for labeling large libraries, adjust volumes as described at the end of the protocol.

1. Grow transformed yeast overnight in SD+CAA. OD_{600} should be greater than 1. As a general approximation, $OD_{600} = 1$ represents 10^7 cells/ml.

2. Inoculate a 5-ml culture of $SG+CAA^8$ (inducing media) with the overnight culture. The final OD_{600} of the new culture should be ~1.

3. Induce at $20°$ with shaking (250 rpm) for at least 18 h. The appropriate induction temperature should be tested for each scFv, from 20, 25, 30, or $37°$.

4. Collect 0.2 OD_{600} ml of induced yeast in a 1.5-ml microcentrifuge tube. Several such aliquots may be necessary to sample the full diversity of the library, as this aliquot will correspond to approximately 2×10^6 cells.

5. Spin down in table-top centrifuge for 30 s at max speed. Discard supernatant.

6. Rinse with phosphate-buffered saline (PBS) plus 0.1% bovine serum albumin (BSA). Centrifuge for 10 s and discard supernatant.

7. Incubate with primary reagents. Add desired concentration of biotinylated antigen and 1 μl 9E10 (1:100, Covance) to a final volume of 100 μl in PBS/BSA. Incubate at desired temperature for 30 min. Larger volumes and longer incubation times are required for very low (<10 nM) antigen concentrations (see notes at end of protocol).

8. Centrifuge, discard supernatant, and rinse with ice-cold PBS/BSA. Centrifuge and discard supernatant from rinse.

9. On ice, incubate with secondary reagents. Add 97 μl ice-cold PBS/BSA, 2 μl goat antimouse FITC conjugate (1:50, Sigma), and 1 μl streptavidin phycoerythrin conjugate (1:100, Molecular Probes). Incubate for 30 min.

10. Centrifuge, discard supernatant, and rinse with ice-cold PBS/BSA. Centrifuge and discard supernatant from rinse.

11. Resuspend cells in 500 μl ice-cold PBS/BSA and transfer to tubes for flow cytometry or FACS sorting.

An important consideration when labeling high-affinity antibodies (<30 nM) is the depletion of antigen from the labeling mixture. This results in a lower than expected concentration of soluble (free) antigen, and hence a lower signal. Sorting libraries under depletion conditions can reduce the difference in signal observed for improved clones compared to their wild-type counterparts. The equivalent concentration of yeast surface-displayed proteins when 0.2 OD_{600} ml of yeast is added to a 100-μl volume is approximately 3 nM or less. To avoid depletion, always use at least a 10-fold excess of antigen by adjusting the total volume and/or reducing the number of yeast added (as little as 0.05 OD_{600} ml can be used).

Note that for labeling large libraries, it is advisable not to scale up directly. Instead use 1 ml volume per 10^8 cells labeled, keeping the reagent dilutions constant. Depletion can be especially severe with such high cell densities, however, and the experiment must be designed to avoid such conditions.

Analyzing Clones and Libraries by Flow Cytometry

Once a yeast population is labeled, it can be analyzed by flow cytometry. This allows quantification of binding affinity by titrating the antigen concentration. In addition to the samples that one wishes to analyze, a negative control (no fluorophores) and two single positive controls (just one fluorophore in each) should be prepared. With standard filters installed, FITC will be detected in the FL1 channel, whereas PE will be detected by FL2 for the settings on most flow cytometers. However, some "bleed over" or spectral overlap will be present in each channel, which must be compensated out. One should use the negative control to set the voltage and gain on each of the detectors so that the negative population has an order of magnitude intensity of 1 to 10. The single positive controls are used to adjust compensation so that no FITC signal is detected in FL2 and no PE signal in FL1.

In a titration, one generally sets a gate on cells that express the antibody (i.e., FITC-positive cells if the preceding labeling protocol is used) to eliminate nonexpressing cells from quantification.

For sorting or analyzing a library, it is helpful to also prepare a labeled sample of the wild-type antibody and saturated library for comparison and to aid in drawing sort windows.

Sorting Yeast Surface Display Libraries by FACS

FACS is the most efficient and accurate way to sort yeast surface display libraries, although magnetic particle strategies have also been employed.[1,2] To sort a library by FACS, one labels cells according to the

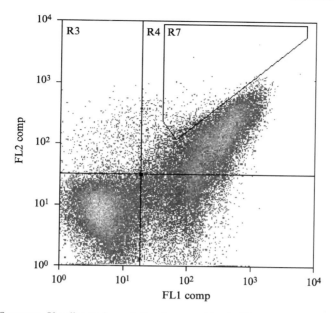

FIG. 3. Sort gate. If a diagonal population is present in the library, a sort gate such as the one labeled R7 should be drawn to take full advantage of expression normalization.

protocol given earlier, taking into consideration the notes that follow the protocol. Equations describing the optimum labeling concentration for a first library sort are available,[3] or one can simply choose a concentration that results in a weak signal (say, one-fourth of the K_d value). One should typically screen 10–100 times the number of independent clones that are in the library. When drawing a gate for collecting cells, it is advisable to use a window with a diagonal edge to normalize for expression if a double-positive diagonal is present (Fig. 3). If no diagonal is observed (little or no binding), the entire double-positive quadrant should be collected. Cells should be sorted directly into SD+CAA with antibiotics such as penicillin and streptomycin to diminish the risk of bacterial contamination. Cells will grow to saturation in 1 (if $>10^5$ cells are collected) or 2 ($<10^5$) days.

The very first time a library is sorted, gates are drawn conservatively (0.5 to 1% of the library is collected) to minimize the likelihood that an improved clone is missed. After the first sort, care should be taken to note the number of cells collected, as this is the maximum number of independent clones remaining in the library. In subsequent sorts, when the library size has been reduced and the amount of sorting time necessary decreases, one should bring several samples labeled under different conditions for sorting. These samples should be sorted at increasing stringency to rapidly isolate the best clones. Sort gates should cover the range of 0.01% of cells collected to 0.5%. All samples should be analyzed, and the one with the greatest improvement should be chosen for further sorting. Typically the single best clone, or clones containing a consensus mutation, is isolated within four sorts.

Cells collected in the final sort are plated out for clonal analysis. The mutant plasmids may be recovered from yeast using the Zymoprep kit (Zymo Research). The following primers may be used for sequencing:

> Forward sequencing primer: gttccagactacgctctgcagg
> Reverse sequencing primer: gattttgttacatctacactgttg

Summary

The protocols and methods described here enable engineering of scFv by yeast surface display. Each protocol is up to date and has been verified and optimized through several years of application. The directed evolution process is often applied iteratively until the desired affinity is achieved. A single round of mutagenesis and screening typically results in 10- to 100-fold improvement in the K_d value, with the largest improvements obtained when the wild-type affinity is low (say, low micromolar binding constant). A complete cycle of mutagenesis and screening, from wild-type clone to improved mutant clone, requires conservatively approximately 3–6 weeks.

Acknowledgments

DWC is the recipient of an NSF Graduate Fellowship, and this work has been funded by a grant from the Hereditary Disease Foundation and NIH CA96504.

[28] A Conditionally Replicating Virus as a Novel Approach Toward an HIV Vaccine

By Atze T. Das, Koen Verhoef, and Ben Berkhout

Introduction

Live-attenuated virus vaccines have proven to be highly successful at inducing protective immunity against pathogenic viruses such as smallpox, polio, and measles. The development of a similar vaccine for HIV is complicated by the enormous flexibility of the viral genome. Reversion of attenuated HIV toward a better replicating, potentially pathogenic virus is due to ongoing low-level virus replication, combined with the error-prone replication machinery of the virus. To prevent this chronic replication and possible restoration of virulence, we generated a conditional live HIV-1 virus of which the replication can be switched on and off at will. We made such an HIV-1 variant by replacing the natural mechanism for gene expression of the virus with the Tet system for tetracycline (Tc)-controlled gene expression. This HIV–rtTA virus replicates exclusively in the presence of the Tc-derivative doxycycline (dox). The concept of this conditional live virus vaccine is that upon vaccination in the presence of dox, the virus will replicate and potently activate all arms of the immune system. Subsequent withdrawal of dox will switch off the virus, abort chronic replication, and thus prevent the generation of virulent virus variants. This article explains in detail how the HIV-1 virus was turned into a dox-dependent variant. A similar strategy can be used to control the replication of other viruses.

Live-Attenuated HIV as a Vaccine

The estimated number of people infected with HIV and dying of AIDS increases each year, even though multiple potent antivirals are currently available. The development of a prophylactic vaccine is needed urgently to block the continuous spread of the virus. The induction of protective immunity most likely requires a vaccine that stimulates both antibody molecules that neutralize the virus and a cellular immune response, particularly cytotoxic T lymphocytes, that kills virus-infected cells.[1,2] Most efforts for developing a vaccine focus on immunization with one or several

[1] B. S. Graham, *Annu. Rev. Med.* **53**, 207 (2002).
[2] N. L. Letvin, D. H. Barouch, and D. C. Montefiori, *Annu. Rev. Immunol.* **20**, 73 (2002).

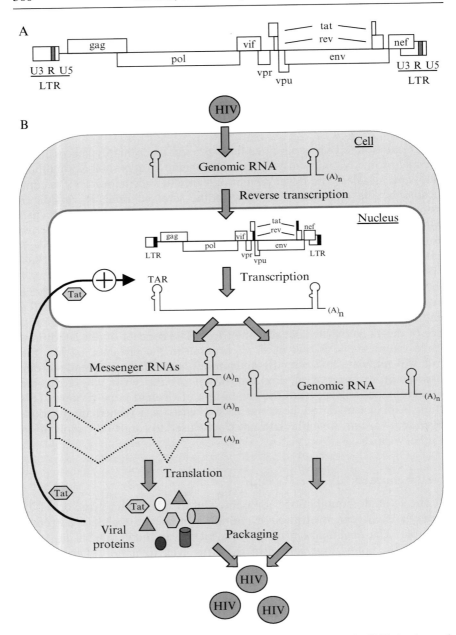

FIG. 1. The HIV-1 virus. (A) The HIV-1 genome is approximately 10 kb in size and encodes nine genes of which five are essential for viral replication: gag, pol, env, tat, and rev. (B) The HIV-1 replication cycle. Upon binding of the virus to specific receptors on the cellular

viral proteins that are either injected directly or delivered by nonreplicating viral vectors or plasmid DNA.[3-12] However, there are serious doubts about the efficacy of such vaccines to induce a sufficiently strong immune response. A live-attenuated virus appears to be a more potent candidate for eliciting both a protective humoral and a cellular immune response. The idea of a live-attenuated virus vaccine is that the nonpathogenic virus replicates to a limited extent and thereby elicits a potent immune response that protects against a new infection with the wild-type pathogenic virus. Several molecular approaches can be used to attenuate a complex retrovirus like HIV-1. The virus encodes nine genes of which five are essential for viral replication (Fig. 1A). This includes the gag gene encoding the structural proteins matrix, capsid, and nucleocapsid; the pol gene encoding the viral enzymes protease, reverse transcriptase, and integrase; and the env gene encoding the envelope proteins SU-gp120 and TM-gp41. Other essential viral functions are the tat gene, encoding the Tat transcriptional activator, and the rev gene, encoding the Rev protein that regulates RNA nuclear export and splicing. In addition, the HIV-1 genome contains four accessory genes encoding the Vif, Vpr, Vpu, and Nef proteins. These viral proteins are not essential in particular experimental settings,[13,14] and these

[3] S. M. Smith, *J. Biomed. Sci.* **9**, 100 (2002).
[4] G. Sutter and J. Haas, *AIDS* **15**(Suppl. 5), S139 (2001).
[5] A. M. Schultz and J. A. Bradac, *AIDS* **15**(Suppl. 5), S147 (2001).
[6] M. J. Schnell, *FEMS Microbiol. Lett.* **200**, 123 (2001).
[7] P. Mooij and J. L. Heeney, *Vaccine* **20**, 304 (2001).
[8] T. Hanke, *Br. Med. Bull.* **58**, 205 (2001).
[9] B. S. Peters, *Vaccine* **20**, 688 (2001).
[10] H. L. Robinson, *Nature Rev. Immunol.* **2**, 239 (2002).
[11] J. R. Mascola and G. J. Nabel, *Curr. Opin. Immunol.* **13**, 489 (2001).
[12] M. Klein, *Vaccine* **19**, 2210 (2001).
[13] C. M. Steffens and T. J. Hope, *AIDS* **15**(Suppl. 5), S21 (2001).
[14] N. J. Deacon, A. Tsykin, A. Solomon, K. Smith, M. Ludford-Menting, D. J. Hooker, D. A. McPhee, A. L. Greenway, A. Ellett, C. Chatfield, V. A. Lawson, S. Crowe, A. Maerz, S. Sonza, J. Learmont, J. S. Sullivan, A. Cunningham, D. Dwyer, D. Dowton, and J. Mills, *Science* **270**, 988 (1995).

---

membrane (CD4 receptor and CCR5 or CXCR4 coreceptor), the virus enters the cell. The viral RNA genome is reverse transcribed into a proviral DNA that integrates into the genome of the cell. The full-length transcript of about 9.5 kb functions both as messenger RNA for the synthesis of Gag and Pol proteins and as genomic RNA that will be packaged into budding virions. This transcript is also the precursor for the more than 30 differentially spliced, subgenomic mRNAs. Translation of these mRNAs results in the production of the other viral proteins. One of these proteins is the Tat protein that subsequently enhances the production of new transcripts by binding to TAR. The RNA genome is packaged into virus particles that bud from the host cell and that subsequently can infect new cells. (See color insert.)

genes have been the major target for attenuation. Alternatively, modulating the function of essential viral proteins or regulatory RNA/DNA signals can be used to attenuate the virus. For instance, the untranslated leader of HIV-1 RNA encodes multiple signals of which the activity can be modulated.[15,16]

Most research on the development of a live-attenuated HIV vaccine has focused on the experimental model system of the pathogenic simian immunodeficiency virus (SIV) and infection of rhesus macaques. Several accessory functions have been deleted from the viral genome, either individually or in combination.[17–20] The majority of monkeys vaccinated with such deletion mutants of SIV can efficiently control the replication of pathogenic challenge virus strains.[21–23] However, there is accumulating evidence that the attenuated virus can revert to virulence and cause disease over time in a minority of the vaccinated animals.[24–27] Similarly, long-term cell culture infections with an HIV-1 variant with deletions in the vpr and nef genes and in the long terminal repeat (LTR) promoter[28] demonstrated that this attenuated virus can regain substantial replication capacity within a few months of viral passage.[29] This revertant virus maintained the introduced deletions, but had acquired compensatory changes elsewhere in the viral genome. These results highlight the genetic instability and evolutionary capacity of attenuated SIV/HIV strains, which poses a serious safety risk for any future experimentation with live-attenuated HIV vaccines in humans.

[15] B. Berkhout, *Adv. Pharmacol.* **48,** 29 (2000).
[16] Y. Guan, J. B. Whitney, C. Liang, and M. A. Wainberg, *J. Virol.* **75,** 2776 (2001).
[17] H. W. Kestler, III, D. J. Ringler, K. Mori, and R. C. Desrosiers, *Cell* **65,** 651 (1991).
[18] J. S. Gibbs, D. A. Regier, and R. C. Desrosiers, *AIDS Res. Hum. Retrovir.* **10,** 607 (1994).
[19] M. S. Wyand, K. H. Manson, M. Garcia-Moll, D. Montefiori, and R. C. Desrosiers, *J. Virol.* **70,** 3724 (1996).
[20] Y. Guan, J. B. Whitney, M. Detorio, and M. A. Wainberg, *J. Virol.* **75,** 4056 (2001).
[21] R. P. Johnson, *Nature Med.* **5,** 154 (1999).
[22] R. C. Desrosiers, *Nature Med.* **4,** 982 (1998).
[23] J. Mills, R. Desrosiers, E. Rud, and N. Almond, *AIDS Res. Hum. Retrovir.* **16,** 1453 (2000).
[24] T. W. Baba, Y. S. Jeong, D. Penninck, R. Bronson, M. F. Greene, and R. M. Ruprecht, *Science* **267,** 1820 (1995).
[25] T. W. Baba, V. Liska, A. H. Khimani, N. B. Ray, P. J. Dailey, D. Penninck, R. Bronson, M. F. Greene, H. M. McClure, L. N. Martin, and R. M. Ruprecht, *Nature Med.* **5,** 194 (1999).
[26] L. A. Chakrabarti, K. J. Metzner, T. Ivanovic, H. Cheng, J. Louis-Virelizier, R. I. Connor, and C. Cheng-Mayer, *J. Virol.* **77,** 1245 (2003).
[27] A. M. Whatmore, N. Cook, G. A. Hall, S. Sharpe, E. W. Rud, and M. P. Cranage, *J. Virol.* **69,** 5117 (1995).
[28] J. S. Gibbs, D. A. Regier, and R. C. Desrosiers, *AIDS Res. Hum. Retrovir.* **10,** 343 (1994).
[29] B. Berkhout, K. Verhoef, J. L. B. van Wamel, and B. Back, *J. Virol.* **73,** 1138 (1999).

Upon vaccination with a live-attenuated SIV/HIV virus, the virus is not cleared, resulting in a persistent, chronic infection. Reversion to an efficiently replicating virus is probably due to ongoing low-level replication of the vaccine strain. Triggered by the error-prone virus replication machinery, fitter virus variants are continuously generated and selected. These variants may eventually become pathogenic. To improve the safety of a live-attenuated virus vaccine, the replication capacity of the virus can be reduced further through a progressive deletion or mutation of (accessory) genes or regulatory elements. This will make it more difficult for the virus to revert to virulence and cause disease. However, reduced replication of the virus also reduces the immunogenicity of the vaccine. The efficacy of vaccination thus depends on the replication capacity of the vaccine virus, and an excessively attenuated virus will not evoke protective immunity.[19,30,31] Instead of attenuating the virus further, we decided to produce a safe replicating virus vaccine by constructing a virus that does not replicate continuously, but exclusively in the presence of an exogenously administered effector. Since replication of such a conditional live virus can be limited to the period of vaccination, restoration of virulence can be prevented. We made such an HIV-1 variant by replacing the natural gene expression mechanism of the virus with an inducible regulatory system.

HIV-1 Replication Is Controlled by a Constitutive Autoregulatory Loop

In infected cells, the double-stranded DNA proviral genome is flanked by long terminal repeats (LTRs; Fig. 1A), of which the 5′ LTR acts as a promoter for the synthesis of viral RNA. The HIV-1 promoter is located mainly in the U3 region of the LTR and contains a large number of cis-acting sequences that control transcription.[32,33] Important regulatory elements in this region are the highly conserved NF-κB and Sp1 sites. Sp1-driven basal transcription at the LTR allows the production of a low level of HIV-1 transcripts.[34–36] Transcription starts at the U3/R border, and the

[30] B. L. Lohman, M. B. McChesney, C. J. Miller, E. McGowan, S. M. Joye, K. K. van Rompay, E. Reay, L. Antipa, N. C. Pedersen, and M. L. Marthas, J. Virol. 68, 7021 (1994).

[31] R. P. Johnson, J. D. Lifson, S. C. Czajak, K. Stefano Cole, K. H. Manson, R. L. Glickman, J. Q. Yang, D. C. Montefiori, R. C. Montelaro, M. S. Wyand, and R. C. Desrosiers, J. Virol. 73, 4952 (1999).

[32] L. A. Pereira, K. Bentley, A. Peeters, M. J. Churchill, and N. J. Deacon, Nucleic Acids Res. 28, 663 (2000).

[33] F. C. Krebs, T. H. Hogan, S. Quiterio, S. Gartner, and B. Wigdahl, in "HIV Sequence Compendium" (C. Kuiken, B. Foley, E. Freed, B. Hahn, P. A. Marx, F. McCutchan, J. W. Mellors, S. Wolinsky, and B. Korber, eds.), p. 29, 2001.

[34] B. Berkhout and K. T. Jeang, J. Virol. 66, 139 (1992).

R sequence at the extreme 5' end of the transcript folds a hairpin structure called the *trans*-acting response region (TAR; Fig. 1B). Upon binding of the viral Tat transactivator protein to this TAR hairpin in the nascent RNA transcript, transcription is enhanced dramatically.[37–44] The transcripts are polyadenylated at the R/U5 border in the 3' LTR, such that the R sequence, including the TAR hairpin, is repeated at the 3' terminus of the RNA. The full-length transcript of about 9.5 kb functions both as messenger RNA for the synthesis of Gag and Pol proteins and as genomic RNA that will be packaged into budding virions. This transcript is also the precursor for the more than 30 differentially spliced, subgenomic mRNAs. Translation of these mRNAs results in the synthesis of other viral proteins. One of these proteins is the Tat protein that subsequently enhances the production of new transcripts by binding to TAR. Thus, virus production and replication are controlled at the transcriptional level by a constitutive autoregulatory loop in which Tat and TAR play a key role. To transform the constitutively replicating HIV-1 virus into a conditional-live variant, the Tat–TAR axis of gene expression was functionally replaced by an inducible regulatory system.

The Tet System

Several artificial gene regulation systems have been developed to control gene expression in a quantitative and temporal way. The most widely used regulatory circuit is the so-called Tet system, which allows stringent control of gene expression by tetracycline or its derivative doxycycline.[45–48]

[35] D. Harrich, J. Garcia, F. Wu, R. Mitsuyasu, J. Gonzalez, and R. Gaynor, *J. Virol.* **63**, 2585 (1989).

[36] K. A. Jones, J. T. Kadonaga, P. A. Luciw, and R. Tjian, *Science* **232**, 755 (1986).

[37] B. Berkhout, R. H. Silverman, and K. T. Jeang, *Cell* **59**, 273 (1989).

[38] B. Berkhout and K. T. Jeang, *J. Virol.* **63**, 5501 (1989).

[39] C. Dingwall, I. Ernberg, M. J. Gait, S. M. Green, S. Heaphy, J. Karn, A. D. Lowe, M. Singh, and M. A. Skinner, *EMBO J.* **9**, 4145 (1990).

[40] A. D. Frankel, S. Biancalana, and D. Hudson, *Proc. Natl. Acad. Sci. USA* **86**, 7397 (1989).

[41] J. W. Harper and N. J. Logsdon, *Biochemistry* **30**, 8060 (1991).

[42] S. Roy, N. T. Parkin, C. Rosen, J. Itovitch, and N. Sonenberg, *J. Virol.* **64**, 1402 (1990).

[43] M. Sumner-Smith, S. Roy, R. Barnett, L. S. Reid, R. Kuperman, U. Delling, and N. Sonenberg, *J. Virol.* **65**, 5196 (1991).

[44] K. M. Weeks and D. M. Crothers, *Cell* **66**, 577 (1991).

[45] M. Gossen and H. Bujard, *in* "Tetracyclines in Biology, Chemistry and Medicine" (M. Nelson, W. Hillen, and R. A. Greenwald, eds.), p. 139. Birkhäuser Verlag, 2001.

[46] U. Baron and H. Bujard, *Methods Enzymol.* **327**, 401 (2000).

[47] S. Freundlieb, U. Baron, A. L. Bonin, M. Gossen, and H. Bujard, *Methods Enzymol.* **283**, 159 (1997).

[48] C. Berens and W. Hillen, *Eur. J. Biochem.* **270**, 3109 (2003).

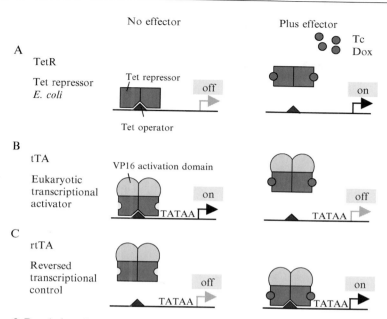

FIG. 2. Regulation of gene expression by TetR and TetR-derived transregulators. (A) The *Tn10* tet operon. In *Escherichia coli* the Tet repressor (TetR) protein binds as a dimer to the tet operator (tetO) sequence in the *Tn10* tet operon, thereby blocking transcription of the downstream-positioned tetA gene.[49] Binding of tetracycline (Tc), or Tc derivatives such as doxycycline (dox), induces a conformational switch in TetR that impedes the interaction with tetO and results in the activation of tetA expression. (B) The Tet–tTA system for Tc-controlled gene expression. In the Tet–tTA system, TetR is fused to the activation domain of the herpes simplex virus VP16 protein, resulting in the transcriptional activator tTA, and the tetO elements are placed upstream of a minimal promoter coupled to the gene of interest. (C) The Tet–rtTA system. A tTA variant with four amino acid substitutions in the TetR moiety exhibits a reverse phenotype. This reverse tTA (rtTA) binds to tetO and activates the expression of downstream-positioned genes in the presence of dox, but not in its absence. The rtTA protein has a low affinity for Tc and is poorly activated by this compound. (See color insert.)

The Tet system is based on the specific, high-affinity binding of the *Escherichia coli* Tet repressor protein (TetR) to the tet operator (tetO) sequence[49] (Fig. 2A). Tc and dox induce a conformational change in TetR, which impedes the interaction with tetO. Fusion of the activation domain of the herpes simplex virus VP16 protein to TetR resulted in the transcriptional activator tTA, which induces gene expression from minimal promoters placed downstream of tetO elements (Ptet) in eukaryotic cells

[49] W. Hillen and C. Berens, *Annu. Rev. Microbiol.* **48**, 345 (1994).

(Fig. 2B). The presence of Tc or dox abolishes this gene expression. A tTA variant with four amino acid substitutions in the TetR moiety exhibits a reverse phenotype.[50] This reverse tTA (rtTA) binds to Ptet and activates gene expression in the presence of dox, but not in its absence (Fig. 2C). Although the Tet system originates from a bacterial gene regulation mechanism, it is now widely applied to control gene expression in eukaryotes, including mammals, plants, and insects.[45] The transactivator components of the Tet system have been improved significantly in recent years. Minimizing the VP16 moiety of tTA and rtTA reduced the cytotoxic squelching of the transcription factor such that higher intracellular concentrations of these transactivators are tolerated.[51] Since prokaryotic DNA sequences can be preferred targets for methylation, which could hinder transcription, the TetR part of the rtTA gene was modified to create a codon composition similar to that found in human coding sequences.[52] Putative splice donor and acceptor sites were eliminated to avoid detrimental splicing of the (r)tTA transcripts. Furthermore, an rtTA protein that requires less dox for full activation and that shows less background activity in the absence of this effector was selected in random mutagenesis studies.[52]

This Tet–rtTA system seems to be the ideal regulatory system to control replication of a conditional live HIV-1 virus vaccine. Controlling virus replication via rtTA instead of tTA will avoid the long-lasting administration of Tc or dox that would be required with tTA. There is ample clinical expertise with dox as an antibiotic since its introduction in 1967. Dox is inexpensive and nontoxic, and blood concentrations higher than that needed for rtTA activation are easily reached in clinical practice by oral administration of the drug.[53,54]

The Conditional Live HIV–rtTA Virus

To transform the constitutively replicating HIV-1 virus into a conditional live variant, the viral genome was mutated to inactivate the Tat–TAR transcription regulation mechanism and to integrate the Tet system (Fig. 3). We used the full-length, infectious HIV-1 molecular clone HIV$_{LAI}$[55] for construction of this HIV–rtTA virus. The viral transcription

[50] M. Gossen, S. Freundlieb, G. Bender, G. Muller, W. Hillen, and H. Bujard, *Science* **268,** 1766 (1995).

[51] U. Baron, M. Gossen, and H. Bujard, *Nucleic Acids Res.* **25,** 2723 (1997).

[52] S. Urlinger, U. Baron, M. Thellmann, M. T. Hasan, H. Bujard, and W. Hillen, *Proc. Natl. Acad. Sci. USA* **97,** 7963 (2000).

[53] N. Joshi and D. Q. Miller, *Arch. Intern. Med.* **157,** 1421 (1997).

[54] D. J. Kelly, J. D. Chulay, P. Mikesell, and A. M. Friedlander, *J. Infect. Dis.* **166,** 1184 (1992).

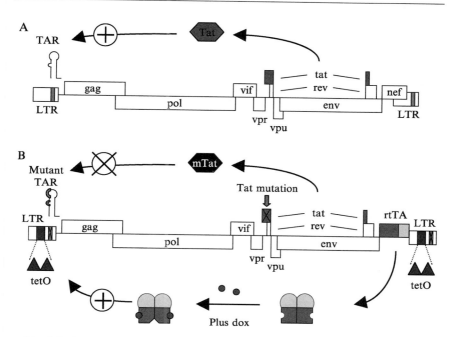

Fig. 3. Inducible gene expression in the dox-dependent HIV–rtTA virus. (A) In wild-type HIV-1, transcription of the viral genome is controlled by the viral Tat protein that binds to the TAR hairpin at the 5′ end of the nascent RNA transcript. Translation of RNA transcripts results in the production of viral proteins, including Tat. Thus, viral gene expression and replication are controlled by a constitutive autoregulatory loop in which Tat and TAR play a critical role. (B) In the HIV–rtTA virus, the Tat protein and its TAR-binding site were inactivated by mutation and functionally replaced by the rtTA transcriptional activator protein and its binding site, the tet operator (tetO). The gene encoding the rtTA protein was inserted in place of the 3′-terminal nef gene, and the tetO-binding sites were inserted in the LTR promoter (see Fig. 6 for details). The activity of rtTA is fully dependent on dox, and the HIV–rtTA virus does not replicate in the absence of this effector. The administration of dox induces transcription of the viral genome and expression of the viral proteins, including rtTA. This rtTA protein subsequently activates transcription, gene expression, and virus replication. Thus, viral gene expression and replication are now controlled by an inducible autoregulatory loop. (See color insert.)

elements TAR and Tat were inactivated by mutation, and the tetO and rtTA elements were inserted into this genome. In general, we took a conservative approach with regard to the type of mutations that

[55] K. Peden, M. Emerman, and L. Montagnier, *Virology* **185,** 661 (1991).

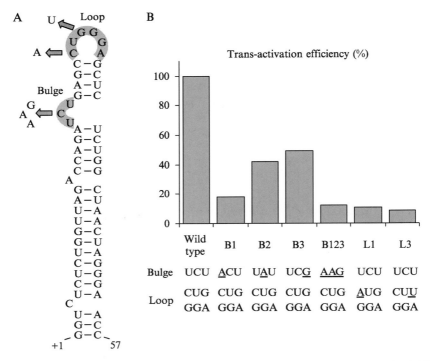

Fig. 4. Inactivation of the TAR hairpin. (A) RNA secondary structure of the TAR element present in the R region of HIV-1 transcripts. The five nucleotide substitutions introduced into the single-stranded bulge and loop sequence to inactivate TAR are indicated. (B) The effect of nucleotide substitutions on TAR activity. The Tat-mediated transactivation efficiency of wild-type (set at 100%) and TAR-mutated LTR-reporter gene constructs was determined in COS cells and described previously.[38] (See color insert.)

were introduced in the HIV-1 genome in order to minimize the risk of unforeseen inactivation of important replication signals.

Inactivation of TAR

The 5′ TAR RNA hairpin is formed by base pairing of nucleotides from positions +1 to +57 relative to the start site of transcription (Fig. 4A). Important features of TAR are the highly conserved 3 nucleotide pyrimidine bulge in the stem and the apical 6 nucleotide loop. Tat binds specifically to the bulge domain of TAR,[56] and mutation of the bulge or the flanking base pairs abolishes Tat-mediated *trans*-activation.[38,42] In the prototype HIV$_{HXB2}$ hairpin, as in most HIV-1 isolates, the bulge is formed by an UCU sequence.[57] Occasionally, like in the HIV_{LAI} isolate, an UUU sequence is present (Fig. 7). Substitution of one of these pyrimidine nucleotides with a purine nucleotide significantly reduces the transactivation efficiency mediated by Tat[38] (Fig. 4B). Although the TAR loop is not directly involved in the Tat–TAR interaction, the integrity of this domain is also important for transactivation. The positive transcriptional elongation factor (pTEFb) binds to the TAR loop through its cyclin T1 subunit in a Tat-dependent manner. Binding of Tat to the bulge facilitates the interaction of cyclin T1 with the TAR loop.[58–61] The kinase component of P-TEFb, CDK9, can then phosphorylate the C-terminal domain of RNA polymerase II, which enhances the processivity of the elongating polymerase.[62–64] Mutation of the loop sequence at position L1 or L3 reduces the Tat-mediated transactivation significantly[38,65] (Fig. 4B).

To disable the Tat–TAR axis of gene expression in our HIV-rtTA variant, we inactivated the TAR function by mutating both the bulge and the loop sequence (Fig. 4A). TAR is encoded by the R region, which is present at both ends of the viral genome (Fig. 1). We did not introduce more gross sequence changes or deletions in TAR because this R region is also essential for virus replication during strand transfer of reverse transcription.[66] Although we demonstrated previously that the TAR element of the 5' LTR is inherited in both LTRs of the viral progeny,[67] we inactivated the TAR motif in both LTRs to minimize the chance of reversion to the wild-type virus by a recombination event.

Inactivation of Tat

The Tat protein of HIV-1 is encoded by two exons and consists of 86 to 101 amino acids, depending on the viral isolate (Fig. 5A). Tat has a modular structure with an RNA-binding and a transcriptional activation domain. The activation domain, which is encoded by the first 48 amino

[56] R. Tan, A. Brodsky, J. R. Williamson, and A. D. Frankel, *Semin. Virol.* **8**, 186 (1997).
[57] B. Berkhout, *Nucleic Acids Res.* **20**, 27 (1992).
[58] S. Richter, Y. H. Ping, and T. M. Rana, *Proc. Natl. Acad. Sci. USA* **99**, 7928 (2002).
[59] S. Richter, H. Cao, and T. M. Rana, *Biochemistry* **41**, 6391 (2002).
[60] P. Wei, M. E. Garber, S.-M. Fang, W. H. Fisher, and K. A. Jones, *Cell* **92**, 451 (1998).
[61] J. Wimmer, K. Fujinaga, R. Taube, T. P. Cujec, Y. Zhu, J. Peng, D. H. Price, and B. M. Peterlin, *Virology* **255**, 182 (1999).
[62] P. D. Bieniasz, T. A. Gardina, H. P. Bogerd, and B. R. Cullen, *Proc. Natl. Acad. Sci. USA* **96**, 7791 (1999).
[63] C. A. Parada and R. G. Roeder, *Nature* **384**, 375 (1996).
[64] R. A. Marciniak and P. A. Sharp, *EMBO J.* **10**, 4189 (1991).
[65] S. Feng and E. C. Holland, *Nature* **334**, 165 (1988).
[66] B. Berkhout, J. van Wamel, and B. Klaver, *J. Mol. Biol.* **252**, 59 (1995).

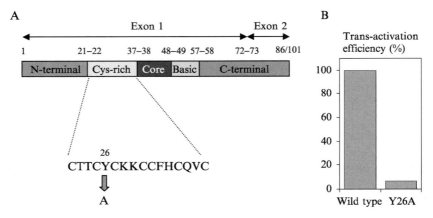

FIG. 5. Inactivation of the Tat protein. (A) The Tat protein of HIV-1 is encoded by two exons and consists of 86 or 101 amino acids, depending on the virus isolate. Tat has a modular structure with an RNA-binding and an activation domain. The activation domain, which is encoded by the first 48 amino acids, can be subdivided into the N-terminal acidic domain, the cysteine-rich domain, and a central core domain. The RNA-binding domain consists of a stretch of positively charged amino acids and is therefore termed the basic domain. The C-terminal part of Tat (amino acids 59–86/101), which is partially encoded by the second exon, contributes marginally to Tat activity. We inactivated Tat by a tyrosine-to-alanine substitution of residue 26 in the cysteine-rich domain. (B) The transactivation efficiency of wild-type (set at 100%) and Y26A-mutated Tat was determined by cotransfection of Tat expression plasmids and an LTR reporter gene construct into SupT1 T cells.[70] (See color insert.)

acids, can be subdivided into three regions. The N-terminal part contains several acidic residues and is termed the acidic domain. Next, the cysteine-rich domain contains seven highly conserved cysteine residues, of which six are critical for Tat function.[68] The central core domain is the most conserved region of Tat and is not only an essential part of the activation domain, but also adds specificity to TAR RNA binding.[69,70] The RNA-binding domain consists of a stretch of positively charged amino acids that interact with the TAR hairpin and is therefore termed the basic domain. It has been described that this motif also functions as a nuclear localization domain that is required to direct Tat to the nucleus[71,72] and as a protein

[67] B. Klaver and B. Berkhout, *Nucleic Acids Res.* **22,** 137 (1994).

[68] M. R. Sadaie, J. Rappaport, and F. Wong-Staal, *Proc. Natl. Acad. Sci. USA* **85,** 9224 (1988).

[69] M. J. Churcher, C. Lamont, F. Hamy, C. Dingwall, S. M. Green, A. D. Lowe, J. G. Butler, M. J. Gait, and J. Karn, *J. Mol. Biol.* **230,** 90 (1993).

[70] K. Verhoef, M. Koper, and B. Berkhout, *Virology* **237,** 228 (1997).

[71] J. Hauber, M. Malim, and B. Cullen, *J. Virol.* **63,** 1181 (1989).

transduction domain that allows the Tat protein to traverse the cell membrane.[73,74] The C-terminal part of Tat (amino acids 59–86/101), which is partially encoded by the second exon, contributes marginally to Tat activity, but may contribute to RNA binding.[69,75]

Since it has been suggested that Tat may play additional roles in the replication cycle in addition to its transcriptional function,[76–80] we decided to minimally mutate the Tat open reading frame and introduced a tyrosine-to-alanine substitution of residue 26 in the cysteine-rich domain. This Y26A mutation results in an almost complete loss of Tat-mediated activation of the LTR promoter (Fig. 5B). Accordingly, the replication of HIV-1 variants carrying this mutation is reduced dramatically.[70] The corresponding codon change (UAU to GCC) was designed to restrict the likelihood of simple reversion to the wild-type amino acid, which requires at least two nucleotide substitutions.[81] Combining the Tat and TAR mutations in the virus will obliterate the Tat–TAR regulation mechanism of gene expression and viral replication.

Substitution of the nef Gene with the rtTA Gene

We wanted to insert the rtTA gene into the HIV-rtTA genome at a position where the gene is expressed properly, without affecting the expression of other essential viral proteins. Our strategy to accomplish this was by replacement of a gene that is not essential for viral replication. Because the nef gene is dispensable for viral replication[13,14,17,27,82] and the nef mRNA is produced and translated early after infection,[83] we decided to substitute the rtTA gene for the nef gene.

In the HIV-1 genome, the nef gene initiates directly downstream of the env gene and the 3' half of the gene overlaps the U3 region of the LTR (Fig. 6A). We introduced two deletions in the nef gene of the HIV$_{LAI}$

---

[72] T. Subramanian, M. Kuppuswamy, L. Venkatesh, A. Srinivasan, and G. Chinnadurai, *Virology* **176,** 178 (1990).

[73] D. A. Mann and A. D. Frankel, *EMBO J.* **10,** 1733 (1991).

[74] E. Vives, P. Brodin, and B. Lebleu, *J. Biol. Chem.* **272,** 16010 (1997).

[75] K. Verhoef, M. Bauer, A. Meyerhans, and B. Berkhout, *AIDS Res. Hum. Retrovir.* **14,** 1553 (1998).

[76] C. Ulich, A. Dunne, E. Parry, C. W. Hooker, R. B. Gaynor, and D. Harrich, *J. Virol.* **73,** 2499 (1999).

[77] D. Harrich, C. Ulich, L. F. Garcia-Martinez, and R. B. Gaynor, *EMBO J.* **16,** 1224 (1997).

[78] L. M. Huang, A. Joshi, R. Willey, J. Orenstein, and K. T. Jeang, *EMBO J.* **13,** 2886 (1994).

[79] A. Apolloni, C. W. Hooker, J. Mak, and D. Harrich, *J. Virol.* **77,** 9912 (2003).

[80] C. W. Hooker, J. Scott, A. Apolloni, E. Parry, and D. Harrich, *Virology* **300,** 226 (2002).

[81] K. Verhoef and B. Berkhout, *J. Virol.* **73,** 2781 (1999).

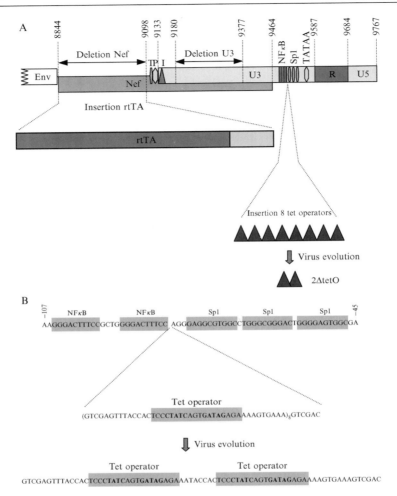

Fig. 6. Introduction of the two components of the Tet system into the HIV-1 genome. (A) The 3′ end of the HIV-1 proviral genome is shown. The nef gene (positions 8844–9464 in the HIV$_{LAI}$ proviral DNA) starts directly downstream of the env gene, and the 3′ half of the gene overlaps the U3 region of the LTR (LTR positions 9133–9767). We introduced two deletions in the nef gene of the HIV$_{LAI}$ genome to create space for insertion of the rtTA gene and tetO elements. We removed a 256-nucleotide nef fragment upstream of the 3′ LTR (positions 8843–9098) and a 198-nucleotide fragment overlapping the U3 region of this LTR (positions 9180–9377). We maintained an 81-nucleotide sequence around the 5′ end of the 3′ LTR that contains important cis-acting sequences for virus replication; the T-rich sequence (T), the 3′ polypurine tract (P), and the sequence essential for integration are indicated (I).[84,85] The rtTA2^s-S2 gene (747 nucleotides) with a translation start codon in an optimized sequence context (CC<u>A</u>UGU[90]) was inserted directly downstream of the env gene, thus in fact replacing the 5′ half of the nef gene. In the rtTA gene, sequences derived from the

genome to create space for the insertion of rtTA gene and tetO elements (Figs. 6A and 7). We removed a 256-nucleotide nef fragment upstream of the 3′ LTR and a 198-nucleotide fragment overlapping the U3 region of the LTR. In the viral progeny this U3 deletion will be inherited in both LTRs. The exact borders of the nef and U3 deletions were chosen carefully such that important *cis*-acting sequences for virus replication were not affected. In particular, we maintained 81 nucleotides around the 5′ end of the 3′ LTR (Fig. 7). This region encodes multiple sequence elements that are critical for reverse transcription[84] and integration.[85] In fact, we tried to mimic spontaneous deletions that have been observed in the nef–U3 region of several HIV and SIV variants in a variety of replication studies.[86–89]

The rtTA gene, with a translational start codon in an optimized sequence context (CCAUGU[90]), was inserted directly downstream of the env gene, thus replacing the 5′ half of the nef gene. We used the novel rtTA2^s-S2 variant that requires less dox for full activation and that shows less background activity in the absence of this effector.[52] Thus, rtTA translation will occur from the subgenomic, spliced mRNA that was originally meant for expression of the Nef protein. As a result, rtTA will be

[82] L. Alexander, P. O. Illyinskii, S. M. Lang, R. E. Means, J. Lifson, K. Mansfield, and R. C. Desrosiers, *J. Virol.* **77**, 6823 (2003).

[83] S. Kim, R. Byrn, J. Groopman, and D. J. Baltimore, *J. Virol.* **63**, 3708 (1989).

[84] P. O. Ilyinskii and R. C. Desrosiers, *EMBO J.* **17**, 3766 (1998).

[85] P. O. Brown, in "Retroviruses" (J. M. Coffin, S. H. Hughes, and H. E. Varmus, eds.), p. 161. Cold Spring Harbor Laboratory Press, Cold Spring Harbor, NY, 1997.

[86] P. O. Ilyinskii, M. D. Daniel, M. A. Simon, A. A. Lackner, and R. C. Desrosiers, *J. Virol.* **68**, 5933 (1994).

[87] F. Kirchhoff, T. Greenough, D. B. Brettler, J. L. Sullivan, and R. C. Desrosiers, *N. Engl. J. Med.* **332**, 228 (1995).

[88] F. Kirchhoff, H. W. Kestler, III, and R. C. Desrosiers, *J. Virol.* **68**, 2031 (1994).

E. coli tet repressor (encoding amino acids 1–207) are fused to sequences derived from the activation domain of the herpes simplex virus VP16 protein (amino acids 208–248). (A and B) We inserted a 342-nucleotide fragment containing eight copies of the tetO-binding site derived from the E. coli Tn10 tet operon between the NF-κB and the Sp1 sites in both the 5′ and the 3′ LTR promoter. Upon culturing of this HIV-rtTA virus, a rapid rearrangement in the LTR–8tetO promoter region was observed in several independent cultures.[91] This rearrangement resulted in a deletion of six of the original eight tetO sequences, followed by a further deletion of 14 or 15 nucleotides in the spacer between the two remaining tetO elements. We made an HIV–rtTA variant in which the LTR–2ΔtetO promoter (with the 15-nucleotide spacer deletion) replaced the LTR–8tetO promoter. This HIV–rtTA 2ΔtetO variant replicated much more efficiently than the original HIV rtTA virus. The nucleotide positions of the sequence shown in B are relative to the transcription start site (−107 to −45). This sequence corresponds to positions 348–410 in the 5′ LTR and to positions 9480–9542 in the 3′ LTR. (See color insert.)

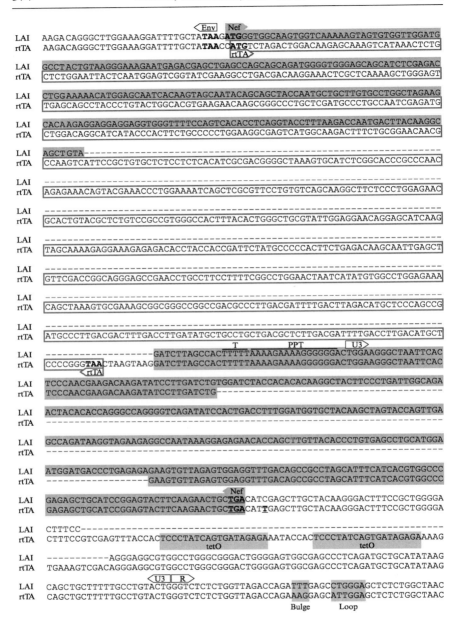

FIG. 7. Introduction of the Tet system into HIV-1. Comparison of the 3′ sequence of HIV$_{LAI}$ (positions 8812–9634) and the HIV–rtTA 2ΔtetO variant. See Fig. 6 for details.

produced early after infection, when it is required for the activation of HIV–rtTA transcription.

Insertion of tetO Elements in the LTR Promoter

We inserted eight copies of the tetO-binding site in between the NF-κB and the Sp1 sites in both the 5′ and the 3′ LTR promoter (Fig. 6B). The introduced tetO sequence is identical to one of the two tetO elements naturally present in the *E. coli* Tn10 tet operon. When the LTR–8tetO promoter configuration was tested in an LTR luciferase construct in transient transfection studies, robust and dox-inducible activation of transcription was observed. Insertion of the tetO elements into the 3′ LTR may have been sufficient to produce a mutant progeny, as the U3 sequences in the 3′ LTR are inherited in both LTRs of the viral progeny. However, we also introduced the tetO motifs in the 5′ LTR to generate molecular clones that are expressed efficiently and in a dox-dependent manner upon the transfection of cells. Most fascinating, a rapid rearrangement in the LTR–tetO promoter region was observed in several independent HIV–rtTA cultures.[91] This rearrangement resulted in a deletion of six of the original eight tetO sequences, followed by a further deletion of 14 or 15 nucleotides in the spacer sequence between the two remaining tetO elements (Fig. 6B). Interestingly, conformation of the tetO elements in this evolved LTR–2ΔtetO promoter resembles the conformation of these elements in the *E. coli* Tn10 tet operon. We made an HIV–rtTA variant in which the LTR–2ΔtetO promoter replaced the LTR–8tetO promoter. This HIV–rtTA 2ΔtetO variant replicated much more efficiently than the original HIV rtTA virus. Strikingly, the transcriptional activity of the LTR–2ΔtetO promoter is lower than that of the original LTR–8tetO promoter and mimics that of the wild-type HIV LTR promoter.[92] These results demonstrate that HIV requires a fine-tuned level of transcription rather than maximal transcription for efficient replication.

Dox-Dependent Replication of the HIV–rtTA Virus

Multiple cloning steps and virus evolution resulted in an efficiently replicating HIV–rtTA virus. In the absence of dox, this virus is inactive. Administration of dox induces transcription of the viral genome and expression of the viral proteins, including rtTA (Fig. 3B). This rtTA protein subsequently activates transcription, gene expression, and virus replication.

[89] S. Pohlmann, S. Floss, P. O. Ilyinskii, T. Stamminger, and F. Kirchhoff, *J. Virol.* **72**, 5589 (1998).

[90] M. Kozak, *J. Cell Biol.* **108**, 229 (1989).

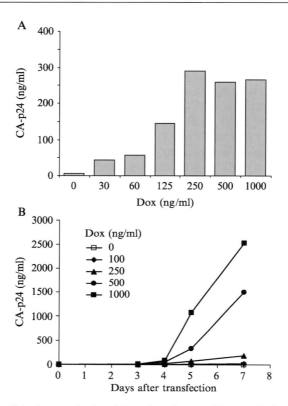

FIG. 8. HIV–rtTA virus production (A) and replication (B) are strictly dox dependent. (A) C33A cervix carcinoma cells (70% confluency in 2-cm² wells) were transfected with 1 μg of the molecular clone encoding HIV–rtTA 2ΔtetO (by Ca-phosphate transfection) and were cultured in the presence of different dox concentrations (0–1000 ng/ml). Virus production was measured by CA-p24 ELISA on culture supernatant samples 2 days after transfection. (B) SupT1 T cells (5 × 10⁶ cells) were transfected with 1 μg of the HIV–rtTA 2ΔtetO molecular clone (by electroporation) and were cultured in 5 ml complete medium in the presence of different dox concentrations (0–1000 ng/ml). Virus replication was monitored by CA-p24 ELISA on culture supernatant samples. Details on cell transfection and virus culture can be found elsewhere.[92a]

Production and replication of this virus are thus strictly dependent on the presence of dox. When cells transfected with the HIV–rtTA molecular clone were cultured for 2 days in the absence of dox, no virus production was observed (Fig. 8A). When these cells were cultured in the presence of

⁹¹ G. Marzio, K. Verhoef, M. Vink, and B. Berkhout, *Proc. Natl. Acad. Sci. USA* **98,** 6342 (2001).

an increasing dox level, the level of virus production increased gradually. Accordingly, when assaying virus replication in replication-competent T cells, the HIV–rtTA virus does not replicate in the absence of dox, and the replication level increases progressively with an increasing dox concentration (Fig. 8B).

Most systems for inducible gene expression, including the original Tet–rtTA system, are known to yield a significant level of leaky expression in the uninduced state. In regular dox-controlled gene expression systems, the transcriptional activator is produced in a constitutive manner from a second locus, e.g., from a plasmid in which rtTA expression is under the control of the constitutive CMV promoter. The resulting high concentration of rtTA in the absence of dox causes a significant level of gene activation when these systems are in the off state.

The HIV–rtTA system differs from regular dox-controlled gene expression systems in that an inducible autoregulatory loop is established that will reduce the level of leaky gene expression drastically.[93] Specifically, we have placed rtTA expression under control of an rtTA-regulated LTR–tetO promoter, a situation that mimics the natural autoregulatory loop of the TAR–Tat axis (Fig. 3). This means that both the activity of rtTA and its synthesis are dox dependent. Thus, only minute amounts of rtTA protein will be present in the absence of dox. Moreover, use of an improved rtTA variant with reduced binding to tetO sequences in the absence of dox[52] may have contributed to the superior performance of HIV–rtTA. As a result, HIV–rtTA demonstrates an extremely low level of gene expression in the absence of dox and, consequently, a more profound dox induction. The lack of virus replication of the HIV–rtTA vaccine candidate in the absence of dox will prevent the evolution toward a pathogenic virus.

Optimization of the HIV–rtTA Virus

The initial HIV–rtTA virus, with eight tetO elements inserted in the LTR promoter, replicated in a dox-dependent manner in cell culture experiments, but replication was much reduced compared to a wild-type HIV-1 strain.[94] We therefore anticipated that viruses with improved replication capacity might evolve during prolonged culturing of the virus. Improved replication might result from repair of the original Tat–TAR system of gene expression by reversion of the introduced mutations. In such viruses the components of the rtTA–tetO system will become

[92] G. Marzio, M. Vink, K. Verhoef, A. de Ronde, and B. Berkhout, *J. Virol.* **76**, 3084 (2002).
[92a] A. T. Das, B. Klaver, B. I. F. Klasens, J. L. B. van Wamel, and B. Berkhout, *J. Virol.* **71**, 2346 (1997).

redundant and are therefore likely to be lost by mutation or deletion. This evolution route is unlikely because we introduced multiple inactivating mutations in the Tat protein and TAR hairpin. Alternatively, the components of the introduced rtTA–tetO system, which are largely derived from *E. coli*, might be optimized to support virus replication in human cells.

We followed the evolution of HIV–rtTA in multiple cell culture infections and analyzed the status of the old Tat–TAR axis and the new rtTA–tetO system. In all viruses examined, the introduced mutations in the Tat and TAR were maintained, demonstrating that the original Tat–TAR system is not repaired. Furthermore, all viruses were found to maintain the introduced rtTA gene and the tetO elements. As anticipated, we did see changes in the tetO–LTR promoter region and the rtTA protein. As mentioned earlier, we observed a rearrangement in the LTR–8tetO promoter that resulted in the LTR–2ΔtetO configuration, which improved the replication potential of HIV–rtTA significantly. We also observed amino acid changes in the rtTA protein that improved the replication capacity of the virus considerably. For instance, a virus was selected with a mutation in the dox-binding site of rtTA. This mutation improves dox sensitivity and maximum activity of the protein greatly, and this HIV–rtTA variant requires less dox for optimal virus replication.[95a]

These results underline the genetic flexibility of HIV-1, which can be exploited for the functional adaptation of the Tet expression system and to improve the HIV–rtTA virus further. For instance, we are currently adapting HIV–rtTA to dox-like compounds that lack antibiotic activity. Selective activation of HIV–rtTA by a compound that is not used in human or veterinarian medicine will reduce the chance of unintentional reactivation of the virus after vaccination. This will be another major step forward in the design of an effective and safe vaccine based on a drug-dependent HIV-1 variant.

The HIV–rtTA Vaccine

The HIV–rtTA vaccine has to be tested vigorously for efficacy and safety. We will therefore construct a similar SIV–rtTA virus that can be tested in rhesus macaques. If the conditional live HIV-rtTA vaccine proves to be safe and effective, vaccination could be performed with the virus or with the proviral DNA, with the latter method being more attractive for logistical reasons.[95] Virus replication can be temporarily activated

[93] A. T. Das, X. Zhou, M. Vink, B. Klaver, and B. Berkhout, *Expert Rev. Vaccines* **1**, 293 (2002).

[94] K. Verhoef, G. Marzio, W. Hillen, H. Bujard, and B. Berkhout, *J. Virol.* **75**, 979 (2001).

and controlled to the extent that is needed for induction of the immune system by transient dox administration. If needed, virus replication can be turned on at a later moment as booster vaccination by the additional administration of dox.[96] Repeated antigenic stimulation may be critical to preserve immunological memory.

Our approach to make a conditional live HIV-1 virus by incorporation of the Tet system seems applicable for the generation of other virus vaccines. Instead of producing a live-attenuated virus vaccine by the deletion of viral genes or by culture on specific cells, a temporally replicating nonpathogenic virus can be made by replacing the natural regulation mechanism with the Tet system. Making the virus dox dependent can also be done as an additional safety measure for live-attenuated virus vaccines, thus reducing the chance that the attenuated virus reverts to a better replicating, pathogenic virus. Other groups have also tried to implement the Tet gene expression system in SIV and HIV,[97,98] but these attempts failed to produce an efficiently replicating virus. These results underscore the notion that it is critical that the mutations, deletions, and insertions, which have to be introduced into the viral genome to construct a conditional live virus, need to be chosen carefully, thus avoiding the risk of unforeseen destruction of important regulatory elements.

Acknowledgments

Vaccine research within the Berkhout laboratory is sponsored by the Technology Foundation STW (applied science division of NWO and the technology program of the Ministry of Economic Affairs, Utrecht, the Netherlands), the Dutch AIDS Fund (AIDS Fonds, Amsterdam, the Netherlands), and the National Institutes of Health (NIH, Bethesda, MD).

[95] S. J. Kent, C. J. Dale, S. Preiss, J. Mills, D. Campagna, and D. F. Purcell, *J. Virol.* **75,** 11930 (2001).

[95a] A. T. Das, X. Zhou, M. Vink, B. Klaver, K. Verhoef, G. Marzio, and B. Berkhout, *J. Biol. Chem.* **279,** 18776 (2004).

[96] B. Berkhout, G. Marzio, and K. Verhoef, *Virus Res.* **82,** 103 (2002).

[97] Y. Xiao, T. Kuwata, T. Miura, M. Hayami, and H. Shida, *Virology* **269,** 268 (2000).

[98] S. M. Smith, M. Khoroshev, P. A. Marx, J. Orenstein, and K. T. Jeang, *J. Biol. Chem.* **276,** 32184 (2001).

Author Index

Numbers in parentheses are footnote reference numbers and indicate that an author's work is referred to although the name is not cited in the text.

A

Abato, P., 252
Abecassis, V., 12, 20(20)
Abulercia, C., 132
Achiwa, K., 250
Adachi, S., 209
Adams, M. W., 8(26), 8
Adhya, S., 128
Admiraal, S. J., 279, 288(50), 299, 314(23)
Affholter, J. A., 36, 43, 48(7), 333
Agarwal, P. K., 250
Aguayo, C., 49
Aguinaldo, A. M., 43
Ahn, N. G., 165
Ait-Abdelkadar, N., 252, 253(47), 254(47)
Akanuma, S., 341, 343(36)
Alber, T., 104
Albrecht, M., 319, 324
Alcade, M., 49
Aldea, M., 276
Alexander, L., 371
Alexander, O. B., 134
Alexeeva, M.,
Allan, F. K., 209
Allawi, H. T., 83
Allison, R. W., 158
Almond, N., 362
Altamirano, M. M., 49
Altenbuchner, J., 200, 201, 205, 254
Alzari, P. M., 23
Amitani, H., 259, 265(7)
Ampofo, S. A., 150, 155(34)
Anderson, A., 132
Andre, C., 49
Anfinsen, C. B., 103
Angelaud, R., 252, 253(52)
Antipa, L., 363
Aparicio, J. F., 269, 292

Apolloni, A., 371
Appel, D., 220
Aragno, M., 342
Arand, M., 254
Arase, A., 158
Arase, M., 209
Archelas, A., 254
Arensdorf, J. J., 43
Arezi, B., 5
Armstrong, G. A., 319
Arndt, K. M., 104
Arndt, M. A. E., 180
Arnold, F. A., 188, 333, 339
Arnold, F. H., 4, 11(5), 12, 17, 35, 36, 37(1; 3),
    39(2; 3; 5), 40(2; 3; 5), 41(2), 42, 43, 48,
    48(7), 49, 134, 145(5), 146, 148, 188, 191,
    199, 201, 211, 213(36), 214, 221, 221(36),
    222(65), 223, 223(36; 50; 61; 62), 229,
    239, 250, 250(10), 254, 317, 318(7),
    325(7), 333, 351
Ashley, G., 274, 281, 282, 283(35),
    284(34; 35)
Asif-Ullah, M., 91, 92(8), 99(8)
Auf der Maur, A., 179
Averhoff, B., 342, 343

B

Baba, T. W., 362
Babik, J. M., 91, 101(3)
Babitzke, P., 276
Bach, H., 209
Back, B., 362
Bai, L., 274
Bailey, M. J., 164
Baker, A., 63
Bakhtina, M., 75, 76(1), 78(1), 90(1)
Balint, R. F., 104
Baltimore, D. J., 371

381

Davies, M. J., 167
Davydov, R., 209
Dawes, G., 11(4; 7), 12, 155
Deacon, N. J., 361, 363, 371(14)
Dean, R. T., 167
de Boer, A. L., 315, 317, 326(2)
De Castro, L., 41
Deege, A., 252(53; 54; 60), 253
de Ferra, F., 310
de Grado, M., 343
Deisenhofer, J., 209, 210(9), 213(9), 214(9; 10)
Delagrave, S., 3
del Cardayré, S. B., 239
de Lemos Esteves, F., 158
Delling, U., 364
Delviks, K. A., 43
Demarest, S. J., 180
Denman, S., 250
Dennis, P. P., 150
de Pedro, M. A., 343
de Prat-Gay, G., 103
Derewenda, U., 23
Derewenda, Z. S., 23
de Ronde, A., 375
DeSantis, G., 4(22), 5, 225, 237(5), 252, 253(44), 254(44)
Deshayes, K., 147
Desrosiers, R. C., 362, 363, 363(19), 371, 371(17), 372(84), 373
Detorio, M., 362
De Voss, J. J., 209
Devreese, B., 158
Diaz, M., 23
DiBlasio-Smith, E. A., 147
Dickins, M., 208
Dijkstra, B. W., 242, 243, 243(17), 244(17), 245(17), 246(17), 247(17), 249, 249(17), 251(17; 24), 252(17)
Dijkstra, R. S., 253
Dingwall, C., 364, 370, 371(69)
Dion, M., 49
Doekel, S., 295, 299, 309(25), 310(25), 311(11), 312(11; 12)
Doi, N., 61, 66(10; 11)
Doi, R. H., 49
Domann, S., 225, 234(7)
Dominguez, R., 23
Donadio, S., 269, 281, 281(2), 283(61)
Donarski, W. J., 257

Dong, S. D., 314
Doran, P. M., 165
Dordick, J. D., 149
Dordick, J. S., 145, 150, 154, 155, 155(34–38)
Dower, M. T., 226
Dower, W. J., 98, 135, 138
Dowton, D., 361, 371(14)
Drauz, K., 238, 239(2), 250(2)
Druzhko, A., 334, 335(26)
Dubendorff, J. W., 136
Dubrawsky, I., 4
Dugaiczyk, A., 98
Dumas, D. P., 256
Dunne, A., 371
Dunster, N. J., 271
Dupont, C., 159
Dwyer, D., 361, 371(14)
Dycaico, M., 119, 132
Dziarnowski, A., 279(58), 280, 288(58), 290(58), 291(58), 292(58)

E

Ebert-Khosla, S., 272, 273, 287
Eggert, T., 247
Egmond, M. R., 201
Ehmann, D. E., 300
Eijsink, V. G. H., 145
Eipper, A., 252, 253(45; 46; 48; 50), 254(45; 46)
Ellett, A., 361, 371(14)
Elovson, J., 279
El-Sayed, M., 345
Emerman, M., 366
Encell, L. P., 43
Endelman, J. B., 35, 37(3), 39(2; 3), 40(2; 3), 41(2)
Endo, Y., 61
Enfors, S. O., 147
England, P. A., 209
English, D. S., 334
Enright, A., 254
Enzelberger, M. M., 200
Eppelmann, K., 295, 299, 307, 308(45), 312(17; 22; 45)
Ernberg, I., 364
Ernst, A., 91
Ernst, S., 169
Escher, D., 179
Estabrook, R. W., 211

Subject Index

A

Aldolases
classification, 235
directed evolution
assays
acetaldehyde-dependent adolase
selection, 231
aldol assay with alcohol
dehydrogenase, 233–234
aldol assay with glycerol 3-phosphate
dehydrogenase, 231–232
aldol assay with lactate
dehydrogenase, 231, 233
caged fluorogenic substrates,
235, 237
retroaldol assay with
α-glycerophosphate
dehydrogenase/triosephosphate
isomerase, 233
retroaldol assay with lactate
dehydrogenase,
230–231, 233
substrate preparation,
227–228
expression and purification, 232
gene disruption of wild-type enzyme,
226–227
hot spots, 238
kinetic parameter modification,
235–236
library construction
DNA shuffling, 229
error-prone polymerase chain
reaction, 228–229
site-directed mutagenesis, 229
library expression, 229–230
materials, 225
plasmid construction, 226
industrial applications, 224–225
Antibody, *see* Immunoglobulin
Assisted protein reassembly, *see* Protein
fragment complementation

B

Bacteriorhodopsin, directed evolution
photonic device utilization, 334
photophysical property optimization
conformational intermediates in
photocycle, 335–337
device testing, 341
M state and O state optimization,
337–338
mutagenesis techniques, 338–339
mutational landscape, 335–336
requirements, 338
screening of variants, 339–341
rationale, 333–334
thermostability optimization
device testing, 345, 347
mutant generation, 342–344
thermoselection of variants, 344–345
Thermus thermophilus heterologous
expression system, 341–344
Biphenyl oxidase, substrate specificity
broadening using staggered extension
process recombination, 48

C

N-Carbamoylase, directed evolution
for functional expression improvement
activity analysis, 190
chaperone coexpression effects on folding,
187–188
fusion protein effects on folding, 187
green fluorescent protein fusion protein
construction, 189, 192
evolved mutant characterization,
193–195
fluorescence correlation with functional
enzyme expression, 192
screening of mutant library, 189–190
industrial applications, 188
materials, 189
purification of enzyme, 190

A For 32-fold degeneracy

```
5' GAT CAG AAC GCT TTC ATC GAG GGT GTG CTC CCG AAA TTC GTC GTC
    D   Q   N   A   F   I   E   G   V   L   P   K   F   V   V
```

```
5' ATCAGAACGCTTTCATCGAGNNKGTGCTCCCGAAATTCGTCGT    3' coding strand
5' ACGACGAATTTCGGGAGCACMNNCTCGATGAAAGCGTTCTGAT    3' non-coding strand
```

B For 64-fold degeneracy

```
5' GAT CAG AAC GCT TTC ATC GAG GGT GTG CTC CCG AAA TTC GTC GTC
    D   Q   N   A   F   I   E   G   V   L   P   K   F   V   V
```

```
5' ATCAGAACGCTTTCATCGAGNNNGTGCTCCCGAAATTCGTCGT    3' coding strand
5' ACGACGAATTTCGGGAGCACNNNCTCGATGAAAGCGTTCTGAT    3' non-coding strand
```

KRETZ *ET AL.*, CHAPTER 1, FIG. 1. Oligonucleotide primer design.

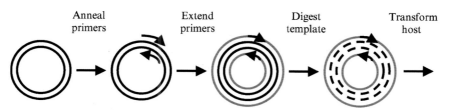

KRETZ *ET AL.*, CHAPTER 1, FIG. 2. Reaction mechanism.

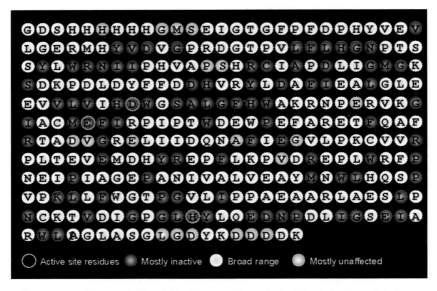

KRETZ *ET AL.*, CHAPTER 1, FIG. 3. Residue-specific analysis of the haloalkane dehalogenase from *Rhodococcus rhodochrous*. The mutant enzymes were tested for dehalogenase activity and compared to the wild-type enzyme.

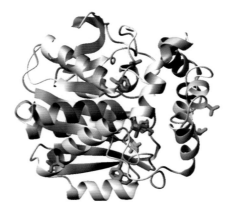

KRETZ *ET AL.*, CHAPTER 1, FIG. 4. Schematic representation of the crystal structure of the *R. rhodochrous* haloalkane dehalogenase. Highlighted in blue are residues that cannot be changed without severely affecting enzyme activity, in white are residues that can accommodate a broad spectrum of amino acids, and in gold are residues that can be replaced with any residue without affecting activity. The ball and stick figures in turquoise are catalytic residues, in red are wild-type residues, and in green are thermostable residues (PDB ref. 1BN7).

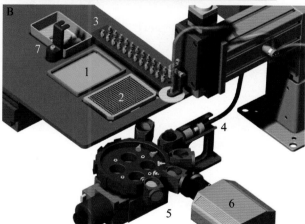

LAFFERTY AND DYCAICO, CHAPTER 11, FIG. 2. (A) The automated detection and recovery system. (B) Exploded three-dimensional model view of the major detection and recovery station components. (1) GigaMatrix plate, (2) recovery microplate, (3) recovery needles, (4) illumination optics, (5) filter wheel, (6) cooled CCD camera with telecentric lens, and (7) needle removal device.

LAFFERTY AND DYCAICO, CHAPTER 11, FIG. 5. Composite false-color image from a GigaMatrix plate in which four control clones encoding different fluorescent proteins were grown. The fluorescent protein encoded by clone A has absorbance/emission peaks at 388/440 nm and is identified here with the color blue. Clones B and C encode proteins with peaks at 463/488 and 488/507 nm, respectively. Clone C is identified by bright green, whereas clone B is a darker combination of blue and green. Clone D encodes a protein with peaks at 530/540 nm and is visualized by the color red in this image.

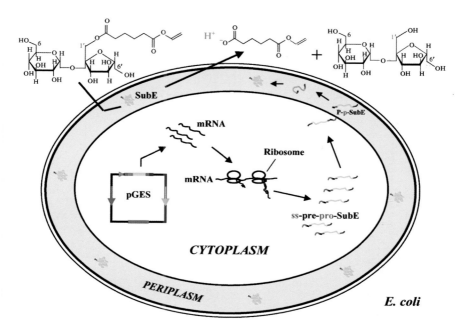

SROGA AND DORDICK, CHAPTER 13, FIG. 1. The principle of periplasmic expression of recombinant proteins in *E. coli*. During expression, subtilisin E is synthesized in its pre-pro form along with the signal sequence of gene III that encodes the minor capsid protein pIII of a filamentous phage fd. The signal sequence directs the recombinant protein to the periplasm, where it matures. Substrates have easy access to the enzyme expressed near the bacterial surface. Hydrolysis of S1'A is shown as an example. Screening is based on the release of vinyl adipate, which causes a pH decrease detected by bromothymol blue.

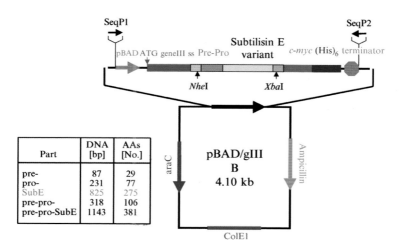

Part	DNA [bp]	AAs [No.]
pre-	87	29
pro-	231	77
SubE	825	275
pre-pro-	318	106
pre-pro-SubE	1143	381

SROGA AND DORDICK, CHAPTER 13, FIG. 2. General plasmid map of DNA libraries carrying subtilisin E variants. The pBAD/gIII B (Invitrogen) vector served as the basis for the generation of libraries of subtilisin E variants.

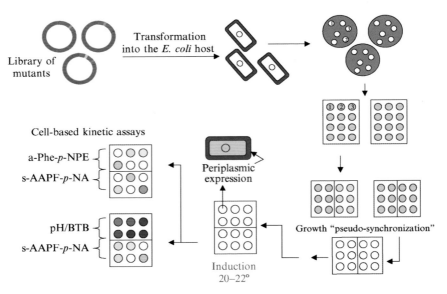

SROGA AND DORDICK, CHAPTER 13, FIG. 3. The principle of pseudo-synchronization prior to the expression of subtilisin E libraries in *E. coli* for cell-based kinetic assays. A library of subtilisin E mutants in plasmid DNA form is transformed into *E. coli*. The resulting single cell colonies are transferred to nutrient-rich medium with an antibiotic. After growth pseudo-synchronization, the synthesis of subtilisin variants is induced at 20–22°. Whole cells are used as biocatalysts in the amidase and esterase kinetic screens.

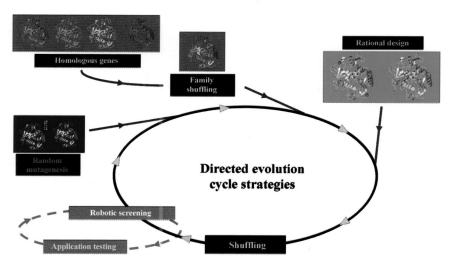

CHERRY AND LAMSA, CHAPTER 15, FIG. 1. Directed evolution is best viewed as an iterative cycle in which gene diversity is generated by one or more methods, variants are screened for improvement in a selected phenotype, and the process is repeated with the best variants resulting from the screening. Here, multiple methods of introducing diversity are selected based on the knowledge of structure/function relationships in the protein, as described in the text. A critical component is the validation of the screening system by testing selected variants ("screening winners") in the intended application.

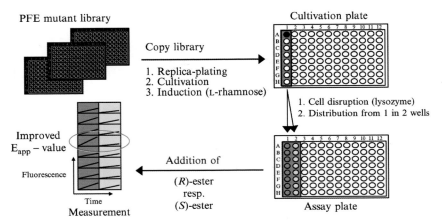

SCHMIDT *ET AL.*, CHAPTER 18, FIG. 1. Principle of the screening of mutant libraries as exemplified for an esterase from *Pseudomonas fluorescens* (PFE). First, clones are transferred from a master plate to a new MTP. After cultivation of *E. coli* harboring the esterase gene and production of the enzyme by induction with L-rhamnose, esterases are isolated by cell disruption using lysozyme and freeze/thaw cycles. The enzyme from the supernatant is then split into two wells of another MTP and optically pure (R)- or (S)-substrate is added. The hydrolytic activity is quantified by measurement of fluorescence. From the initial rate, apparent enantioselectivity (E_{app}) can be calculated.

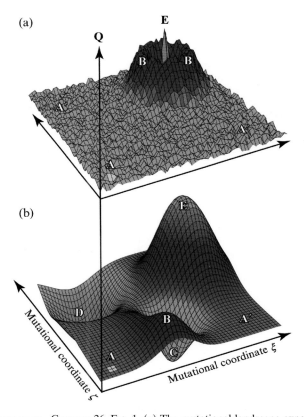

HILLEBRECHT *ET AL*., CHAPTER 26, FIG. 1. (a) The mutational landscape expected for more variable characters such as photophysical optimizations. (b) The mutational landscape expected for optimization of a fluid trait, such as temperature stability. Starting at mutational coordinate A, troughs and local maxima must be examined (A′, B, C, D) before reaching an optimization peak (E). A large number of variants must be screened to ensure that a global peak has been reached.

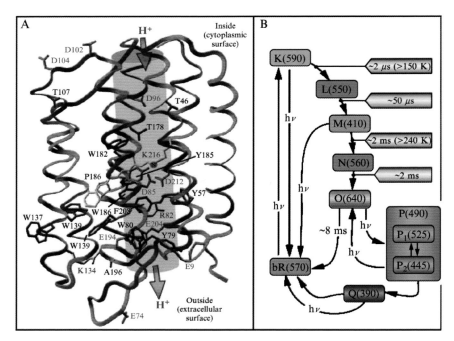

HILLEBRECHT ET AL., CHAPTER 26, FIG. 2. Bacteriorhodopsin structure and photocycle. (A) BR is an ion pump that results in the net transfer of a proton from the intracellular to the extracellular surface. Key residues are highlighted, and the proton transfer channel is shown. (B) The main and branched photocycle in BR used for three-dimensional memory device applications. The branching reaction involves the O → P transition that is optimized using directed evolution. M and O state lifetimes, and yields can also be optimized using directed evolution.

DAS ET AL., CHAPTER 28, FIG. 1. The HIV-1 virus. (A) The HIV-1 genome is approximately 10 kb in size and encodes nine genes of which five are essential for viral replication: gag, pol, env, tat, and rev. (B) The HIV-1 replication cycle. Upon binding of the virus to specific receptors on the cellular membrane (CD4 receptor and CCR5 or CXCR4 coreceptor), the virus enters the cell. The viral RNA genome is reverse transcribed into a proviral DNA that integrates into the genome of the cell. The full-length transcript of about 9.5 kb functions both as messenger RNA for the synthesis of Gag and Pol proteins and as genomic RNA that will be packaged into budding virions. This transcript is also the precursor for the more than 30 differentially spliced, subgenomic mRNAs. Translation of these mRNAs results in the production of the other viral proteins. One of these proteins is the Tat protein that subsequently enhances the production of new transcripts by binding to TAR. The RNA genome is packaged into virus particles that bud from the host cell and that subsequently can infect new cells.

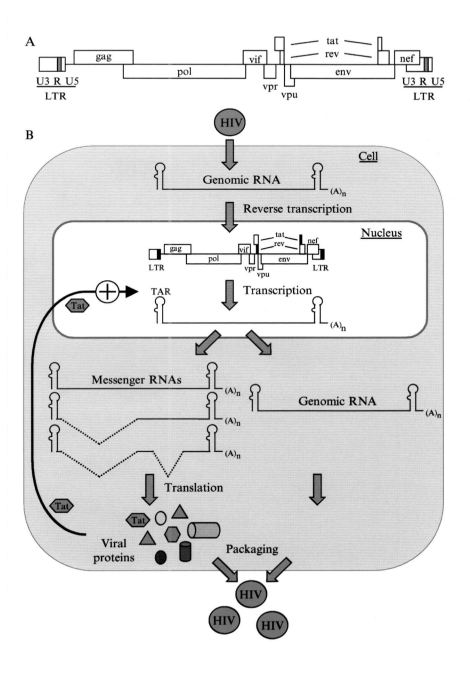

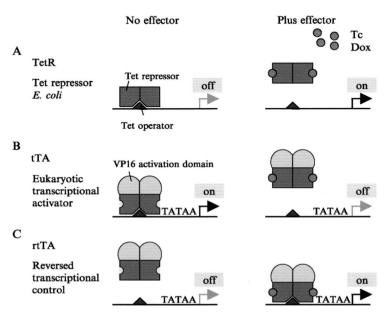

DAS *ET AL.*, CHAPTER 28, FIG. 2. Regulation of gene expression by TetR and TetR-derived transregulators. (A) The *Tn*10 tet operon. In *Escherichia coli* the Tet repressor (TetR) protein binds as a dimer to the tet operator (tetO) sequence in the *Tn*10 tet operon, thereby blocking transcription of the downstream-positioned tetA gene. Binding of tetracycline (Tc), or Tc derivatives such as doxycycline (dox), induces a conformational switch in TetR that impedes the interaction with tetO and results in the activation of tetA expression. (B) The Tet–tTA system for Tc-controlled gene expression. In the Tet–tTA system, TetR is fused to the activation domain of the herpes simplex virus VP16 protein, resulting in the transcriptional activator tTA, and the tetO elements are placed upstream of a minimal promoter coupled to the gene of interest. (C) The Tet–rtTA system. A tTA variant with four amino acid substitutions in the TetR moiety exhibits a reverse phenotype. This reverse tTA (rtTA) binds to tetO and activates the expression of downstream-positioned genes in the presence of dox, but not in its absence. The rtTA protein has a low affinity for Tc and is poorly activated by this compound.

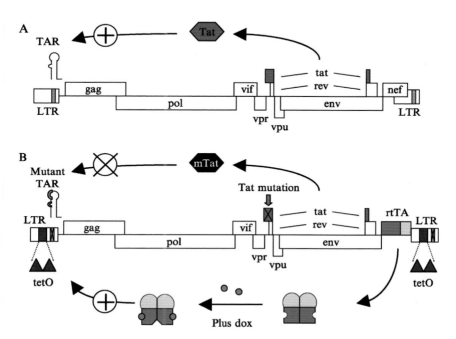

DAS *ET AL.*, CHAPTER 28, FIG. 3. Inducible gene expression in the dox-dependent HIV–rtTA virus. (A) In wild-type HIV-1, transcription of the viral genome is controlled by the viral Tat protein that binds to the TAR hairpin at the 5′ end of the nascent RNA transcript. Translation of RNA transcripts results in the production of viral proteins, including Tat. Thus, viral gene expression and replication are controlled by a constitutive autoregulatory loop in which Tat and TAR play a critical role. (B) In the HIV–rtTA virus, the Tat protein and its TAR-binding site were inactivated by mutation and functionally replaced by the rtTA transcriptional activator protein and its binding site, the tet operator (tetO). The gene encoding the rtTA protein was inserted in place of the 3′-terminal nef gene, and the tetO-binding sites were inserted in the LTR promoter. The activity of rtTA is fully dependent on dox, and the HIV–rtTA virus does not replicate in the absence of this effector. The administration of dox induces transcription of the viral genome and expression of the viral proteins, including rtTA. This rtTA protein subsequently activates transcription, gene expression, and virus replication. Thus, viral gene expression and replication are now controlled by an inducible autoregulatory loop.

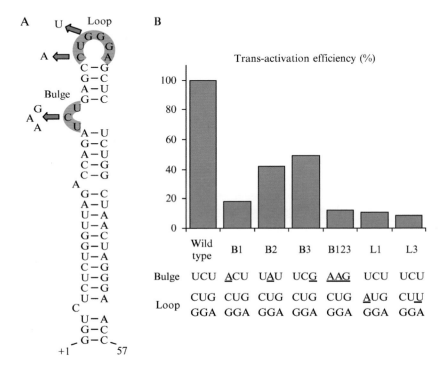

DAS *ET AL.*, CHAPTER 28, FIG. 4. Inactivation of the TAR hairpin. (A) RNA secondary structure of the TAR element present in the R region of HIV-1 transcripts. The five nucleotide substitutions introduced into the single-stranded bulge and loop sequence to inactivate TAR are indicated. (B) The effect of nucleotide substitutions on TAR activity. The Tat-mediated transactivation efficiency of wild-type (set at 100%) and TAR-mutated LTR-reporter gene constructs was determined in COS cells and described previously.

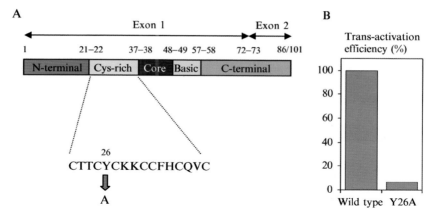

Das *et al.*, Chapter 28, Fig. 5. Inactivation of the Tat protein. (A) The Tat protein of HIV$_{LAI}$ is encoded by two exons and consists of 86 or 101 amino acids, depending on the virus isolate. Tat has a modular structure with an RNA-binding and an activation domain. The activation domain, which is encoded by the first 48 amino acids, can be subdivided into the N-terminal acidic domain, the cysteine-rich domain, and a central core domain. The RNA-binding domain consists of a stretch of positively charged amino acids and is therefore termed the basic domain. The C-terminal part of Tat (amino acids 59–86/101), which is partially encoded by the second exon, contributes marginally to Tat activity. We inactivated Tat by a tyrosine-to-alanine substitution of residue 26 in the cysteine-rich domain. (B) The transactivation efficiency of wild-type (set at 100%) and Y26A-mutated Tat was determined by cotransfection of Tat expression plasmids and an LTR reporter gene construct into SupT1 T cells.

Das *et al.*, Chapter 28, Fig. 6. Introduction of the two components of the Tet system into the HIV-1 genome. (A) The 3′ end of the HIV-1 proviral genome is shown. The nef gene (positions 8844–9464 in the HIV$_{LAI}$ proviral DNA) starts directly downstream of the env gene, and the 3′ half of the gene overlaps the U3 region of the LTR (LTR positions 9133–9767). We introduced two deletions in the nef gene of the HIV$_{LAI}$ genome to create space for insertion of the rtTA gene and tetO elements. We removed a 256-nucleotide nef fragment upstream of the 3′ LTR (positions 8843–9098) and a 198-nucleotide fragment overlapping the U3 region of this LTR (positions 9180–9377). We maintained an 81-nucleotide sequence around the 5′ end of the 3′ LTR that contains important *cis*-acting sequences for virus replication; the T-rich sequence (T), the 3′ polypurine tract (P), and the sequence essential for integration are indicated (I). The rtTA2^s-S2 gene (747 nucleotides) with a translation start codon in an optimized sequence context (CC<u>AUG</u>U) was inserted directly downstream of the env gene, thus in fact replacing the 5′ half of the nef gene. In the rtTA gene, sequences derived from the *E. coli* tet repressor

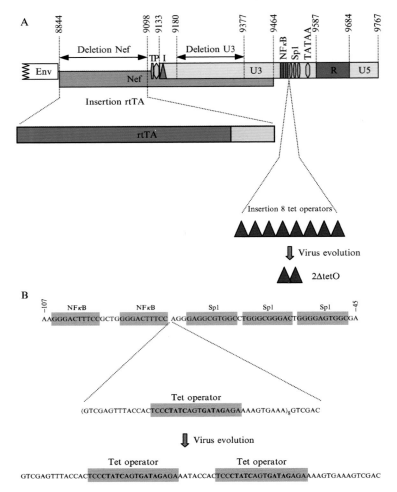

(encoding amino acids 1–207) are fused to sequences derived from the activation domain of the herpes simplex virus VP16 protein (amino acids 208–248). (A and B) We inserted a 342-nucleotide fragment containing eight copies of the tetO-binding site derived from the *E. coli* Tn10 tet operon between the NF-κB and the Sp1 sites in both the 5′ and the 3′ LTR promoter. Upon culturing of this HIV-rtTA virus, a rapid rearrangement in the LTR–8tetO promoter region was observed in several independent cultures. This rearrangement resulted in a deletion of six of the original eight tetO sequences, followed by a further deletion of 14 or 15 nucleotides in the spacer between the two remaining tetO elements. We made an HIV–rtTA variant in which the LTR–2ΔtetO promoter (with the 15-nucleotide spacer deletion) replaced the LTR–8tetO promoter. This HIV–rtTA 2ΔtetO variant replicated much more efficiently than the original HIV rtTA virus. The nucleotide positions of the sequence shown in B are relative to the transcription start site (−107 to −45). This sequence corresponds to positions 348–410 in the 5′ LTR and to positions 9480–9542 in the 3′ LTR.